GOVERNORS STATE UNIVERSITY LIBRARY
3 1611 00343 1415

W9-BCA-873

Global Warming and Climate Change

Ten Years after Kyoto and Still Counting

Global Warming and Climate Change

Ten Years after Kyoto and Still Counting

Volume 2

Editor

Velma I. Grover
United Nations University
International Network on Water,
Environment and Health
Hamilton
Ontario
Canada

GOVERNORS STATE UNIVERSITY
UNIVERSITY PARK
IL 60466

Science Publishers

Enfield (NH) Jersey Plymouth

QC
981.8
.C5
G6638
2008
v. 2

Science Publishers
www.scipub.net
234 May Street
Post Office Box 699
Enfield, New Hampshire 03748
United States of America

General enquiries : *info@scipub.net*
Editorial enquiries : *editor@scipub.net*
Sales enquiries : *sales@scipub.net*

Published by Science Publishers, Enfield, NH, USA
An imprint of Edenbridge Ltd., British Channel Islands
Printed in India

© 2008 reserved

ISBN (Set) 978-1-57808-539-2
ISBN (Vol. 1) 978-1-57808-540-8
ISBN (Vol. 2) 978-1-57808-541-5

Library of Congress Cataloging-in-Publication Data

Global warming and climate change : ten years after Kyoto and still counting/editor, Velma I. Grover. -- 1st ed.
p. cm.
Includes bibliographical references and index.
ISBN 978-1-57808-540-8 (v. 1 : alk. paper)--ISBN 978-1-57808-541-5 (v. 2 : alk. paper)--ISBN 978-1-57808-539-2 (set : alk. paper)
1. Climatic changes. 2. Climatic changes--Government policy--International cooperation. 3. Climatic changes--Effect of human beings on. 4. Global warming--Economic aspects. 5. Global warming--Social aspects. I. Grover, Velma.
QC981.8.C5G6638 2008
363.738'74--dc22

2008035763

All rights reserved. No part of this publication may be reproduced, stored in a retrieval system, or transmitted in any form or by any means, electronic, mechanical, photocopying or otherwise, without the prior permission of the publisher, in writing. The exception to this is when a reasonable part of the text is quoted for purpose of book review, abstracting etc.

This book is sold subject to the condition that it shall not, by way of trade or otherwise be lent, re-sold, hired out, or otherwise circulated without the publisher's prior consent in any form of binding or cover other than that in which it is published and without a similar condition including this condition being imposed on the subsequent purchaser.

UNITED NATIONS ENVIRONMENT PROGRAMME

Programme des Nations Unies pour l'environnement   Programa de las Naciones Unidas para el Medio Ambiente

Программа Организации Объединенных Наций по окружающей среде   برنامج الأمم المتحدة للبيئة

联合国环境规划署

Foreword

By Achim Steiner, UN Under-Secretary General and UNEP Executive Director

Ten years after the signing of the Kyoto Protocol the momentum on the issue of climate change has never been higher.

During this tenth anniversary year, several key outstanding questions have been resolved and a new quality of consensus has been built.

Earlier in 2007 the Intergovernmental Panel on Climate Change (IPCC), established by UNEP and the World Meteorological Organisation, concluded that it was "unequivocal' that the global warming being witnessed right now is linked to human activity.

This full stop behind the scientific debate has been some time in coming and others may say 'too long'.

But anyone who now believes that causes other than human activity are behind climate change is now in the ideological camp of the "Flat Earthers' who continue to view the world as flat rather than round.

Indeed one of the advancements in 2007 has been the rhetoric by world leaders with none now disputing the fact that climate change is a serious and urgent challenge.

The other feature, ten years after Kyoto, is the quality of the research on the likely impacts. Again the IPCC has produced sobering reports that underline that climate change will affect every community and every country on this planet—in many cases with profound economic and social impacts.

Executive Office

P. O. Box 30552, Nairobi, Kenya • Tel: (254 20) 762 3386/3416/3652 • Fax: (254 20) 762 4275/4608

E-mail: executiveoffice@unep.org

www.unep.org

From the loss of Arctic livelihoods and traditions to the melting of glaciers in the Himalayas and from sea level rise in Africa to storm surges in New York, many of the these impacts may well take place in time frames that will impact people alive today—not at some distant point in the future.

And finally we have the optimism provided by the IPCC—combating climate change is do-able, it will not cost the Earth. This expert panel suggests that 0.1 per cent of global GDP per year for the next 30 years might be enough to stabilize the atmosphere and avoid the worst.

Indeed according to the IPCC and studies by other bodies, emissions savings in some sectors like the building sector can be achieved with net benefits to the global economy.

Other signs of hope are also glimpsed in this tenth year after Kyoto. A report by the Sustainable Energy Finance Initiative in which UNEP plays a key role, claims that investments in renewable energies have grown over 40 per cent and that many of the technologies are no longer subject to the vagaries of fossil fuel prices.

Renewables are no longer the sole preserve of industries in developed countries either, with two of the world's biggest companies now based in China and India.

The carbon markets, triggered by the coming into force of Kyoto, are partly to explain. The Clean Development Mechanism (CDM) of the protocol is set to generate some $100 billion to finance projects—funds that are flowing from North to South, and, in part, assisting in fulfilling the aid promises made by developed countries in 1992 at the Rio Earth Summit.

Meanwhile action on climate change has fairly and squarely moved from being the business of government to being the business of business, as well as, local authorities and individual citizens.

In the United States, which decided not to ratify the Protocol, states like California and over 300 cities have signed up for, or are considering emission controls, in the spirit of Kyoto.

Developing countries, often perceived as waiting and watching for developed countries to take their responsibilities seriously, are in fact already contributing to the climate change challenge.

The IPCC concludes that rapidly developing economies like Brazil, China, India, Mexico and others have over recent years reduced greenhouse gas emissions—albeit and often for reasons other than climate change—reduced greenhouse gas emissions by some 500 million tonnes, an amount bigger than Annex I countries.

Meanwhile Brazil has reduced the rate of deforestation in the Amazon by over 50 per cent in the past three or four years. Indeed another defining feature of 2007 is the growing understanding that standing forests, which sequestrate carbon gases need to be part of the solution.

Deforestation may be adding 20 per cent of the greenhouse gas burden to the atmosphere with other wide ranging impacts, including impacts on freshwaters, to increased risks from natural hazards.

Various bodies including the G8, and institutions like the World Bank, are backing funding support for standing forests, and perhaps, this may be a prelude to including them in a post 2012 emission reduction regime.

Meanwhile other multilateral treaties are looking at how they may be able to assist in the climate change challenge. The Montreal Protocol, the agreement to protect and restore the ozone layer, has already contributed greatly to mitigate climate change by the phasing out of chlorflurocarbons (CFCs) and other controlled ozone depleting substances.

Governments which met in Montreal in September, just days before the UN Secretary-General Ban Ki Moon hosted a heads of state meeting on climate change in New York, agreed to an accelerated freeze and phase-out of chemicals known as HCFCs specifically for their global warming impacts, and also, for their ozone layer-damaging potential.

This was a decision taken under a separate treaty—the Montreal Protocol—but one that supports the aims and objectives of the Kyoto Protocol.

All this represents good news but it is not the final chapter in the history of the Kyoto Protocol and the climate change story.

Despite all the momentum, we still do not have a truly global deal on emission reductions beyond Kyoto which expires in 2012.

However, the climate convention meeting in Bali, at the end of 2007, produced a 'Road Map' to be concluded in Copenhagen in 2009. If this can be followed without too many detours and dead ends, it is possible that the geopolitical landscape on climate change will evolve into something quite radically different to what we see today.

Let us hope this leads us to a point in, perhaps 2009, when a successor to Kyoto can be in place that delivers deep and meaningful emission reductions so urgently needed and so manifestly possible.

Since being agreed in 2007, there have been many quick and eager to bury the Kyoto Protocol on the grounds that is more dead than alive, or, was a failed agreement based on even more flawed science.

But that did not happen, the world community has not only signed but ratified Kyoto, bought into being completely novel and operational markets that are starting to have the desired impacts and scientific debate has been put to bed.

Ten Years and still Counting for sure—but also counting a mounting level of positive developments that bode well for the next ten years.

Achim Steiner
Executive Director

Preface

As the tenth of anniversary of the Kyoto Protocol was approaching the first thing which came to my mind was "ten years and still counting" - when we should be talking about the results achieved and the progress made we are still defining the targets to be achieved. This prompted me to go for another book as a sequel to the one produced on the fifth anniversary of the Kyoto Protocol. Just like the long drawn and complex multi-lateral negotiations of the Protocol itself, it has been a challenging task to put together over fifty chapters by approximately seventy authors from nearly forty countries. In the process (just like the protocol negotiations) some authors came on board while others left-although the effort was to include the perspective from all major players but Russian, Indian and Chinese points of views are not that well represented as I would have liked to. Having said that it has been an interesting process to communicate with all the different authors (from different countries and different continents - some of whom I have never met) and in the end producing this interesting collection of chapters (which I must agree would not have been possible without full cooperation from the authors as well as from the publisher). In this long drawn process there have been a few roadblocks such as a car accident I was involved in, which has changed my perspective on life, and how things do not always happen as planned. Although the plan was to launch this book in 2007, one thing or the other delayed the project but it will finally see the light of the day in 2008 and I am sure my colleagues from different disciplines like academia, policy makers, professionals and scientists will find it a useful book.

Everyone is getting a taste of global warming by extreme weather conditions due to climate change (even if it is a cyclical phenomenon). Global warming is shifting rainfall patterns, causing extended drought in some areas and excessive rainfall in others. The weather conditions thus vary from severe droughts in Africa leading to food insecurity to intensive hurricane season causing havoc and devastation even in one of the most developed countries, the USA. As the climate or weather is changing differently in different parts of the world, so is the impact being felt/observed is different in different parts of the world. In some places global warming will initially boost agriculture productivity (while in others it will lead to droughts and loss of agriculture leading to food security issues), in some regions it will reduce demand for energy for heating in winter (as the temperatures are expected to rise) while in other regions demand for cooling in summers will increase (as temperatures get unbearable). The main issue is inability of our society to deal with these extreme conditions making it a natural disaster. For example, the summer of 2003 was extremely hot in Europe. Prolonged and intense heat led to about 22,000 – 35,000 deaths (mainly elderly) and an economic loss of US $ 13 billion.

The socio-economic impact of climate change on tourism, freshwater availability and quality, aquaculture, human settlements and human health will be immense, devastating and negative. A rise of average sea level by one meter, when superimposed on storm surges, could easily submerge low-lying islands. Some small islands are concerned that their entire culture, and perhaps the lives of their citizens, are at a risk. As discussed by Kolbert (in the second chapter of the book) change in temperature, moisture and seasonality has a big impact on plant species (based on a study done by Webb), some of them survive all the changes and adapt to the changes while others perish. All of this also influences eco-systems and the services they offer. Climate change stress on organisms and ecosystems is thus providing opportunities to invasive species and disease vectors, which could not survive in earlier climate, putting a strain on capacity and limits on the system in place to deal with it.

This book combines philosophical approach to climate change (including a development debate and a discussion on need for equitable approach to climate change negotiations), with scientific facts and its impact on human health. Adaptation is one of the important issues and its impact on human health is equally important issue – a number of chapters address these issues in the book. At least two chapters in the book have described impact of climate change on human health in the Arctic area of Northern (Canada) and another one near equator (Cuba) – in both cases temperature change is changing the local dynamics and health issues in the regions. Some authors have tried to illustrate how to deal with climate change (under Kyoto implementation mechanisms) such as land use changes.

There is also a section devoted to gender issues and lack of female involvement in the discussions of climate change. The book also looks at the Kyoto Protocol (legal challenges; policies and national, international institutional structures required), way forward from Bali and beyond

2012. It has not been easy to put such a book together because of such diversified topics, but I feel satisfaction in having achieved such a diversified and not so easy task successfully.

Now as we stand at the crossroads, the decision made by our National Governments, business communities and others involved in the negotiations will determine which path we will take as a world community, to mitigate and to adapt to climate change. As discussed in the book, in the coming years it is more politics than policies that will determine the way forward to climate change negotiations, Kyoto protocol and beyond regime. I am sure I will have the honour to produce another report in another five years time to see which path we follow from this crossroads on our way to Copenhagen (and probably commitments till 2020) and how far have we succeeded. Do not hesitate to contact me with your suggestions and ideas.

Velma I. Grover
Hamilton, ON

Contents

Volume 2

Section V: Legal Issues

Section VI: Impact of Climate Change and/or Kyoto (non) Implementation of Different Regions or Countries

Section VII: Gender and Climate Change

Section VIII: Safe Landing: Protecting the Climate for Future Generations and Health

Section IX: Beyond 2012: The Next Phase

Section X: Kyoto Protocol: Bali and Beyond

List of Contributors

Acosta-Michlik, Lilibeth

Département de Géologie et de Géographie, Université Catholique de Louvain (UCL), Place Louis Pasteur 3, B-1348 Louvain-la-Neuve, Belgium.

Aerts, Jeroen

Institute for Environmental Studies, Vrije Unversiteit, Amsterdam, De Boelelaan 1087, 1081 HV Amsterdam, The Netherlands.

Ali, Saleem H.

University of Vermont and Brown University, George D. Aiken Center, 81 Carrigan Drive, Burlington, Vermont, USA.

Baumert, Kevin A.

World Resources Institute, 9106 Woodland Drive, Silver Spring, MD 20910, USA.

Birkel, Kathrin

Vakgroep Milieu & Beleid, Department of Political Sciences of the Environment, Nijmegen University, P.O. Box 9108, 6500 HK Nijmegen, The Netherlands.

Bodansky, Daniel

University of Georgia, School of Law, Athens, GA 30602, USA.

Bressers, Hans Th.A.

Centre for Clean Technology and Environmental Policy, University of Twente, PO Box 217, 7500 AE Enschede, The Netherlands.

Bultó, Paulo Lázaro Ortíz

Climate Center, Meteorological Institute, Havana, Cuba.

Busby, Joshua W.

Lyndon B. Johnson School of Public Affairs, The University of Texas at Austin, P.O. Box Y, Austin, TX 78713-8925, USA.

Chen, Deliang

Earth Science Centre, Göteborg University, PO Box 460, 405 30 Göteborg, Sweden.

Chen, Yibing

Soil and Fertilizer Institute, Sichuan Academy of Agricultural Sciences, 20 Jingjusi Road, Chengdu, Sichuan 610066, P.R. China.

Chunling, Liu

Graduate School of Geography, Clark University, 950 Main Street, Worcester, MA 01610, USA.

Coelho, Suani Teixeira

Av-Prof. Luciano Gualberta 1289-05508-010, Sao Paulo/SP-Brazil.

Crête, Jean

Department of Political Science, Université Laval, Québec G1K 7P4, Québec, Canada.

Crowley, Paul

Senior attorney, Climate Law and Policy Project, Washington D.C., USA.

Daibes-Murad, Fadia

Palestinian Authority, P.O. Box 51875, East Jerusalem.

Erion, Graham

512 Whitmore Ave., Toronto, ON, M6E 2N8, Canada.

Estela, Luis Lecha

Centre for Environmental Research and Services of Villa Clara, Cuba.

Furgal, Christopher

Trent University, Environment and Resource Studies Program, 1600 West Bank Drive, Peterborough ON, Canada K9J7B8.

Goldberg, Donald M.

Climate Law & Policy Project, Washington D.C., USA.

Gosselin, Pierre

Laval University, Québec, Canada G1K 7P4.

Grover, Velma I.

United Nations University–International Network on Water, Environment & Health 916-981 Main St. West Hamilton, ON, L8S 1A8 Canada.

Guardabassi, Patricia

Av-Prof. Luciano Gualberta 1289-05508-010, Sao Paulo/SP-Brazil.

Gupta, Joyeeta

De Boelelaan 1087, 1081 HV Amsterdam.

Hamilton, Paul

Department of Political Science, Brock University, St Catharines, Ontario, Canada L2S-3A1.

Hanschel, Dirk

Senior Research Assistant, Chair of German and Comparative Public Law, International Law and European Law, University of Mannheim, Germany.

Hemmati, Minu

Ansbacher Str. 45, 10777 Berlin, Germany.

Hoffman, John S.

President, WorkSmart Energy Enterprises Inc., Washington DC 20008, USA.

Hotimsky, Samy

Rva Estevão de Almeida 74 apt 31, São Paulo-SP/05014-010 Brasil.

Huang, Jingjing

Soil and Fertilizer Institute, Sichuan Academy of Agricultural Sciences, 20 Jingjusi Road, Chengdu, Sichuan 610066, P.R. China.

Huitema, Dave

Institute for Environmental Studies, Vrije Universiteit, Amsterdam, De Boelelaan 1087, 1081 HV Amsterdam, The Netherlands.

Hussainy, Syed U.

Institute for Sustainability and Innovation, Werribee Campus, Victoria University, PO Box 14428, Melbourne City, MC 8001, Australia.

Kanie, Norichika

Associate Professor, Department of Value and Decision Science, Tokyo Institute of Technology, Tokyo, Japan.

Kelkar, Ulka

The Energy and Resources Institute (TERI), Darbari Seth Block, IHC Complex, Lodhi Road, New Delhi-110 003, India.

Kolbert, Elizabeth

Staff writer, The New Yorker, 326 Oblong Rd., Williamstown, MA 01267, USA.

Kubursi, Atif
Economics Dept., 1280 Main St W, McMaster University, Hamilton, ON, Canada.

Kumar, Santosh
School of Computer Science and Mathematics, Victoria University, Footscray Park Campus, PO Box 14428, Melbourne City, MC 8001, Australia.

Leone, Roberto P.
Department of Political Science, McMaster University, 1280 Main St. W., Hamilton, ON, L8S 4M4, Canada.

Lin, Chaowen
Soil and Fertilizer Institute, Sichuan Academy of Agricultural Sciences, 20 Jingjusi Road, Chengdu, Sichuan 610066, P.R. China.

Liptow, Holger
Director Energizing Africa, Deutsche Gesellschaft fur Technische Zusammenarbeit (GTZ), Postfach 5180, 65726 Eschborn, Germany.

Lucon, Oswaldo
Av-Prof. Luciano Gualberta 1289-05508-010, Sao Paulo/SP Brazil.

Martins, Osvaldo Stella
Av-Prof. Luciano Gualberta 1289-05508-010, Sao Paulo/SP Brazil.

McBean, G.A.
Institute for Catastrophic Loss Reduction, Departments of Geography and Political Science, The University of Western Ontario, 1491 Richmond Street, London, ON, N6G 2M1, Canada.

Mercier, Jean
Department of Political Science, Université Laval, Québec G1K 7P4, Québec, Canada.

Montiel, Alcira Noemí Perlini
Av. Montes de Oca 1408 5°B (1271), Buenos Aires, Argentina.

Ostwald, Madelene
Earth Science Centre, Göteborg University, PO Box 460, 405 30 Göteborg, Sweden.

Philibert, Cédric
Principal Administrator, Energy Efficiency and Environment Division, International Energy Agency, 9 rue de la Féddération, 75739 Paris, Cedex 15, France.

Read, Peter

Massey University, Palmerston North 4442, New Zealand.

Rei, Fernando

Av-Prof. Luciano Gualberta 1289-05508-010, Sao Paulo/SP Brazil.

Riedacker, Arthur

INRA Unité Mona, 63 Bd de Brandebourg, 94205 Ivry Cedex, France.

Robbins, Mike

School of Development Studies, University of East Anglia, Norwich, NR4 7TJ, UK.

Rodríguez, Antonio Pérez

Tropical Medicine Institute "Pedro Kouri" (IPK). Havana, Cuba.

Röhr, Ulrike

Genanet - focal point gender justice and sustainability, LIFE e.V., Dircksenstr. 47, D-10178 Berlin, Germany.

Rosales, Jon

Assistant Professor, Environmental Studies, St. Lawrence University, 23 Romoda Drive, Canton, NY 13617, USA.

Salamat, Mohammad Reza

Officer-in-Charge, Energy and Transport Branch, Division for Sustainable Development, Department of Social and Economic Affairs (DESA), United Nations, DC2-2224, New York, NY 10017, USA.

Scheffran, Jürgen

Adjunct Associate Professor, Political Science and Atmospheric Sciences, ACDIS, University of Illinois, 359 Armory Building, MC 533, 505 East Armory Ave., Champaign, IL 61820, USA.

Schönfeld, Martin

Philosophy FAO 226, University of South Florida, Tampa, FL 33620 USA.

Skutsch, Margaret M.

Technology and Sustainable Development Group, Centre for Clean Technology and Environmental Policy, University of Twente, PO Box 217, 7500 AE Enschede, The Netherlands.

Stokes, Alexia

INRA, AMAP, TA-A51/PS2 Boulevard de la Lironde, 34398 Montpellier Cedex 5, France.

Tladi, Dire

PO Box 13139, Hatfield, Pretoria, 0028, South Africa.

Valencia, Alina Rivero

Climate Center, Institute of Meteorology, Havana, Cuba.

Van Asselt, Harro

Institute for Environmental Studies, Vrije Universiteit, Amsterdam, De Boelelaan 1087, 1081 HV Amsterdam, The Netherlands.

Van Laake, Patrick E.

Department of Natural Resources, International Institute for Geo-information Science and Earth Observation (ITC), P.O. Box 6, 7500 AA Enschede, The Netherlands.

Vézeau, Nicolas

Université Laval, Québec, Canada, G1K7P4.

Verplanke, Jeroen J.

Department of Urban Planning and Geo-information Management International Institute for Geo-information Science and Earth Observation (ITC), P.O. Box 6, 7500 AA Enschede, The Netherlands.

Wagner, Martin

Director and Managing attorney International Program Earthjustice Oakland, CA, USA.

Willace, Simon

C/o 48 Second Ave, Fortsville SA 5035, Australia.

Section V

Legal Issues

28

CHAPTER

Petition to the Inter-American Commission on Human Rights Seeking Relief from Violations Resulting from Global Warming Caused by Acts and Omissions of the United States

Martin Wagner[1], Paul Crowley[2] and Donald M. Goldberg[3]

[1]Director and Managing Attorney, International Program, Earthjustice, Oakland, CA, USA

[2]Senior Attorney, Climate Law & Policy Project, Washington D.C., USA

[3]Executive Director, Climate Law & Policy Project, Washington D.C., USA

SUMMARY OF THE PETITION

In this petition, Sheila Watt-Cloutier, an Inuk woman and Chair of the Inuit Circumpolar Conference, requests the assistance of the Inter-American Commission on Human Rights in obtaining relief from human rights violations resulting from the impacts of global warming and climate change caused by acts and omissions of the United States. Ms. Watt-Cloutier submits this petition on behalf of herself, 62 other named individuals, and all the Inuit of the arctic regions of the United States of America and Canada who have been affected by the impacts of climate change described in this petition.

Global warming refers to an average increase in the Earth's temperature, causing changes in climate that lead to a wide range of adverse impacts on plants, wildlife, and humans. There is broad scientific consensus that global warming is caused by the increase in concentrations of greenhouse gases in the atmosphere as a result of human activity. The United States is, by any measure, the world's largest emitter of greenhouse gases, and thus bears the greatest responsibility among nations for causing global warming.

The Inuit, meaning 'the people' in their native Inuktitut, are a linguistic and cultural group descended from the Thule people whose traditional range spans four countries - Chukotka in the Federation of Russia, northern and western Alaska in the United States, northern Canada, and Greenland. While there are local characteristics and differences within the broad ethnic category of 'Inuit,' all Inuit share a common culture characterized by dependence on subsistence harvesting in both the terrestrial and marine environments, sharing of food, travel on snow and ice, a common base of traditional knowledge, and adaptation to similar arctic conditions. Particularly since the World War II, the Inuit have adapted their culture to include many western innovations, and have adopted a mixed subsistence- and cash-based economy. Although many Inuit are engaged in wage employment, the Inuit continue to depend heavily on the subsistence harvest for food. Traditional 'country food' is far more nutritious than imported 'store-bought' food. Subsistence harvesting also provides spiritual and cultural affirmation, and is crucial for passing skills, knowledge and values from one generation to the next, thus ensuring cultural continuity and vibrancy.

Like many indigenous people, the Inuit are the product of the physical environment in which they live. The Inuit have fine-tuned tools, techniques and knowledge over thousands of years to adapt to the arctic environment. They have developed an intimate relationship with their surroundings, using their understanding of the arctic environment to develop a complex culture that has enabled them to thrive on scarce resources. The culture, economy and identity of the Inuit as an indigenous people depend upon the ice and snow.

Nowhere on Earth has global warming had a more severe impact than the Arctic. Building on the 2001 findings of the Intergovernmental Panel on Climate Change, the 2004 Arctic Climate Impact Assessment - a comprehensive international evaluation of arctic climate change and its impacts undertaken by hundreds of scientists over four years - concluded that:

> The Arctic is extremely vulnerable to observed and projected climate change and its impacts. The Arctic is now experiencing some of the most rapid and severe climate change on Earth. Over the next 100 years, climate change is expected to

accelerate, contributing to major physical, ecological, social, and economic changes, many of which have already begun.

As the annual average arctic temperatures are increasing more than twice as fast as temperatures in the rest of the world, climate change has already caused severe impacts in the Arctic, including deterioration in ice conditions, a decrease in the quantity and quality of snow, changes in the weather and weather patterns, and a transfigured landscape as permafrost melts at an alarming rate, causing slumping, landslides, and severe erosion in some coastal areas. Inuit observations and scientific studies consistently document these changes. For the last 15 to 20 years, Inuit, particularly hunters and elders who have intimate knowledge of their environment, have reported climate-related changes within a context of generations of accumulated traditional knowledge.

One of the most significant impacts of warming in the Arctic has been on sea ice. Commonly observed changes include thinner ice, less ice, later freezes and earlier, more sudden thaws. Sea ice is a critical resource for the Inuit, who use it to travel to hunting and harvesting locations, and for communication between communities. Due to the loss in the thickness, extent and duration of the sea ice, these traditional practices have become more dangerous, more difficult or, at times, impossible. In many regions, traditional knowledge regarding the safety of the sea ice has become unreliable. As a result, more hunters and other travellers are falling through the sea ice into the frigid water below. The shorter season for safe sea ice travel has also made some hunting and harvest activities impossible, and curtailed others. For the Inuit, the deterioration in sea ice conditions has made travel, harvest, and everyday life more difficult and dangerous.

The quality, quantity and timing of snowfall have also changed. Snow generally falls later in the year, and the average snow cover over the region has decreased ten per cent over the last three decades. The spring thaw comes earlier and is more sudden than in the past. As with decreased ice, the shorter snow season has made travel more difficult. In addition, the deep, dense snow required for igloo building has become scarce in some areas, forcing many travellers to rely on tents, which are less safe, much colder and more cumbersome than igloos. The lack of igloo-quality snow can be life threatening for travellers stranded by unforeseen storms or other emergencies. These changes have also contributed to the loss of traditional igloo building knowledge, an important component of Inuit culture.

Permafrost, which holds together unstable underground gravel and inhibits water drainage, is melting at an alarming rate, causing slumping, landslides, severe erosion and loss of ground moisture, wetlands and lakes. The loss of sea ice, which dampens the impact of storms on coastal areas, has resulted in increasingly violent storms hitting the coastline, exacerbating erosion and flooding. Erosion in turn exposes coastal permafrost to warmer air and water, resulting in faster permafrost melts. These transformations have had a devastating impact on some coastal communities, particularly in Alaska and the Canadian Beaufort Sea region. Erosion, storms, flooding and slumping harm homes, infrastructure, and communities, and have damaged Inuit property, forcing relocation in some cases and requiring many communities to develop relocation contingency plans. In addition, these impacts have contributed to decreased water levels in rivers and lakes, affecting natural sources of drinking water, and habitat for fish, plants, and game on which Inuit depend.

Other factors have also affected water levels. Changes in precipitation and temperature have led to sudden spring thaws that release unusually large amounts of water, flooding rivers and eroding their streambeds. Yet, after spring floods, rivers and lakes are left with unusually low levels of water further diminished by increased evaporation during the longer summer. These changes affect the availability and quality of natural drinking water sources. The fish stocks upon which Inuit rely are profoundly affected by changing water levels. Fish sometimes can not reach their spawning grounds, their eggs are exposed or washed ashore, or northward moving species compete with the native stocks for ecological niches.

The weather has become increasingly unpredictable. In the past, Inuit elders could accurately predict the weather for coming days based on cloud formations and wind patterns, allowing the Inuit to schedule safe travel. The changing climate has made clouds and wind increasingly erratic and less useful for predicting weather. Accurate forecasting is crucial to planning safe travel and hunting. The inability to forecast has resulted in hunters being stranded by sudden storms, trip cancellations, and increased anxiety about formerly commonplace activities.

Observers have also noted changes in the location, characteristics, number, and health of plant and animal species caused by changes in climate conditions. Some species are less healthy. In the words of the Arctic Climate Impact Assessment, "[m]arine species dependent on sea ice, including polar bears, ice-living seals, walrus, and some marine birds, are very likely to decline, with some facing extinction."

Other species are becoming less accessible to the Inuit because the animals are moving to new locations, exacerbating the travel problems resulting from climate change. Still others cannot complete their annual migrations because the ice they travel on no longer exists, or because they cannot cross rivers swollen by sudden floods. More frequent autumn freeze-thaw cycles have created layers of solid ice under the snow that makes winter foraging more difficult for some game animals, including caribou, decreasing their numbers and health. These impacts on animals have impaired the Inuit's ability to subsist.

Increased temperatures and sun intensity have heightened the risk of previously rare health problems such as sunburn, skin cancer, cataracts, immune system disorders and heatrelated health problems. Warmer weather has increased the mortality and decreased the health of some harvested species, impacting important sources of protein for the Inuit. Traditional methods of food and hide storage and preservation are less safe because of increased daytime temperatures and melting permafrost.

The current impacts in the Arctic of climate change are severe, but projected impacts are expected to be much worse. Using moderate - not worst case - greenhouse gas emission scenarios, the Arctic Climate Impact Assessment finds that:

- "Increasing global concentrations of carbon dioxide and other greenhouse gases due to human activities, primarily fossil fuel burning, are projected to contribute to additional arctic warming of about 4-7°C, about twice the global average rise, over the next 100 years."
- "Increasing precipitation, shorter and warmer winters, and substantial decreases in snow and ice cover are among the projected changes that are very likely to persist for centuries."
- "Unexpected and even larger shifts and fluctuations in climate are also possible."
- "Reductions in sea ice will drastically shrink marine habitat for polar bears, ice-inhabiting seals, and some seabirds, pushing some species toward extinction."
- "Caribou/reindeer and other animals on land are likely to be increasingly stressed as climate warming alters their access to food sources, breeding grounds, and historic migration routes."
- "Species ranges are projected to shift northward on both land and sea, bringing new species into the Arctic while severely limiting some species currently present."
- "As new species move in, animal diseases that can be transmitted to humans, such as West Nile Virus, are likely to pose increasing health risks."
- "Severe coastal erosion will be a growing problem as rising sea level and a reduction in sea ice allow higher waves and storm surges to reach shore."

- "Along some Arctic coastlines, thawing permafrost weakens coastal lands, adding to their vulnerability."
- "The risk of flooding in coastal wetlands is projected to increase, with impacts on society and natural ecosystems."
- "In some cases, communities and industrial facilities in coastal zones are already threatened or being forced to relocate, while others face increasing risks and costs."
- "Many indigenous peoples depend on hunting polar bear, walrus, seals, and caribou, herding reindeer, fishing, and gathering, not only for food and to support the local economy, but also as the basis for cultural and social identity."
- "Changes in species' ranges and availability, access to these species, a perceived reduction in weather predictability, and travel safety in changing ice and weather conditions present serious challenges to human health and food security, and possibly even the survival of many cultures."

Noting the particular impact these changes will have on the Inuit, the ACIA states: "For Inuit, warming is likely to disrupt or even destroy their hunting and food sharing culture as reduced sea ice causes the animals on which they depend on to decline become less accessible, and possibly become extinct."

Several principles of international law guide the application of the human rights issues in this case. Most directly, the United States is obligated by its membership in the Organization of American States and its acceptance of the American Declaration of the Rights and Duties of Man to protect the rights of the Inuit described above. Other international human rights instruments give meaning to the United States' obligations under the Declaration. For example, as a party to the International Covenant on Civil and Political Rights ("ICCPR"), the United States is bound by the principles therein. As a signatory to the International Covenant on Economic, Social, and Cultural Rights ("ICESCR"), the United States must act consistently with the principles of that agreement.

The United States also has international environmental law obligations that are relevant to this petition. For instance, the United States also has an obligation to ensure that activities within its territory do not cause transboundary harm or violate other treaties to which it is a party. As a party to the UN Framework Convention on Climate Change, the United States has committed to developing and implementing policies aimed at returning its greenhouse gas emissions to 1990 levels. All of these international obligations are relevant to the application of the rights in the American Declaration because, in the words of the Inter-American

Commission, the Declaration "should be interpreted and applied in context of developments in the field of international human rights law … and with due regard to other relevant rules of international law applicable to [OAS] member states."

The impacts of climate change, caused by acts and omissions by the United States, violate the Inuit's fundamental human rights protected by the American Declaration of the Rights and Duties of Man and other international instruments. These include their rights to the benefits of culture, to property, to the preservation of health, life, physical integrity, security, and a means of subsistence, and to residence, movement, and inviolability of the home.

As the Inuit culture is inseparable from the condition of their physical surroundings, the widespread environmental upheaval resulting from climate change violates the Inuit's right to practice and enjoy the benefits of their culture. The subsistence culture central to Inuit cultural identity has been damaged by climate change, and may cease to exist if action is not taken by the United States in concert with the community of nations.

The Inuit's fundamental right to use and enjoy their traditional lands is violated as a result of the impacts of climate change because large tracks of Inuit traditional lands are fundamentally changing, and still other areas are becoming inaccessible. Summer sea ice, a critical extension of traditional Inuit land, is literally ceasing to exist. Winter sea ice is thinner and unsafe in some areas. Slumping, erosion, landslides, drainage, and more violent sea storms have destroyed coastal land, wetlands, and lakes, and have detrimentally changed the characteristics of the landscape upon which the Inuit depend. The inability to travel to lands traditionally used for subsistence and the reduced harvest have diminished the value of the Inuit's right of access to these lands.

The Inuit's fundamental right to enjoy their personal property is violated because climate change has reduced the value of the Inuit's personal effects, decreasing the quality of food and hides, and damaging snowmobiles, dog sleds and other tools. Their right to cultural intellectual property is also violated, because much of the Inuit's traditional knowledge, a formerly priceless asset, has become frequently unreliable or inaccurate as a result of climate change.

The Inuit's fundamental rights to health and life are violated as climate change exacerbates pressure on the Inuit to change their diet, which for millennia has consisted of wild meat and a few wild plants. Climate change is accelerating a transition by Inuit to a more western store-bought diet with all of its inherent health problems. Life-threatening accidents are increasing because of rapid changes to ice, snow, and land. Traditional food preservation methods are becoming difficult to practice safely.

Natural sources of drinking water are disappearing and diminishing in quality. Increased risks of previously rare heat and sun related illnesses also implicate the right to health and life.

The Inuit's fundamental rights to residence and movement, and inviolability of the home are likewise violated as a result of the impacts of climate change because the physical integrity of Inuit homes is threatened. Most Inuit settlements are located in coastal areas, where storm surges, permafrost melt, and erosion are destroying certain coastal Inuit homes and communities. In inland areas, slumping and landslides threaten Inuit homes and infrastructure.

The Inuit's fundamental right to their own means of subsistence has also been violated as a result of the impacts of climate change. The travel problems, lack of wildlife, and diminished quality of harvested game resulting from climate change have deprived the Inuit of the ability to rely on the harvest for year-round sustenance. Traditional Inuit knowledge, passed from Inuit elders in their role as keepers of the Inuit culture, is also becoming outdated because of the rapidly changing environment.

The United States of America, currently the largest contributor to greenhouse emissions in the world, has nevertheless repeatedly declined to take steps to regulate and reduce its emissions of the gases responsible for climate change. As a result of well-documented increases in atmospheric concentrations of greenhouse gases, it is beyond dispute that most of the observed change in global temperatures over the last 50 years is attributable to human actions. This conclusion is supported by a remarkable consensus in the scientific community, including every major US scientific body with expertise on the subject. Even the Government of the United States has accepted this conclusion.

However, and notwithstanding its ratification of the UN Framework Convention on Climate Change, United States has explicitly rejected international overtures and compromises, including the Kyoto Protocol to the U.N. Framework Convention on Climate Change, aimed at securing agreement to curtail destructive greenhouse gas emissions. With full knowledge that this course of action is radically transforming the arctic environment upon which the Inuit depend for their cultural survival, the United States has persisted in permitting the unregulated emission of greenhouse gases from within its jurisdiction into the atmosphere.

Protecting human rights is the most fundamental responsibility of civilized nations. Because climate change is threatening the lives, health, culture and livelihoods of the Inuit, it is the responsibility of the United States, as the largest source of greenhouse gases, to take immediate and effective action to protect the rights of the Inuit.

As this petition raises violations of the American Declaration of the Rights and Duties of Man by the United States of American, the Inter-American Commission on Human Rights has jurisdiction to receive and consider it. The petition is timely because the acts and omissions of the United States that form the basis for the petition are ongoing, and the human rights violations they are causing is increasing. Since there are no domestic remedies suitable to address the violations, the requirement that domestic remedies be exhausted does not apply in this case.

The violations detailed in the petition can be remedied. As such, the Petitioner respectfully requests that the Commission:

1. Make an onsite visit to investigate and confirm the harms suffered by the named individuals whose rights have been violated and other affected Inuit;
2. Hold a hearing to investigate the claims raised in this Petition;
3. Prepare a report setting forth all the facts and applicable law, declaring that the United States of America is internationally responsible for violations of rights affirmed in the American Declaration of the Rights and Duties of Man and in other instruments of international law, and recommending that the United States:
    a. Adopt mandatory measures to limit its emissions of greenhouse gases and cooperate in efforts of the community of nations - as expressed, for example, in activities relating to the United Nations Framework Convention on Climate Change - to limit such emissions at the global level;
    b. Take into account the impacts of U.S. greenhouse gas emissions on the Arctic and affected Inuit in evaluating and before approving all major government actions;
    c. Establish and implement, in coordination with Petitioner and the affected Inuit, a plan to protect Inuit culture and resources, including, *inter alia*, the land, water, snow, ice, and plant and animal species used or occupied by the named individuals whose rights have been violated and other affected Inuit; and mitigate any harm to these resources caused by US greenhouse gas emissions;
    d. Establish and implement, in coordination with Petitioner and the affected Inuit communities, a plan to provide assistance necessary for Inuit to adapt to the impacts of climate change that cannot be avoided;
    e. Provide any other relief that the Commission considers appropriate and just.

Section VI

Impact of Climate Change and/or Kyoto (non) Implementation of Different Regions or Countries

29

CHAPTER

Responding to Climate Change and Its Impact on Water Resources: A Case Study from the Middle East

Fadia Daibes-Murad
Palestinian Authority
P.O. Box 51875, East Jerusalem

INTRODUCTION

The area of the Middle East (ME) within this chapter encompasses those countries that have significantly interconnected water resources and that do not yet have alternative water sources such as desalination and wastewater reuse. That is: Lebanon, Syria, Iraq, Israel, Jordan, and the Palestine (West Bank and Gaza Strip including East Jerusalem).

Water availability is a major concern in most countries of the ME region, which is particularly sensitive to droughts, that occur approximately every 10 years with very low water input. Although the effects and extent of climate change are as yet uncertain and cannot be easily quantified or foreseen, there is however, a certain consensus regarding a presumed increase in climate contrast. Climatic change combined with population growth and the conflicting interests in the social and economic uses of shared water resources will increase the pressure on the available water resources and may cause social instability in the area. The impacts on the water resources themselves will affect surface and groundwater supplies for all beneficial uses.

Also in the ME, the quality of knowledge in the field of water availability, climate trends and the impact of climate change on them, is mainly dependent on the availability of historical data sets and therefore on the continuity of monitoring and data collection. In the case of Palestine, although meteorological monitoring and data collection started early in the 19th century, hydrological and hydrogeological monitoring is fairly recent. Uptodate, the Palestinians have limited access to water-related data, and the possibilities to conduct relevant research is fairly limited, due to the political control over the area by Israel. Furthermore, in the areas where the Palestinians have control, there seems to be no direct relation between the data sets in the meteorological departments and how they relate to the water-related information and data. These institutions lack the vision for coordination and cooperation and therefore their efforts are not complimentary.

This chapter endeavours to identify the major challenges facing managers, policy and decision makers in the field of water with regard to climate change. It addresses the need for developing capacities to consider how these global change processes will affect the water resources availability in Palestine, how to minimize adverse impacts, how to improve monitoring and research, and how to facilitate all such actions through coordination, sharing of information, and cooperation in a wider regional context. This will ultimately lead to decreasing uncertainties and to improving the scientific basis for decision-making with regard to the global climate change. The chapter confirms the need for an integrated approach in managing water resources and in responding to the effects of characteristics, better information and knowledge base which links meteorology and climatology with water-related monitoring and assessment.

CLIMATE AND CLIMATE CLASSIFICATION IN PALESTINE

The climate in Palestine is traditionally described as Mediterranean, which is characterized by winter rain and summer drought. However, there is great diversity in this climate which ranges from extremely arid to humid, according to the De Martonne aridity index classification for arid areas.[1]

[1]Dudeen, B. (2000). Soil Contamination in the West Bank and Gaza Strip – Palestine, Proceedings of the First Environmental Scientific Symposium, Hebron – Palestine, October, 2000. found at http://ressources.ciheam.org/om/pdf/b34/01002095.pdf#search=%22Basim%20Dudeen%20climate%20change%22 (*last visited September 2006*).

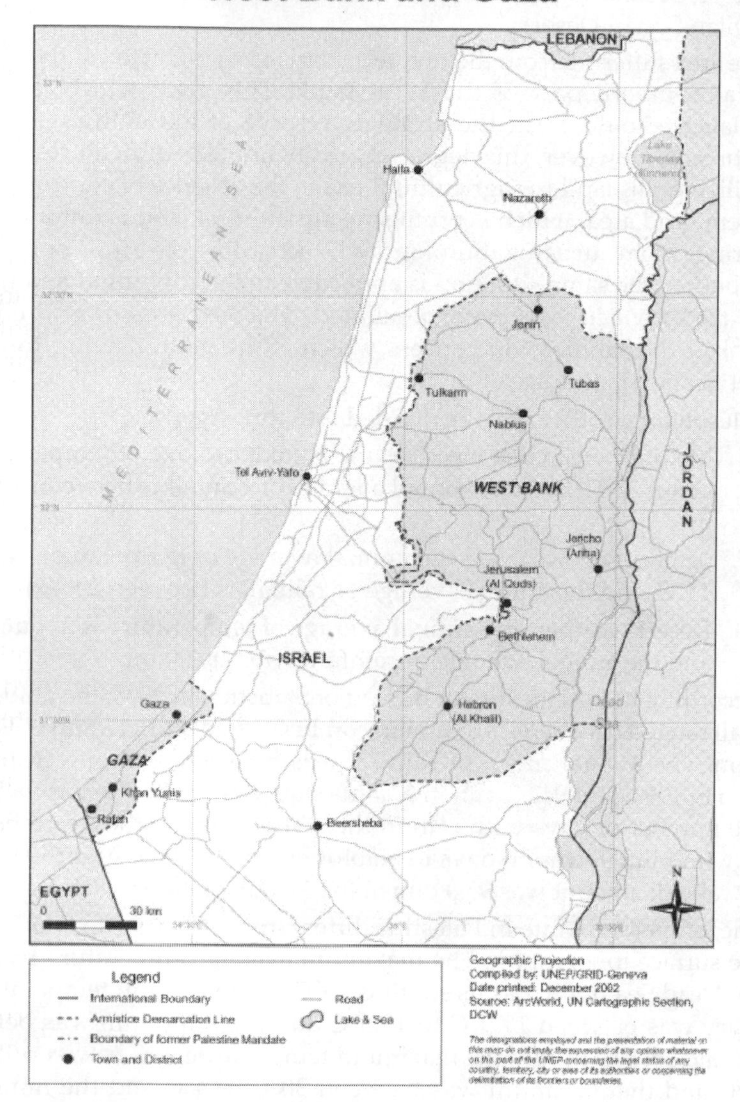

Fig. 1 General Location Map of the West Bank and Gaza Strip

Colour image of this figure appears in the colour plate section at the end of the book.

Palestine is one of the most ancient regions in the Middle East, it is located in the northern moderate site of the eastern coast of the Mediterranean Sea. Palestine's climate is affected by three factors: First, the mountain series which extends from the North to the South, parallel to the coast. Second is the Sinai and North Africa Desert and third, is the Jordanian-Syrian Desert.[2]

The area suffering from the severe aridity (44%) is located at the eastern and far southern part of the West Bank. This area, which is lightly populated, should form the strategic reserve of agricultural land for Palestinians. However, this degree of aridity imposes difficult restrictions on utilizing this land for agricultural use in the absence of control over it. The semi-arid area, which is promising agricultural land is unfortunately suffering from urbanization sprawl, according to the population distribution; the same situation is applicable to the subhumid and humid areas (26%) which is heavily populated. The vast area of arid climate provoked the salinization process which is the main driving force for desertification in this area.[3]

Palestinian weather may be divided into three types:

1. Mediterranean Sea weather: the annual average of temperature is about 22°C, and the annual average or rainfall is between 400-500 ml.
2. Semi-desert weather: the annual average of temperature is about 18°C, and the annual average or rainfall is between 200-350 ml.
3. Desert weather: the annual average of temperature is about 22°C, and the annual average or rainfall is about 200 ml.

According to the Palestinian Bureau of Statistics, in 2005, the amount of rainfall ranged between 790.5 mm in Nablus station and 117 mm in Jericho Station. The annual mean of rainfall was between 48.7 mm in Jericho station in 1999, and 942.7 mm in Nablus station in 2003. There is variation in the number of days with rain in the stations, the highest number of l days with rainfall was 56 days in Nablus station in 2005. And the highest daily rainfall amount was 95.0 mm in the Nablus station in February 2005.

The temperature rate in Palestine differs from one place to another due to the surface topography. The maximum average temperature recorded in the Dead Sea Basin and its south is 25°C, the maximum temperature in January was between 11.9°C to 19.5°C and the minimum was between 4,4°C and 39,4°C. In July the maximum temperature was between 19.5°C-11.9°C and the minimum was between 30°C-39.4°C and the minimum between 15.9°C-24°C.[4]

[2]Palestinian Bureau of Statistics Press Release, March 2006, found at http://www.pcbs.gov.ps/Portals/_pcbs/PressRelease/arsad23E.pdf#search=%22climate%20in%20Gaza%20temperature%22 (last visited September 2006)

[3]See supra note 1.

[4]Palestinian National Information Center, http://www.pnic.gov.ps/english/information/fact4.html (last visited September) 2006.

Main Improtant Meteorological Indicators in the Palestinian Territory by the Indicator and Station Location, 2005

Month	*Station location*							
	Jenin	*Meithalun*	*Tulkarm*	*Nablus*	*Ramaliah*	*Jericho*	*Hebron*	*Gaza*
Annual Mean of Air Temperature (C°)	20.3	20.5	23.1	18.0	16.5	23.4	16.7	21.0
Annual Mean of Maximum Air Temperature (C°)	25.6	25.3	26.3	22.9	20.8	30.3	21.0	23.6
Annual Mean of Minimum Air Temperature (C°)	16.0	12.9	16.0	14.3	13.3	16.2	12.3	17.7
Annual Rainfall Quantity (mm)	431.1	519.2	585.8	790.5	711.6	117.0	475.9	260.5
Total Number of Rainfall Days	55	50	51	56	47	36	46	38
Annual Mean of Relative Humidity (%)	65.1	59.3	60.3	60.2	68.1	52.5	60.0	65.6
Annual Evaporation Quantity (mm)	1,932.2	-	-	1,991.3	2,282.2	2,085.3	2,047.0	1,542.8

Source: Palestinian Bureau of Statistics Press Release, March 2006, found at http://www.pcbs.gov.ps/Portals/_pcbs/PressRelease/arsad23E.pdf#search=%22climate%20in%20Gaza%20temperature%22

THE WATER SITUATION IN PALESTINE

Palestine is a semi-arid area and is characterized by limited surface and groundwater water resources. Groundwater resources, however, are very important in this region, surface water resources are often unreliable, poorly distributed, and large amounts of its water are lost due to evaporation. The main surface water system in the region is the Jordan River Basin which begins in three headwaters. The Jordan River Basin is considered under international law as an international river with waters shared by Israel, Jordan, Syria, Lebanon and the Palestine.[5]

As Palestine depends mostly on groundwater, the main source of replenishment is precipitation. Recharge to groundwater aquifers may occur directly from rain, or indirectly from surface drainage following concentration by runoff. The degree of aridity of the area plays an important role in defining the type of recharge. The relation is one that as aridity increases, it is more likely that recharge from indirect sources is dominant. Recharge by the ephemeral and intermittent streams in these areas is very common.

As semi-arid and arid environments, rainfall is short-lived and often very intense. As soils tend to be thin, much of the rainfall runs directly off the surface, only to infiltrate into deeper soils down slope or along river beds. Also in semi-arid and arid areas, where groundwater recharge occurs after flood events, changes in the frequency and magnitude of rainfall events will alter the number of recharge events. Non-renewable aquifers are important water sources in many deserts; they will not be affected on a time scale relevant to humans.

In Palestine, the collection of meteorological data started in the 19th century. However, hydrological and hydrogeological measurements are more recent, and sensitive to political change. Israel, which is the power in control since the 1967 occupation of Gaza and the of the West Bank including East Jerusalem, held control over meteorological, hydrological and hydrogeological monitoring. Therefore, large historical data sets are either rare or unreliable on the Palestinian side.

Finally, almost all groundwater resources in Palestine are transboundary with Israel, a fact that complicates the task of developing and implementing a harmonized and comprehensive policy at the

[5]Shuval, H.I. 1996. Towards Resolving Conflicts Over Water between Israel and Its Neighbours: The Israeli Palestinian Shared Use of the Mountain Aquifer as a Case Study. In: J.A. Allan (ed.), Water, Peace and the Middle East: Negotiating Resources in the Jordan River Basin, (Tauris Academic Studies: London, UK 1996. See also Gvirtzman, H. 1994. Groundwater Allocation in Judea and Samaria. In: J. Issac and H. Shuval (eds.) Water and Peace in the Middle East (Elsevier: Amsterdam, The Netherlands).

Watersheds

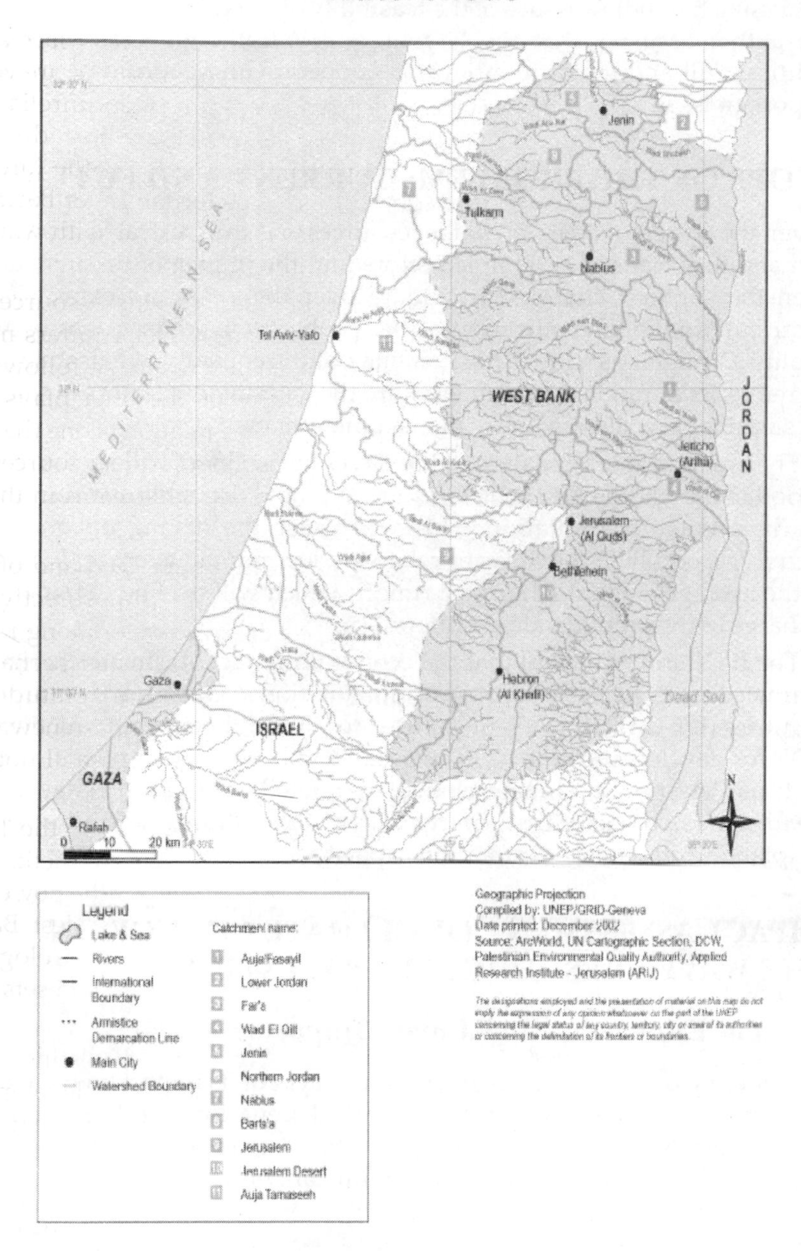

Fig. 2 Watersheds in Israel and Palestine

Colour image of this figure appears in the colour plate section at the end of the book.

international level. To that end, and in order to avoid the dilemma of discussing the endless issues of the feasibility of international cooperation, the author assumes that the Israelis and Palestinians shall have the political will and good intentions to cooperate and coordinate on this important issue.

SCOPE OF THE PROBLEM: CURRENT AND FUTURE

Given the current scarcity of water resources and the inevitable increased demand in social and economic sectors and the impact of urbanization, even the slightest changes in climate over the years are expected to contribute to the intensification of the problems in water quantity and quality. Droughts as natural phenomena occur frequently in Palestine and therefore; its duration and frequency are the determining factors in setting up sustainable water resources management plans.

The expected intensification of the Greenhouse Effect will spell a dryer period in the Middle East which results in longer droughts and declining yearly average precipitation. This will cause the drying up of small springs especially in the areas of relatively low rainfall average. In terms of the average amount of surface runoff, which will decline causing less recharge to the groundwater aquifers.

The IPCC in 1997 stated that the economic impact of climate change is estimated at 2 to 9 percent of annual national GDP for developing countries (IPCC-WGII 1997) in contrast to a 1 to 1.5 per cent reduction in GDP for developed countries. The greater vulnerability of poor regions, such as Palestine to climate change is related to its high reliance on weather-related activities, particularly agriculture, and the low adaptation and damage restoration capacity.[6]

IMPACT ASSESSMENT OF CLIMATE CHANGE ON THE WATER RESOURCES IN PALESTINE

Current Knowledge on Climate Impacts

Considering the climate change, the assessment of its impact on the water resources are mostly based on hypothetical scenarios and data obtained under controlled conditions. Thus predictions are rather more qualitative than quantitative and could contain significant bias.

---

[6]Watson, R.T., M.C. Zinyowera, R.H. Moss, and D.J. Dokken (eds). 1997. Intergovernmental Panel for Climate Change, Working Group II, Geneva.

Results correlated with historical data led to the conclusion that in the case of global warming, the Mediterranean region will become drier. This will involve the movement northward of the westerlies belt, while at the same time the Sahara belt will move northwards. Thus lower rates of precipitation and longer periods of droughts are the forecast for the countries bordering the Mediterranean, especially those in its southern domain. Various expected climatic changes in Palestine by 2020, such as the mean temperature increase (0.3-0.4°C), reduction in precipitation (-2 to -1%), increase in evapo-transpiration, increased rain intensity and shortened rainy season, greater temperature variability, etc. would have negative impacts on the quality and quantity of water resources. Also, it will reduce the water-holding capacity and soil permeability. Soil salinity will also increase because of high evapo-transpiration and the lower leaching effect of the reduced rain.

Also, in their general study of the climate change and water resources in the Middle East, Bou-Zeid and El Fadel (2001) realized the danger of using complicated hydrological models in cases where long-term rainfall-runoff data and basin water balance are either not available or unreliable. They resorted to using simulated data produced by the so called 'General Circulation Models' (GCM). Their results show minor changes in mean precipitation for the region, while temperatures are projected to increase in all seasons. Mean summer temperatures, already high in the region, will rise significantly (0.8-2.1°C). Areas bordering the Mediterranean (Lebanon, Israel, Palestinian Authority, coastal Syria) would be the least affected. They also demonstrated that groundwater aquifers in these areas will be under the hazard of increased seawater intrusion due to higher sea levels.

In another attempt to estimate regional impacts of climate change on Palestine, Al Kadi (2005)[7] conducted a study on Gaza temperature variations between 1975 and 1995. He concluded that the mean temperature in Gaza showed an upward trend in most of the months. This upward trend was very clear in the seasonal values, especially in autumn, spring and summer. The increase intensified in the mid-1980s onwards. The annual warming may be estimated to be 0.04°C/year, which is amounted to about 0.4°C from 1976 to1995: This warming is close to the estimated values of the Northern Hemisphere global warming. The projected rises in the mean temperature for the years 2005, 2010, and 2015 would be 0.72°C and 0.92°C and 1.12°C respectively. He also showed that

[7] Al Kadi, J. 2006. Global Warming: A Study of the Gaza Temperature Variations in the Period 1976-1995, found at http://www.iugaza.edu/ara/research/articles/D8%AF.%20%D8%A3%D8%AD%D9%85%D8%AF%20%D8%A7%D9%84%D9%82%D8%A7%D8%B6%D9 %8A.pdf#search=%22climate%20in%20Gaza%20temperature%22 (last visited September 2006)

the trend is clearly towards increasing mean temperatures. This will increase irrigation water demand due to higher evaporation. Extreme temperatures are predicted to increase more than mean temperature values. Al Kadi also concluded that increased temperature and evapo-transpiration coupled with constant precipitation are highly associated with desertification. Mean winter temperatures will also increase; however, the rise is lower than for the summer season. Higher winter temperatures will enhance evapo-transpiration and reduce potential groundwater recharge. If the increased runoff due to sharper precipitation patterns is also considered, the net effect will be a reduction in groundwater recharge and hence in the baseline renewable water resources.

The Need for Integrated Impact Assessment

In the future it is recommended that impacts assessment must follow an integrated approach, thus considering the fact that climate change is only one of many issues. Therefore, decision-making needs to consider climate change in conjunction with other issues affecting the same decision strategies. Adaptation to, and mitigation of, climate change are both necessary complementary strategies, so it may be advantageous to consider both in any integrated assessment. Integrated assessments are essential to gauge the benefits and possible clashes of interest to be identified. The choice of implementation strategies will be a combined outcome of the most beneficial and the most reasonable cost. This requires wide understanding of natural and human systems, and consultation with stakeholders so that the human element can be included and stakeholders can identify with strategies to be adopted.

Climate variability is also a major factor in the Palestinian economy, principally through the effects of major droughts on agriculture. Agriculture will be increasingly affected if inter-annual droughts occur more frequently or are more intense in the future. Less secure water supplies would exacerbate competition between users and uses and threaten allocations for environmental flows and future economic growth. Salinity in the Gaza Aquifer is already of increasing concern for Palestine. Any increase in flood frequency would adversely affect housing and other aspects of the built environment, such as industry and the communication networks in the Jordan Valley region. Water quality may also be affected due to increased soil erosion following drought, lower flows and higher water temperatures, leading to more eutrophication and algal blooms.

While there are many pressing problems regarding water supply in Palestine including the political one, climate change is likely to add to them, making solutions more difficult. Given the above, there is an obvious need for properly assessing the impact of climate change and in light of that to develop policies, strategies and mechanisms for implementation in order to confront this challenge.

Given the above, any assessments must take into account 'uncertainty'. This requires that assessments be set in a risk- management framework, where risk is seen as the product of the probability of a climatic effect and the consequences of that effect. Climate change, and our understanding of it, is evolving rapidly in the real world, and on the scientific, technological and policy fronts, so policies and decisions need to be decided on the basis of the best current information, but in the knowledge that they will need to be adjusted with time.

TOWARDS A COMPREHENSIVE POLICY

There is need for a comprehensive policy in order to respond to the impacts of climate change, in light of the increasing stress on water resources in the study region and the potential adverse impacts of climate change. In 1996, the Palestinian Authority adopted a set of policy elements that aims at protecting the water resources from pollution and depletion.[8] Although this policy provides an adequate framework for managing and protecting the water resources of Palestine, it does not specifically incorporate climate change and its impact on the water resources. The Strategy for Water Management in Palestine of 1998, translated the requirements of the policy into a set of practical strategies that must be pursued by the relevant stakeholders in the water sector in Palestine. Among other things, the strategy called for strengthening national policies and regulations in order to "develop a consistent and integrated water legislation with the intention to avoid unnecessary gaps and contradictions in the legal framework" and to "develop and manage water more efficiently and flexibly in order to respond to the changing demands and costs of water for the different uses"[9]. Therefore, there is a basis for revisiting the current policy to assess the modifications and adaptations needed.

As a young government, there is a great opportunity for Palestine to start with the right steps towards developing comprehensive policies and strategies and the proper legal and institutional framework which addresses climate change impacts in an integrated manner. This challenge will require sustained effort over the decades - on the part of

[8]Elements of the National Water Policy, 1996 (*on file with author*)

[9]Excerpts from the Strategic Intervention No. 2 from the Strategy for Water Management in Palestine for 1998 (*on file with author*)

governments, who must establish the rules and modify them as we learn more of the uncertainties and risks.

The modification to the existing policy and strategies should therefore address technical (scientific and technological), legal, institutional and educational aspects of the problem. There is no universally applicable best set of policies; rather, it is important to consider both the robustness of different policy measures against a range of possible future worlds, and the degree to which such climate-specific policies can be integrated with broader sustainable development policies.

Technical Aspects

From the technical perspective, there is a need for better predications and analysis of the current and future impact of climate change on Palestine. Natural, technical, and social sciences can provide essential information and evidence needed for decisions on what the impacts of climate change on the hydrological regime in Palestine, are. Such decisions must be determined through socio-political processes, taking into account considerations such as development, equity, and sustainability, as well as uncertainties and risk.

Sophisticated analysis of climate change and/or its impacts requires continuous data sets provided through a well-designed monitoring system. Monitoring should be enhanced from a strategic perspective in order to integrate key unknowns, particularly groundwater conditions and surface water quality. There are critical gaps in the existing monitoring system in Palestine whereby most of the monitoring is only done by Israel.

As a first step, there is a priority need to assess the availability of such technical information and knowledge and identify gaps in order to decide what researches and studies are needed in the future to bridge the gap. There is also a need for a scientific and programmed mechanism for decision makers for a better and sustainable management to preserve the land cover by a comprehensive long term monitoring of the Palestinian area. This mechanism will also promote and develop the awareness, knowledge and comprehension of the implementation of the technologies relevant to space sciences in order to follow up dynamically the desertification and drought processes.

Legal Aspects

At the domestic level, there is a need to review the legal frameworks and make sure that climate change is being incorporated in the water sector laws and regulations. Law No. 3 for 2002 stipulates that the Water Authority should in coordination with the relevant actors participate in preparing mechanisms for crisis management when drought or floods occur.[10] The said obligation is only a general call to react when and if droughts or floods or any emergency situation appear. There is an urgent need to revise the law or enact the appropriate regulations that ensures preparedness of the Water Authority for emergencies resulting from climate change. This preparedness could take the form of having drought and flood predictions and plans that provide the mechanisms for implementation.

As for the international level, and as most of the groundwater in Palestine is transboundary, it is essential that the procedural rules in any future agreement over the shared water resources incorporate climate change. The basis for setting these procedures is the equitable and reasonable utilization rule that govern the utilization of international watercourse including groundwater.

Institutional Aspects

The institutional frameworks of each country will influence the adaptation process in the water sector. There are many players and stakeholders in the water sector. At the national level, the Palestinian Water Authority is the regulator of the water sector, mainly responsible for implementing the national water policy that is adopted by the National Water Council. According to the law, the latter comprises the main ministries and utilities in the water and water-related sectors including but not limited to the ministries of agriculture, planning, health and representatives from the water utilities and research centres. The roles and responsibilities of these organizations were described in Law No. 3 in 2002, however, up to date the National Water Council has not been able to fulfill its mandate. As Palestine follows a centralized way of governance, more responsibility will lie with the Palestinian Water Authority for long-term strategic planning, creation of formal channels for information flow, providing an adequate level of budget funding, and for making sure that regional and local priorities are incorporated into national plans.

Developmental Aspects

Development activities in the water and environmental sectors must take climate change into account. After reviewing the Palestinian Agricultural

[10] Article 42, para 3 of the Palestinian Water Law, No. 3 for 2002. (on file with author).

Development Plan,[11] and the Palestinian National Water Master Plan (2005-2010), not all climate risks are being incorporated in decision making, even with regard to natural weather extremes. Moreover, practices that take into account historical climate are not necessarily suitable under climate change. Many planning decisions focus on shorter timescales and tend to neglect the longer-term perspective. An analysis of national development plans, poverty reduction strategy papers, sectoral strategies and project documents in climate-sensitive sectors indicates that such documents generally pay little or no attention to climate change, and often pay only limited attention to current climate risk. Even when climate change is mentioned, specific operational guidance on how to take it into account is generally lacking.

There is a general lack of awareness of climate change within the development and donor communities in Palestine. This is due to the fact that climate change expertise is typically the domain of environment and water departments in governments and donor agencies, who normally have limited influence over sectoral country policies programmes. Therefore, sectoral managers find that integrating climate change in development is an extra burden to their ongoing efforts to integrate issues such as gender, governance and environment in development activities. Moreover, as many development projects are funded over three to five years, it would be rather difficult to incorporate within them long-term climate risk reduction. The second reason is that development activities are dependent on a broad range of climate variables, only some of which can be available. Temperature, for example, is typically easier to project than rainfall. Climate extremes, which are often critical for many development-related decisions, are much more difficult to project than mean trends.

PROPOSED POLICY ORIENTATION: NATIONAL AND INTERNATIONAL PERSPECTIVES

The comprehensive policy must respond to the need for incorporating climate change impacts in the water resources management practices at the national and international level. To that end, and given the above needs and knowledge gaps, it is essential for any policy to answer the following questions:

- Does the current Palestinian National Water Policy frameworks deal with water resources and existing stresses and how?

[11] http://www.pnic.gov.ps/english/agriculture/agr112.html#The%20Problems%20and%20Obstacles%20Facing%20the%20Agricultural%20Sector:

- Are the current policy frameworks flexible to respond to possible future climatic conditions?
- What are the main gaps in the policy frameworks in the context of adaptation to climatic change?
- What elements of the current framework might become an obstacle for adaptation?
- Who are the main stakeholders in the water-related sectors and who should participate in adaptation?
- Is there a generic model or some best practices for a policy framework in the water sector to make it more prepared for handling adaptation to climate change?

At the National Level

Ownership Rights

Traditionally, and in accordance with basic principles of Roman law, groundwater has been regarded at law as the property of the owner of the land above (*nuda proprietas*).[12] Countries following the Napoleonic Code tradition, as well as countries following the Anglo-Saxon Common law tradition, equally subscribe to the same principle.[13] The Moslem tradition, by comparison, regards water as a public or communal commodity, 'a gift of God which cannot be owned.[14] Only wells can be owned, whereby

[12]Caponera, D.A. and D. Alheritiere. 1978. Principles for International Groundwater Law. In: L. Teclaff and A. Utton (eds). 25 IGL (1981) also found in 18 NRJ 589-618.

[13]*Ibid.* See also Malanczuk, P. 1997. Modern Introduction to International Law. Oxford University Press, Routledge, p. 6.

[14]Water Codes in Moslem Countries recognize that water is a public property to be administered by the States for the benefit of the Citizens. In Oman, for example, the 'Shari's principles are the governing legal framework of the State including those aspects dealing with water. In 1988 the Royal Decree No. 82 was issued declaring that 'water resources were a national wealth'. In Jordan, for example, Law No. 18 of 1988 (Jordan Water Law 1988) declares water as a public property. Article 25 states:

"Any water resources available within the boundaries of the Kingdom, whether they are surface or groundwater, ... are considered State owned property and shall not be used or transported except in compliance with this law".

In Yemen, water ownership, diversion, use are governed by a mixture of laws and customary practices depending on the water source land use economic activities and prevailing social and religious customs. Article 8 of the Constitution of Yemen states

"all types of natural resources and sources of energy, whether above ground, under ground, in territorial water, ... are the property of the State which assures their exploitation for public welfare".

In Lithuania and according to Article 2 of the Underground Law of 1995 No. I-1034 groundwater is owned by the State, "The underground shall be the exclusive ownership of the State. The basis of the exploitation of the underground resources shall be the right to

Footnote contd.

exclusive or priority user rights in the water accrue to the well-owners.[15] Furthermore, the ownership of wells entails ownership of an area around the well in which new wells cannot be dug (known as *harim*, or forbidden area).[16] However, economic development, the increasing need for water, the introduction of modern extraction methods, and the uncontrolled use of groundwater have compelled States to introduce groundwater regulations in order to replace the traditional ones.[17] The trend is in the

---

exploitation, which in the procedure established by this and other laws can be granted to legal and natural persons of the Republic of Lithuania by the Government of the Republic of Lithuania or the public institution authorised by the Government, and to foreign legal and natural persons - by the Government of the Republic of Lithuania. Actions which directly or indirectly violate the right of State ownership of the underground shall be prohibited". *found at* http://www.litlex.lt/Litlex/Eng/Frames/Laws/Documents/369.HTM (last visited May 2004).

[15]Caponera, D. *supra* note 96.

[16]Bogdanovic, S. *supra* note 13 p. 332, and Caponera, D. *supra* note 96. Caponera and Alheritiere explain that the codified Moslem law is one important source of law in Moslem countries, whereby groundwater belongs to the community. The definition of water as a non-saleable publicly owned commodity applies to all water dug by unknown persons.

[17]In Yemen the proposed new law (on file with the author) contains many articles dealing with groundwater resource ownership, investigation, development, monitoring and Management. Article 3 states:

"All water resources which exist within the boundaries of the Republic, whether on the land surface or underground, are considered natural resources owned as public property, and the State shall orient and organize their exploitation so as to serve public welfare".

Article 3 of the 1996 Albanian Law on Water Resources, *supra* note 87, grants the government ownership rights to all water resources: "The state ownership comprised: a) All water resources of the Republic of Albania, as defined in the Article 2, paragraph 1 of this law. b) All river beds and banks, torrents and other natural streams either temporary or permanent, channels, lakes, ponds, natural or artificial lagoons and reservoirs, islands: end accumulations of sand, stones and silt on river beds, lakes and reservoirs as well as geological formations of underground waters. c) All objects and hydrotechnic works accomplished by the state such as dams, irrigation, drainage and navigation systems, potable water stations and their respective channels and works. d) The land obtained from withdrawal of water or extended land toward the water. The ownership right of the state as specified under paragraph 1 of this article, is hereby unchangeable and irrevocable."

In Lithuania Article 4 of the Water Law of 1997, *supra* note 101, states that:

"Water resources comprise a component part of the natural resources of the Republic of Lithuania. The use thereof shall comply with the Constitution of Republic of Lithuania, the Laws on Environmental Protection, the Law on the Earth Entrails, as well as this and other laws and legal acts" *found at* http://www.dundee.ac.uk/law/iwlri/Research_Documents_National.php (last visited December 2003).

Also see the 'African Water Act' *supra* note 77, which describes the National Government as the public trusteeship of nations water resources' in Chapter 1, paragraph 3:

"(1) As the public trustee of the nation's water resources the National Government, acting through the Minister, must ensure that water is protected, used, developed, conserved, managed and controlled in a sustainable and equitable manner, for the benefit of all persons and in accordance with its constitutional mandate. (2) Without limiting subsection (1), the Minister is ultimately responsible to ensure that water is allocated equitably and used

Footnote contd.

direction of severing water rights from the property law concept of ownership in the classical strict sense.[18] There is a major movement in the law of national and sub-national levels, in recent decades toward limiting and regulating groundwater withdrawals and especially mandating pollution prevention.[19] Much legislation have tended toward the limitation of exclusive private ownership in favor of a form of central administrative control over groundwater uses, thereby creating a formal separation between ownership (*nuda proprietas*) and use rights.[20]

Monitoring and Assessment

A proper monitoring and assessment of the meteorological and hydro meteorological data is essential for effective water resources management and regulation. Modern national water codes include provisions on the monitoring[21] and assessment[22] of the water related data, the assessment of

beneficially in the public interest, while promoting environmental values. (3) The National Government, acting through the Minister, has the power to regulate the use, flow and control of all water in the Republic".

In Israel there is no private ownership of the water resources Section 1 of the 1959 Water Law (on file with the author) states that:

" The water resources of the State are public property; they are subject to the control of the State and are destined for the requirements of its inhabitants and for the development of the country".

Article 3 paragraph (1) of the Palestinian Water Law No 3/2002 (on file with the author) states:

" All water resources available in Palestine are considered public property".

[18]Ibid.

[19]*See* African Act No. 36 for 1998, *supra* note 77.

[20]Caponera, D. *supra* note 96.

[21] " 'Monitoring' is:

"the process of repetitive observing, for defined purposes of one or more elements of the environment according to pre-arranged schedules in space and time and using comparable methodologies for environmental sensing and data collection. It provides information concerning the present state and past trends in environmental behaviour." See UNECE Task Force on Monitoring and Assessment *supra* note 52. *Also see* Article 38 of the Lithuania Water Law of 1997, *supra* note 101, provides:

"1. Users of water who extract water from water bodies or discharge waste water into the natural environment by using their own facilities and equipment and subscribing users of water shall supply statistical data on the use of water to the institution authorized by the Ministry for Environmental Protection according to the established procedure. 2. The Ministry for Environmental Protection in agreement with the Department of Statistics shall establish the procedure for submitting of statistical data, 3. The Ministry for Environmental Protection shall establish the procedure for accounting for the amount of water extracted and used as well as the amount of waste water discharged into the natural environment."

[22]"Assessment" means:

"The evaluation of the hydrological, chemical and/or micro-biological state of groundwaters in relation to the background conditions, human effects, and the actual or intended uses, which may adversely affect human health or the environment." UNECE Task Force on Monitoring and Assessment, See generally *supra* note 52

the effectiveness of measures taken to prevent, control and reduce pollution, and the establishment of a credible data base on water and effluent monitoring.[23] Other relevant aspects deal with the rules for setting up and operating monitoring programmes, which includes measurement systems and devices, analytical techniques, data processing and evaluation techniques. Monitoring includes also the development of systems for surveillance and early-warning systems to identify outbreaks or incidents of water-related diseases or significant threats of such outbreaks or incidents, including those resulting from water pollution or extreme weather.[24]

Permitting System

Countries in the Middle East that rely heavily on groundwater, have issued directives aimed at regulating groundwater extraction through well drilling permits, and the formulation of drinking and renovated waste water standards. For example, the oldest water legislation was established in 1925 in the Syrian Arab Republic and Lebanon governing

[23]For example the UNECE Charter on Groundwater management, *supra* note 76 includes provisions on the monitoring and control of groundwater aquifers. Article XIII provides:

"(1) Monitoring programmes for ground-water protection should be set up and applied. These programmes should include monitoring at the source of potential pollution which could pose a serious or chronic threat to an aquifer. There should be regular inspections to ensure compliance with protection requirements imposed. Attention should also be paid to the monitoring of ground-water quality changes brought about by air-borne pollution. (2). Systematic monitoring should be carried out for all aquifers found to be vulnerable to pollution and/or over-use, as well as for those whose particular importance has been recognized for public water supply, mineral water supply and industry. (3) Monitoring and control should be considered a public-service activity. Facilities should be set up for coordinating the assessment and availability of monitoring data and information on aquifers. The resulting collections of data should be related to information on ground-water quantity and quality characteristics of aquifers as well as details of their location, use, and exposure to various impacts from land uses such as agriculture, industry and urban development. Information should be readily available to those interested. (4) The data from monitoring should make it possible, *inter alia,* to revise periodically plans and forecasts of groundwater use, taking into account actual evolution of aquifers, and to determine measures necessary to ensure the sustainable use of ground-water resources in the long term. Legislative provisions and regulations should, as appropriate, allow for the revision of protection requirements imposed depending on the measures thus determined. (5) Monitoring programmes should be periodically reviewed to ensure that they are achieving their stated aims and that the results have been used effectively."

[24]This strategy aims at collecting information on whether and where accidental spills may affect the drinking water supply, to determine public health hazards of 'abandoned or illegal' land disposal sites, or to determine actual sources of groundwater quality deterioration. For early warning, special wells may be drilled, whereas for surveillance production wells can often be used.

See generally, UNECE Task Force on Monitoring and Assessment, *supra* note 52.

public ownership of water and licensing.[25] In Oman, which depends almost entirely on groundwater, there is a water deficit of about 275 mcm due to over pumping and recurrent drought events.[26] This situation was a consequence of the use of the random and unregulated use of modern machinery and equipment in drilling and pumping in 1970 onwards. In 1975 and as a response to this deteriorated situation the government issued a series of directives to regulate and develop these resources.[27] Most recently - in 1995 - a new Ministerial Directive (No. 13/95) was[28] issued. This directive concerns the regulation of groundwater wells and irrigation and it includes a whole chapter on the permitting of wells including all details that are needed to ensure control over the groundwater utilization and development.[29]

---

[25]In Syria the public property regulation No. 144 of 1925 addresses water rights in some of its articles. Another law was issued in 1926 addressing the protection of public water sources including rules for regulating licenses and water concessions. See Al-Safady, M.S. 1986. Water Legislation in the Arab World. Arab Organization for Education Culture and Science Publication No. 002/03/1986 Tunisia. In Israel the Water Law of 1959 stipulates the need for acquiring a license for four areas: i) the Production License ii) the Recharging License; iii) the Drilling License; and iv) The Construction License (Chapter 2, Articles 3 & 5). Among the responsibilities entrusted to the Palestinian Water Authority is the licensing of the exploitation of water resources as stated in Article 7 paragraph (5) of the Palestine Water Law No. 3/2002:

"Licensing the exploitation of water resources including the construction of public and private wells, drilling, and monitoring and production wells." Within the Palestine Water Law 3/2002 Article 31 states that: "the Authority, in coordination with the other relevant parties, may consider any area that contains groundwater a protected area, if the quality or quantity of water is in danger of pollution, or if achieving the objectives of the water policy requires such action, on condition that the Authority provides alternative water resources."

[26]Khameisi, S. 1996. Water Legislation in Sultanate of Oman. E/ESCWA/ENR/1996/WG.1/CP.8 (in Arabic).

[27]In 1975 Directive No. 45/75 which establishes the Water Resources Wealth Council, and in Directive No. 76/77 of 1977 on Water Resources Development and Directive No. 63/79 of 1979 concerning the Establishment of a Public Authority for Water Resources. In 1989 the Sultan Decree No. 100/89 established the Ministry of Water Resources. The responsibilities of this Ministry is to develop, regulate and further explore the groundwater resources. This decree was followed by Ministerial Decision No. 2/90 of 1990 which regulates the registry of existing wells and regulate the licensing of new wells. In terms of Article 7 of Ministerial Decision No. 2/1990, groundwater exploration activities require a permit. A permit is necessary for: the construction of a new well, the modification of an existing well (by deepening or increasing the diameter of the well, cleaning or maintenance of the well), for the change in use of an existing well and for the installation of a pump in a borehole (Article 8). Permits are granted only to landowners (Article 9 of Ministerial Decision No. 2/1990). However, this provision does not apply to wells excavated for emergency purposes or in case of temporary water shortage; to wells constructed solely for exploration and monitoring purposes and to wells constructed for petroleum exploration activities. (Arabic texts available on file with the author).

[28](text of the law found on file with the author).

[29]Chapter 3 of Directive 13/95 of 1995 includes 18 articles on the procedures for obtaining a permit and the organization concerned and the conditions upon which a permit can be granted.

In Jordan, as well, Law No. 26 for 1968 mandates the competent authority to issue drilling permits. Also in Jordan Resolution Number 26 of 1977 (on file with the author) concerning the 'Supervision of Groundwater Regulation' prohibits any drilling, abstraction and utilization of groundwater without a prior permit.

Groundwater utilization in Yemen was regulated by the issuance of drilling permits according to rules set forth in Law No. 320 of 1926. The decree emphasized the issuing of drilling permits, and that individuals must either obtain prior permission to drill or exemption from a permit. The permit contained specific information on the well for which it was issued including location, type and method of drilling, as well as an estimate of the volume of discharged water.

In Palestine water became a public property by Law No. 3 for 2002. This implies that only the user or usufructuary type rights accrued to the owners of overlying land or to the developer of the resources other than the landowners. These rights are granted by the Palestinian Water Authority subject to terms and conditions. However, in practice, the enforcement procedures of these laws are not effective. There is a need for more elaborated regulations that outline the duties and obligations of water users and specify the sanctions on violators.

Protection and Conservation Areas

There are situations where accelerated depletion of the resources exists and adequate management and legal responses established. The declaration of special zones is invoked by the fact that control of abstraction and of recharge is rendered indispensable as a consequence of increased water use and recurrent drought events. In Israel, for example, the competent authority is mandated to declare 'Rationing Areas' where the water resources are not sufficient for the maintenance of existing consumption water.

In modern codes measures relevant to the control of water utilization to preserve groundwater include well spacing, regulation of pumping rate, monitoring and regulation practices, establishment of minimum flows in surface streams interconnected with transboundary aquifers and control of surface water divisions and declaration of protection zones.

Groundwater protection zoning also has a key role in setting priorities for groundwater quality monitoring, environmental audit of industrial premises, pollution control within the agricultural advisory system, determining priorities for the clean-up of historically-contaminated land, and in public education generally. All of these activities are essential components of a sustainable strategy for groundwater quality and quantity protection. Rationalization and seasonal limitations on groundwater utilization is another means for resources protection and conservation. This measure is particularly important in Palestine as a semi-arid area which is subject to drought periods. In dry years, when the flow rate in aquifers is lower than estimations made for the most likely reference year, the use of the water resources is limited.

Economic and Financial Regulatory Instruments

Economic measures such as fees on the use of water and the charges for waste-disposal are increasingly applied in coordination and have sufficient impact to constitute an effective incentive to use groundwater rationally or be a disincentive to polluting aquifers. The abstraction of groundwater is in some instances subject to differentiated fees in proportion to the volume abstracted, in relation to the available resource or according to the anticipated use of the abstracted groundwater, while complying with legal provisions and regulations governing the applied permit system. Costs attributable to pollution are to be borne by the polluter whenever the latter can be identified. Serious consideration should be given to all possible economic measures which could have an influence on preventing, mitigating and counteracting damage as well as those bearing on remedying critical situations caused by pollution or over-exploitation of aquifers.

At the International Level[30]

There are some crossing cutting elements between the policy to be pursued at the national level and those at the international level. However, in the latter case, there will be two States involved with the two different legal regimes. In this case the harmonization of policies and legal framework becomes essential, especially with regard to that part of the watercourse along the border region.

The Duty to Exchange Information

Israel and Palestine are urged to establish and maintain reliable data and information concerning transboundary waters in order to use and protect these waters in a rational and informal manner. To that end, it is essential to establish and manage a database. A special Joint Commission could be charged with the creation and maintenance of a comprehensive and unified database pertaining to transboundary waters. Greater knowledge about the quality and quantity of transboundary groundwater is essential

[30]This analysis is based on the Bellagio Draft Treaty prepared by Hayton, R.D. and A.E. Utton. 1989. Transboundary Groundwaters: the Bellagio Draft Treaty, 29NRJ [hereinafter referred to as the Bellagio Treaty].

for the improvement of the management of such resources. Regular and systematic collection of hydro meteorological, hydrological and hydrogeological data needs to be promoted and be accompanied by a system for processing quantitative and qualitative information for various types of water bodies.

Monitoring and Assessment

In order to fulfill the objectives of data and information exchange, there is a need to establish observation networks and strengthen existing systems and facilities for measurements and recording fluctuations in water quality and level; organize the collection of all existing data on groundwater (borehole logs, geological structure, and hydrogeological characteristics, etc.); systematically index such data, and attempt a quantitative assessment so as to determine the present status of and gaps in knowledge; increase the search for, and determination of, the variables of aquifers.

Establishment of Joint Commissions

Palestine and Israel are encouraged to establish a 'Joint Commission' responsible for the implementation of an agreement that includes responsibility for all transboundary waters related matters. The joint Commission is mandated to undertake a variety of tasks and functions to fulfill the requirement of the Treaty.

The Duty of Prior Notification

Among the tasks entrusted to this Commission is groundwater quality and quantity protection. This is ensured through the prompt notification on any actual or planned activity that may cause "appreciable harm" to the transboundary groundwaters or recharge areas.[31]

Declaration of Transboundary Groundwater Conservation Areas

Also as part of the quality and quantity conservation plan the Joint Commission shall determine the desirability of declaring any area within the border region containing transboundary groundwaters to be a Transboundary Conservation Area. This is done to protect and to improve, the quality of transboundary waters. In making its determination of these Conservation Areas, the Commission shall consider whether: a) groundwater withdrawals exceed or are to exceed recharge even so as to endanger yield or water quality or are likely to

[31]*Ibid*, Art. VI paras 1-3:1.
"The Parties undertake cooperatively"

diminish water, the quantity or quality of interrelated surface waters; b) recharge has been or may become impaired; c) the use of the included aquifer(s) as an important source of drinking source of water has been or may become impaired water; d) the aquifer(s) have been or may become contaminated; and e) recurring or persistent drought conditions necessitate management of all or some water supplies in the particular area. In making its determination, the Commission shall take into account the impact of implementation of the declaration under consideration on the sources and uses of water previously allocated by agreement between the Parties or under the Drought Management Plan. The Joint Commission is also encouraged to prepare a Comprehensive Management Plan for each declared Transboundary Groundwater Conservation Area to ensure the rational development, use, protection and control of the waters in the Transboundary Groundwater Conservation Area. Allocations of uses under this plan should take the following factors into account (a) hydrogeology and meteorology; (b) existing and planned uses; (c) environmental sensitivity; (d) quality control requirements; (e) socio-economic implications (including dependency); (f) water conservation practices (including efficiency of water use); (g) artificial recharge potential; and (h) comparative costs and implications of alternative sources of supply. The Master Plan is to include a description of what measures are needed (i) to prevent, eliminate or mitigate degradation of transboundary water quality and quantity; (ii) to allocate the uses of groundwaters and interrelated surface waters taking into account the other allocations previously made applicable within the transboundary groundwater conservation area; and (iii) to limit pumping, set criteria for well placement and number of wells, decisions on retirement of existing wells, imposition of extraction fees, planned depletion regimes or reservations of groundwaters for future use.

CONCLUSIONS AND RECOMMENDATIONS

The impact of climate change in Palestine is still uncertain. In the context of water resources there is an inherent lack of proper information and knowledge, particularly in relation to the long term hydrometeorological data and time series, which complicates the task of assessing impact of climate change on the water resources in Palestine. To remedy this situation, there is a need to identify and implement research priorities at the national and international levels.

There is a specific need to identify the key uncertainties such as climate change and attribution, future emissions of greenhouse gases and future changes in global and regional climate. Additionally uncertainties regarding regional and global impacts of changes in average climate and climatic extremes, and lastly, on the costs and benefits of mitigation and adaptation must be identified.

In the absence of adequately reliable information, it is a very challenging to assess the impact of climate change on the water resources availability in Palestine and consequently on the Palestinian economy. Without this knowledge base, policymaking regarding adaptation and mitigation cannot be soundly based on economic considerations and may not be effective in avoiding significant damages to the economy, ecology and people.

In the mean time and until the relevant knowledge is built, there is a need for implementing innovative techniques to undertake the assessment. The future challenge is to develop a reliable knowledge base that enables the use and implementation of proper models for evaluation.

In the context of climate change impact on the water resources of Palestine, this chapter recommends four different areas of interventions. First, there is a need to bridge the gap in the scientific knowledge. At the national level, this requires the creation and maintenance of a credible monitoring and assessment systems for the most relevant parameters. At the international level, it is recommended here that Israel, which is information-rich on climate parameters and trends, shares with the Palestinians as they are information-starved.

Second the policy and legal frameworks have to be revised, modified and adapted to incorporate climate change. The National Water Policy in Policies at the national level could include licensing of drilling and abstractions, developing management and drought plans, declaring conservation areas and imposing economic measures on the use and consumption— are all policies responding to the increasing stress on the water supply to the possible deterioration of the climatic problem. As for the international level, and assuming there is a political will for cooperation Israel and Palestine must establish a joint monitoring and assessment systems and maintain a joint knowledge data base. In this context prior notification and consultation on any planned measure on the shared resources is obligatory. Finally the declaration of protection zones in the border areas is advisable whereby both parties develop a joint management plan including a drought and emergency plan which could incorporate climate change impacts.

Third, this chapter finds that there numerous institutions involved in the water sectors at the different levels. The challenge at the national level is to coordinate the efforts between these institutions identifying who is doing what in terms of the implementation of the national policies. And as institutions are the vehicle for successful cooperation, it is recommended that Israel and Palestine develop and operationalize a Joint Commission at the international level, responsible for implementing the desired cooperation.

Finally, the integration of climate change in development programmes is the outcome of proper implementation of the previously mentioned recommendations. As the work progresses and the knowledge is built, it becomes easier for donors' representative to incorporate responses to them within the development programmes.

References

Applied Research Institute 2001. Localizing Agenda 21 in Palestine. Found at *http://www.arij.org/Agenda-21/l_o_c_a_l_i_z_i_n_g.htm (last visited August 2006).*

Bou-Zeid, E. and M. El-Fadel. 2002. Climate change and water resources in the Middle East: A vulnerability and adaptation assessment. Journal of Water Resources Planning and Management (ASCE) 128 (5): 343–355.

Bou-Zeid, E. and M. El-Fadel, Elie R., Climate change and water resources in the Middle East: Vulnerability, Socio-Economic Impacts and Adaptation, June 2001. FEEM Working Paper No. 46. 2001.

Caponera, D.A. and D. Alheritiere. 1978. Principles for International Groundwater Law. In: L. Teclaff and A. Utton. (eds.). 1981. 25 International Groundwtaer Law. also found in 18 Natural Resources Journal 589-618 (1978).

Daibes-Murad, F. 2005., A New Legal Framework for Managing the World's Shared Groundwater, International Water Association Publishing, UK.

Dudeen, B. 2000. Soil Contamination in the West bank and Gaza Strip—Palestine. proceedings of the First Environmental Scientific Symposium. Hebron—Palestine. October, 2000 at http://resources.ciheam.org/om/pdf/b34/01002095.pdf#search=%22Basim%20Dudeen%20climate%20change%22 (last visited September 2006)

FAO. 2000. Aquastat information system. The land and water development division. Found at www.fao.org/landandwater/aglw/aquastat/dbase/index.stm (last visited August 2006).

Gvirtzman, H. 1994. Groundwater Allocation in Judea and Samaria. In: J. Issac and H. Shuval (eds.). Water and Peace in the Middle East. Elsevier: Amsterdam, The Netherlands.

IPCC Second Assessment Report, 1995.

J. Al Kadi. 1995. Global Warming: A Study of the Gaza Temperature Variations in the Period 1976 - found at http://www.iugaza.edu/ara/research/articles/%D8%AF.%20%D8%A3%D8%AD%D9%85%D8%AF%20%D8%A7%D9%84%D9%82%D8%A7%D8%B6%D9%8A.pdf#search=%22climate%20in%20Gaza%20temperature%22 (last visited September 2006)

Khameisi, S. 1996 Water Legislation in Sultanate of Oman, E/ESCWA/ENR/1996/WG.1/CP.8 (in Arabic).

Levina E. and H. Adams. 2006. Domestic Policy Frameworks for Adaptation to Climate Change in the Water Sector Draft for Review OECD/IEA Project for the Annex I Expert Group on the UNFCCC Paris, found at http://www.oecd.org/dataoecd/17/57/36294928.pdf#search=%22comprehensive%20policy%20to%20cope%20with%20climate%20change%20and%20impact%20on%20water%20resources%20legal%20framework%22 (last visited, September 2006).

Pe'er, G. and U. Safriel. 2000. Climate Change, Israel National Report, the United Nations Framework Convention on Climate Change, Impact, Vulnerability and Adaptation, found at http://www.bgu.ac.il/BIDR/rio/Global91-editedfinal.html last visited September 2006.

Ragab, R. and C. Prudhomme. 2001. Climate Change and Water Resources Management. In: The Southern Mediterranean And Middle East Countries, Centre for Ecology & Hydrology, CEH, Wallingford, OX10 8BB, UK, found at http://www.irncid.org/workshop/pdf/w22/Ragab%20Climate%20change.pdf (last visited August 2006)

OECD, Policy Brief 2006, http://www.oecd.org/dataoecd/57/55/36324726.pdf#search=%22comprehensive%20policy%20to%20cope%20with%20climate%20change%22 Putting Climate Change Adaptation in the Development Mainstream.

Malanczuk, P. 1997. Modern Introduction to International Law, Oxford University Press, Routledge.

Palestinian Bureau of Statistics Press Release, March 2006, found at http://www.pcbs.gov.ps/Portals/_pcbs/PressRelease/arsad23E.pdf#search=%22climate%20in%20Gaza%20temperature%22 (last visited September 2006).

Palestinian National Information Center, http://www.pnic.gov.ps/english/information/fact4.html, (last visited September) 2006.

Palestinian Water Law No. 3 for 2002 (On file with Author).

The Palestinian Agricultural Development Plan 1999 http://www.pnic.gov.ps/english/agriculture/agr112.html#The%20Problems%20and%20Obstacles%20Facing%20the%20Agricultural%20Sector:

Shuval, H.I. 1996. Towards Resolving Conflicts Over Water Between Israel and Its Neighbours: The Israeli Palestinian Shared Use of the Mountain Aquifer a Case Study. In: J.A. Allan (ed.). Water, Peace and the Middle East: Negotiating Resources in the Jordan River Basin, Tauris Academic Studies, UK.

30

CHAPTER

Differential Vulnerability to Climate Change in Asia and Challenges for Adaptation within the Kyoto Context

Lilibeth Acosta-Michlik[1], Liu Chunling[2] and Ulka Kelkar[3]

[1]Département de Géologie et de Géographie
Université Catholique de Louvain (UCL)
Place Louis Pasteur 3, B-1348 Louvain-la-Neuve, Belgium
E-mail: acosta@geog.ucl.ac.be

[2]Graduate School of Geography, Clark University
950 Main Street, Worcester, MA 01610, USA
E-mail: cliu@clarku.edu

[3]The Energy and Resources Institute (TERI), Darbari Seth Block
IHC Complex, Lodhi Road, New Delhi - 110 003, India
E-mail: ulkak@teri.res.in

1. INTRODUCTION

The Kyoto Protocol provides legally binding measures to implement the agreements developed in the United Nations Framework Convention on Climate Change (UNFCCC), which aims to reduce global warming and to cope with its inevitable impacts. But "developing countries at large would suffer the negative effects of any slowdown in global growth caused by implementing the Kyoto Protocol" (Oxley and Macmillan 2004). Similar reports from other studies have caused many developing countries to be cautious in signing and ratifying the Kyoto Protocol. Moreover, they argue that the "[u]nsustainable consumption patterns of the rich industrialised nations are responsible for the threat of climate change" (Parikh and Parikh 2002). Referring to the study of Parikh et al., (1991), Parikh and Parikh mentioned that only 25% of the global population lives in these countries, but they emit more than 70% of the total global CO_2 emissions. Moreover, per capita emissions measured in metric tons are much higher in developed countries (e.g. United States 20.2, Germany 10.3, United Kingdom 9.2, etc.) than in developing countries (e.g. China 2.7, India 1.2, Philippines 0.9, etc.) (WB 2006). The interests of the latter have been defended with success in the ratified Kyoto Protocol, in which they are not required to have binding emissions targets. In addition, the Protocol offers them three market-based mechanisms to support reduction of emissions in a more flexible manner. These mechanisms are the joint implementation mechanism (JIM), clean development mechanism (CDM), and emissions trading mechanism (ETM). Among these mechanisms, CDM has a direct impact not only on promoting mitigation, but also adaptation. "A levy from each CDM project – known as a 'share of the proceeds' – will help finance adaptation activities in particularly vulnerable developing countries..." (UNFCCC, 2002). Because it is designed to promote sustainable development, CDM should help decrease vulnerability to the impacts of climate change in developing countries.

Although existing global climate models (GCMs) require improvements to gain further confidence in climate model projections (IPCC 2001), climate change is believed to contribute to the increase in frequency and intensity of climatic variability causing floods and droughts among others. Developing and less developed countries are more vulnerable to the impacts of these climate extremes because they have less economic and institutional capacity to cope and adapt. For the affected people, projects that promote adaptation to short-term impacts are thus as important as, if not more important than, mitigation of long-term effects of climate change. Mechanism such as CDM that aims to link mitigation and adaptation is thus fundamental to an equitable distribution of benefits and costs of emission production. That is, the major polluters (i.e. industrialized countries) should transfer part of the national benefits (i.e. income growth) to the minor polluters (i.e. developing countries), which must share the global costs of pollution (i.e. climate change). Vulnerability to the impacts of climate change differs across regions, countries, and communities. So does CDM address the differential vulnerability and adaptation needs in developing countries? This chapter aims to provide answers to this question by discussing the sources of vulnerability in Asia (section 2) and presenting selected case studies in most vulnerable Asian countries (section 3). In section 4, the relevance of current and proposed CDM projects to the adaptation needs of vulnerable people in these countries will be discussed. The chapter is concluded in section 5.

2. CURRENT VULNERABILITIES IN ASIA

Many countries in Asia have experienced impressive growth in industry, commerce and services in the past decades. While the share of agriculture to GDP has significantly decreased, the sector continues to play a key role in the Asian economies. Agricultural sector contributes substantial foreign exchange earnings for the government, employs a large portion of labour force, provides source of income for many households and produces food for the population. However, the increase in the frequency and intensity of climatic variability is a great challenge for maintaining a high level agricultural productivity. Asia is vulnerable to many natural hazards including floods, droughts, windstorms, volcanic eruptions, earthquakes, and others that damage not only agricultural production, but also property and life. Figure 1 shows that, for the period 1900-2000, Asia has had the highest number of people reported affected by natural hazards among the different continents. Droughts and floods have affected the highest number of people globally. The most pronounced year-to-year variability in climate features including droughts and floods is influenced by E1 Niño-Southern Oscillation (ENSO) (IPCC 2001). ENSO plays a key role in determining yearly agricultural production across the entire region of South and Southeast Asia (Amadore et al. 1996). Because of this, agriculture of the region would be affected by the changes in frequency and severity of ENSO events.

The IPCC (2001) reported a number of studies that used GCMs to assess the changes that might occur in ENSO in connection with future climate warming and in particular, those aspects of ENSO that may affect future climate extremes. Whilst the models show conflicting results, recent studies on intensification of the Asian summer monsoon and enhancement of summer monsoon precipitation variability with increased greenhouse gases confirmed possible relationship between ENSO and climate warming (IPCC 2001). In addition to the impacts of climate change on long-term agricultural productivity[1], its impacts on climate extremes

---

[1]Crop yield simulation results vary widely (e.g., +20% changes in yield) for specific sites, countries and GCM scenarios (Iglesias et al. 1996). According to Iglesias et al., South and Southeast Asia agriculture sector appears to be among the more vulnerable, whilst East Asia appears to be relatively less vulnerable. However, summarizing various studies, Amadore et al. indicated that crop production appears to be more vulnerable not only in South and Southeast Asia, but also in China.

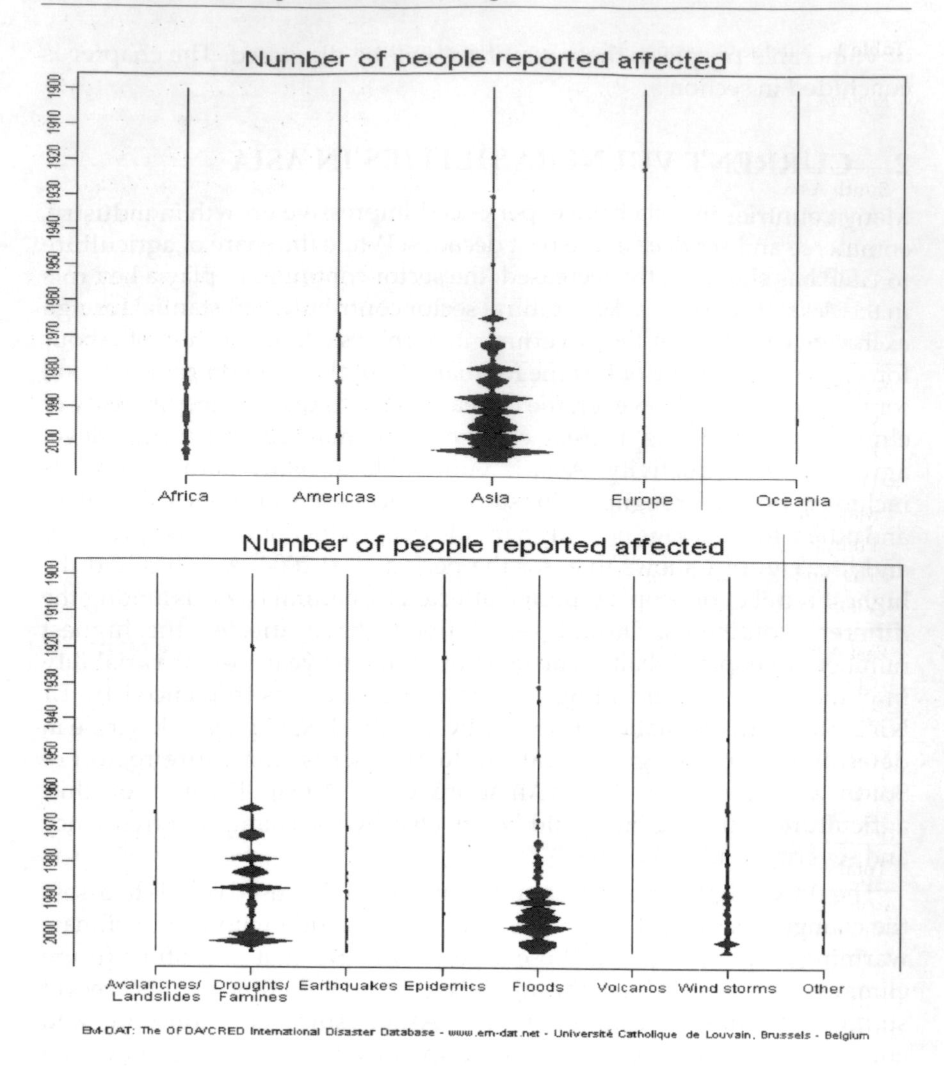

Fig. 1 Number of people affected by natural disasters, 1900-2000
Source: EM-DAT

affect many people in Asia. More than half of the 664 million people affected by droughts were reported in South Asia and about a quarter in East Asia in the last decade (Table 1). At the country level, India (59%) accounted for the highest number of affected people in the entire Asia from 1990 to 2000, followed by China (26%). Compared to South and East Asia, the statistics for Southeast Asian countries are low and for West Asia rather insignificant. In Southeast Asia region, Thailand, Vietnam and the

Table 1 Number of people affected by droughts in Asia, 1990-2000.*

Region/Countries	*People affected in thousand*	*Percentage share*	
		to region	*to continent*
South Asia	**463,381**		**69.78**
Afghanistan	6,380	1.38	0.96
India	391,175	84.42	58.91
Pakistan	2,269	0.49	0.34
Sri Lanka	1,557	0.34	0.23
Iran	62,000	13.38	9.34
Southeast Asia	**27,336**		**4.12**
Cambodia	1,550	5.67	0.23
Indonesia	1,170	4.28	0.18
Laos	20	0.07	0.00
Malaysia	5	0.02	0.00
Philippines	3,981	14.56	0.60
Thailand	13,500	49.38	2.03
Vietnam	7,110	26.01	1.07
East Asia	**172,340**		**25.95**
China	171,890	99.74	25.89
Mongolia	450	0.26	0.07
West Asia	**988**		**0.15**
Jordan	330	33.40	0.05
Syria	658	66.60	0.10
Total Asia	**664,046**		**100.00**

*Asian countries not included in the table did not have any reports on drought events or affected people.
Source: EM-DAT

Philippines experienced the highest number of affected people. The number of affected people, and thus the magnitude of damage, differs across Asia. But does the degree of vulnerability to drought increases with the magnitude of damage? To answer this question, it is important not only to quantify, but also to qualify vulnerability. In this case, evidences on the ground are useful to describe the character and peculiarities of vulnerability of people from different region with different economic, social and cultural backgrounds. The next section discusses selected country case studies in South Asia (India), East Asia (China) and Southeast Asia (the Philippines) and presents recent evidences of differential vulnerabilities and adaptive responses of farmers to droughts. These provide valuable information on the relevance of Kyoto mechanisms for supporting adaptation goals.

3. EVIDENCES ON THE GROUND

India

India is highly vulnerable to climate change, not only because of high physical exposure, but also because livelihoods and economic activities are closely tied to the natural resource base. Agriculture and aquaculture will be threatened by a combination of thermal and water stresses, sea level rise, increased flooding, and strong winds associated with intense tropical cyclones. Freshwater availability and biodiversity, which are already under pressure due to population growth and land use change, will be further impacted by climate change. Finally, warmer and wetter conditions will increase the potential for a higher incidence of heat-related and infectious diseases. By adversely impacting sectors like agriculture, water resources, and health, climate change presents a formidable challenge for efforts to reduce poverty and achieve the Millennium Development Goals.

Of particular relevance to India is the fact that climate change will lead to increased variability in summer monsoon precipitation. Given that even today rain-dependent agricultural area constitutes about 60% of the net sown area of 142 mha, Indian agriculture continues to be fundamentally dependent on the weather. In addition, glacial melt is expected to increase under changed climate conditions, which would lead to increased summer flows in glacier fed river systems for a few decades, followed by a reduction in flow as the glaciers disappear.

Although people in India are already coping with current levels of climate variability, there are wide disparities in the capacity to adapt. Access to adaptation options is severely constrained by economic resources, technological factors, access to information and skills, infrastructure, and institutions. In a detailed study of district-level vulnerability of Indian agriculture (O'Brien et al. 2004), adaptive capacity was mapped as a composite of biophysical, socioeconomic, and technological factors, and juxtaposed against a map of sensitivity to climate change (using output from the HadRM2 downscaled general circulation model). The study revealed higher degrees of adaptive capacity in districts falling in the Indo-Gangetic plains (except for Bihar) and lower degrees of adaptive capacity in the interior regions of the country, including districts in Bihar, Rajasthan, Madhya Pradesh, Maharashtra, Andhra Pradesh, and Karnataka. Community-level case studies were carried out in highly vulnerable districts (including Jhalawar in Rajasthan, Ananthapur in Andhra Pradesh, Raipur in Chattisgarh, and Jagatsingpur in Orissa). These brought out the wide disparities in adaptive capacity across villages, across communities in villages, and specifically across individuals depending on land holding size, education, caste, etc. While larger farmers are able to benefit from government subsidies (e.g. for drip irrigation), formal bank credit, crop insurance, and access to larger markets, smaller farmers are disadvantaged due to lack of information and dependence on local merchants for credit.

In the hilly regions of India, agriculture is impacted by the loss of rainfall due to surface runoff, drying up of streams and springs, and decline in soil moisture. This is particularly a concern in Uttarakhand State, where irrigated area is merely 14% in the hilly districts, and the primary means of irrigation for the terrace fields is small gravity flow channels which divert water from snow-fed streams. Narula and Bhadwal (2003) estimate that about 1500 km^2 of the 4000 km^2 of the Lakhwar sub-basin, part of the Upper Yamuna sub-basin in the state of Uttarakhand, receives an annual runoff of less than 1250 mm, and is hence highly sensitive to increased water stress due to climate change. A decrease of 20% to 30% in total flows on account of climate change alone was estimated, indicating that the amount of water available for usage in the future would be reduced substantially. Under the HadRM2 scenario, there is a net decrease in the volume of rainfall as well as the intensity of rainfall, leading to decrease in the total availability of water in the region including groundwater recharge. The potential impacts could be reduction in ground and surface water availability, level of crop yields and water quality.

In a case study of Lakhwar sub-basin, Kelkar et al. (2005) explored the vulnerability of farm households to water stress and climate variability through a participatory approach. Community-level interactions were carried out in two villages - Lakhwar which is characterized by purely rainfed farming, and Chhotau which has a mix of irrigated and rainfed farming (Figure 2). As compared to Lakhwar, Chhotau has poorer basic amenities (e.g. road connectivity, electricity, drinking water, health care, schools, etc.) and smaller average farm size. As in many villages in the Uttarakhand, however, both villages have experienced massive out migration. In response to increasing drought events and migration pressures, farmers have changed cropping patterns from food grains to feed (i.e. maize) and cash crops (e.g. potatoes, peas, ginger, etc.). However, such adaptive measures without reliable sources of water can further increase the farmers' vulnerability to droughts. Moreover, with lack of transport infrastructure, storage facilities and market opportunities, cultivation of cash crops such as fruits and vegetables is not a lucrative alternative to food grains.

The stratified caste system in India continues to influence people's access to education and land. For example, the Joshi and Rajput castes own

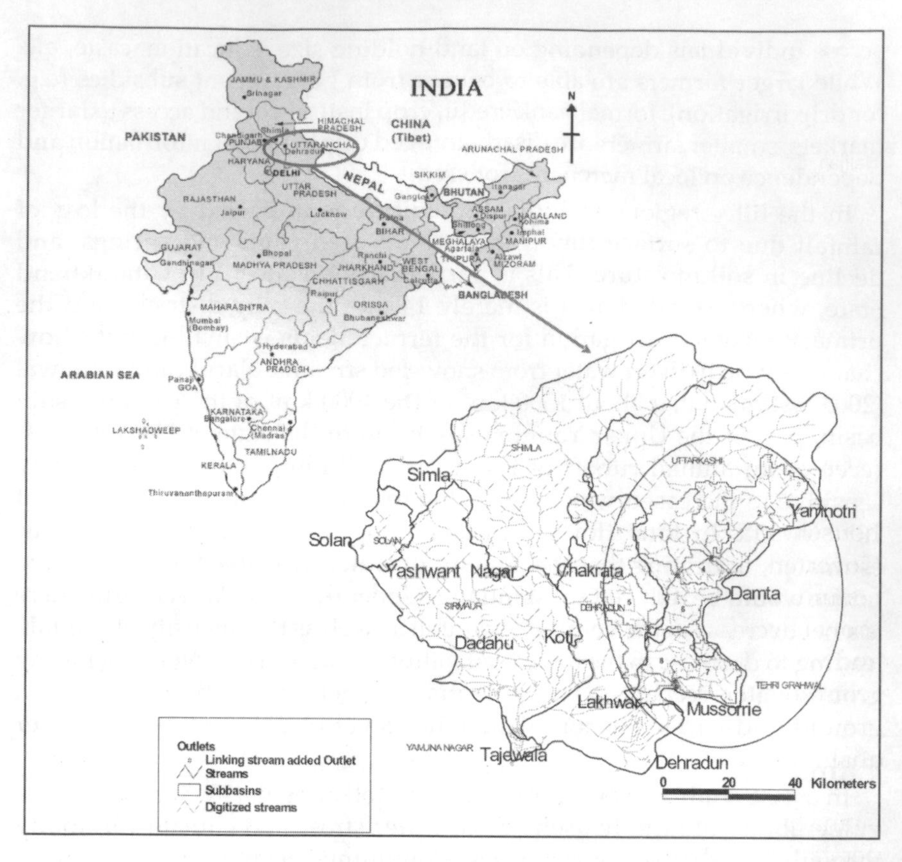

Fig. 2 Uttarakhand State: Case study area in India

The above is a rough sketch map which does not attempt to depict political boundaries

Colour image of this figure appears in the colour plate section at the end of the book.

a large part of landholdings in the two villages (Figure 3). These two upper castes also benefitted most from the government policy of promoting education and employment for tribal communities. The lower caste—Harijans—continues to live in impoverished conditions. None from this caste has completed college education and very few have reached high school. In contrast, many of the interviewed households from the upper castes have family members with either Bachelor or Master degrees.

Through a participatory approach, the farm households in the two villages were asked to identify adaptation measures to overcome increasing drought problems. Most farm households in both villages believe that their current agricultural activities are not longer viable and secure sources of livelihood. Among the adaptation measures suggested by the farm

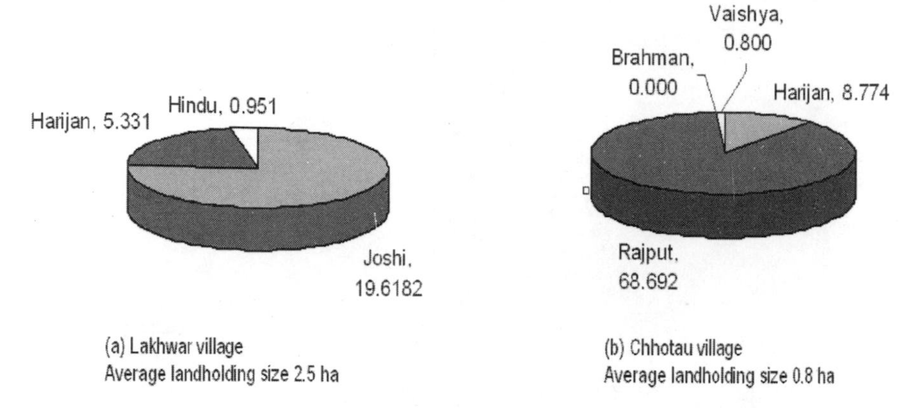

Fig. 3 Distribution of landholdings by caste in (a) Lakhwar and (b) Chhotau villages.

households, only rainwater harvesting is currently supported by the Government of Uttarakhand. Many innovative measures identified by the community for adding value to traditional agricultural products (e.g. grow reetha for producing soaps and shampoos, mundhwa for baby food and wine, jhangura for pillows, etc.) would need technical support and a reliable raw material sourcing and marketing chain.

China

China has a large water resource endowment on an absolute level, but on a per capita basis, it has among the lowest in the world. While the nation's water resources are overwhelmingly concentrated in southern China, northern China, in particular the area north of the Yangtze River Basin, has only one-fourth the per capital water availability of the South and one-tenth the world average (Ministry of Water Resources 2000). As one of the most densely populated regions in the world and an important agricultural and industrial region in China, the areas in northern China have long suffered from water stress in the process of their rapid economic development, and it became more severe in recent years with growing water demand. Furthermore, the lower levels of precipitation in these areas are highly irregular and mostly occur in the flood season. Based on the different scenarios used for Global Circulation Models (Watson et al. 1997), changes in runoff could range greatly between –16 and +7%, which means that climate change in the dry years would exacerbate the water shortage problem in these areas. Water deficiency has been increasingly evident in northern China, with the signs of falling water table and receding surface supplies. This trend is particularly evident in the Yellow River, which supports the irrigation in the North China Plain that

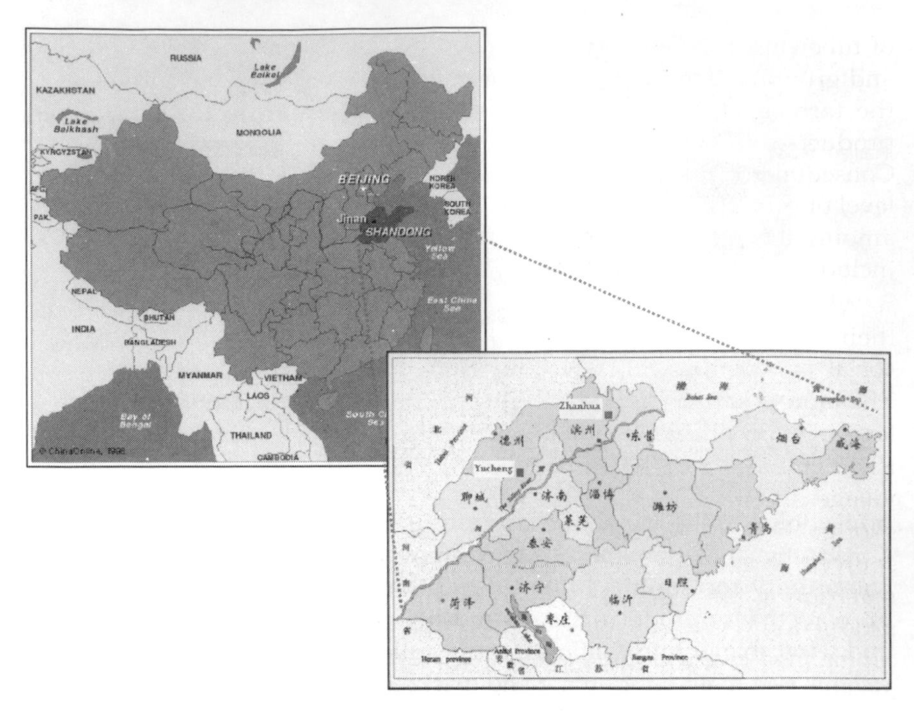

Fig. 4 Shandong Province: Case study Area in China
Colour image of this figure appears in the colour plate section at the end of the book.

produces about half of China's wheat and one third of its corn production. Due to the climatic variation and excessive withdrawal of irrigation water in the upper streams, the lower Yellow River has frequently dried up (i.e. "low-flow" events) during the 1990s and caused severe water shortages to the irrigation agriculture in the lower areas of the river.

Case studies in three villages (i.e. Ma, Xing and Ding) in Shandong (Figure 4), which is most heavily hit province by low-flow events in the Yellow River, were recently carried out to assess the vulnerability of farmers in these areas (Liu 2005). Because these villages are at lower part of the river, they are very much affected by the competition for water supply. Farm income in these villages has been affected by frequent occurrence of low-flow, with surface irrigation becoming increasingly unreliable and costly source of water. To avoid the use of irrigation during peak season when low-flow occurs, farmers adopted either early or delayed irrigation resulting in 10-30% yield loss. Others have adopted farming practices that maintain soil moisture such as the use of plastic film and mulch. However, with the further drying up of the river, sometimes resulting to no-flow events, the farmers have to adopt other measures including the installation of tubewells and pumps. The average irrigation costs (for surface water and groundwater, including pumping cost) have increased to 10-20% of the farmers' total production costs. Notwithstanding the increase in production costs, yield losses continued to increase between 20% and 60%. Consequently, many farmers limit the use of irrigation in lands with high level of productivity to reduce the costs. A more recent adaptive strategy among the farmers is the diversification of agricultural production including livestock, fruits and vegetables.

Whilst there are alternative measures available to adapt to droughts, their accessibility and affordability depend on the financial resources, social connections and farm locations of the farmers. Ma village in Zhanhua County, which is close to the mouth of the river, has the highest level of exposure to drought because the groundwater is deep and Yellow River is the only source of drinking and irrigation water. Xing and Ding villages in Yucheng County, which are in the upper stream, have not only shorter and less frequent no-flow events, but also alternative groundwater sources. Farmers from Xing village have higher income per capita and material assets and thus more capacity to diversify agricultural production. Despite having better farm location and larger farms per capita, the farmers in Ding village have gained relatively lower per capita income than those in Ma village in 1999 (Figure 5). This is attributed to the influence of community leadership and innovators, which are lacking in Ding village. In contrast, both leadership and innovators' influence have helped some farmers in Xing and Ma villages identify alternative

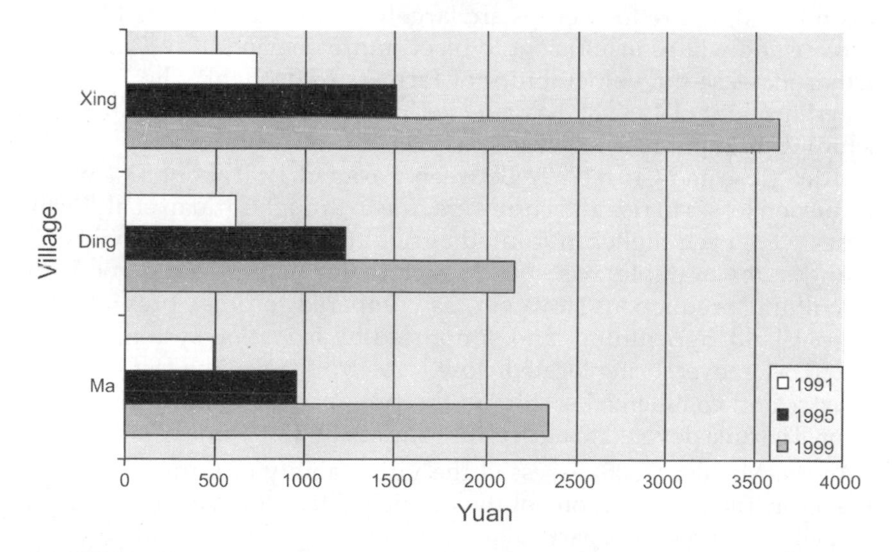

Fig. 5 Changes in per capita income in selected villages in Shandong Province

livelihood options, modern farm practices, better market access and transaction deals and, in few cases, source of loans. Promoting not only the role of these actors in farm villages, but also the link of the government to them will help improve the adaptive capacity of the farmers, particularly the poor. The latter is particularly important because farmers have complained that local governments have promoted technologies not suited to the physical characteristics of the farms (e.g. soil type, water supply) and cash crops without available market, not to mention the lack of technology extension services. Liu (2005) emphasized the need to promote the use of effective water-saving technologies that are affordable to farmers (e.g. drip or sprinkler system) and to implement water reforms that can provide incentives for farmers to conserve water and improve irrigation efficiency.

The Philippines

Many regions in the Philippines have been adversely affected by extreme climatic events during the past few decades. The current vulnerability of farmers to climatic variability is "due to the high frequency of tropical cyclones and floods, the seasonal occurrence of droughts, especially during El Niño episodes, and the incidence of saline water intrusion in coastal areas due to coastal inundation caused by storm surges" (Buan et al. 1996). Results of GCM showed that changes in rainfall and temperature in the future will be critical to future inflow in major reservoirs in major agricultural areas in the Philippines. For example, the Angat reservoir in Luzon Island, where food crops are largely produced, runoff is likely to decrease and will be insufficient to meet future demands. Thus, this will further increase the vulnerability of farmers to drought. The El Niño-related drought of 1982-83 has affected thousands of agricultural areas, including multipurpose reservoirs, where very low water reservoirs were recorded (Jose and Cruz 1999). Between 1968 and 1990 about 33% of the production losses in rice and corn were due to droughts (Buan et al. 1996). Losses were even higher in rainfed agricultural areas. The province of Batangas, for example, has soil characteristics that are very good for agricultural production. However, as compared to other provinces in Luzon Island, agricultural land supported by irrigation system is very limited. Moreover, being located close to Metro Manila, the fast growing industry and commercial sectors in the province affect the agricultural sector. The area devoted to agriculture shrank by 15% from 1993 to 2001.

Acosta-Michlik (2005) assessed the vulnerability of farmers in three villages in Tanauan City, one of the municipalities of Batangas Province. Each village represents a particular ecosystem: Cale (agricultural), Natatas (urban) and Gonzales (coastal). Cale and Natatas have the same number of farmers (138), whilst Gonzales has less number because fishing and eco-tourism are alternative sources of income for the villagers. Only 33% of the total land area in Natatas is devoted to agriculture, hence the farm sizes are smaller than in the other villages. The farms in Natatas are however closest to the largest market in the province. A common characteristic of these three villages is the lack of irrigation system. Due to the increase in frequency and intensity of drought as well as increase in the costs of labour and other production inputs, the productivity and profitability of staple food crops (i.e. rice and corn) and traditional commodity exports (i.e. sugar and coconut) have declined causing many farmers to shift land use to cash crops such as vegetables and fruits. Some farmers adopted farming practices such as the use organic fertilizers, drought-resistant varieties and plastic cover to cut the use of water for producing vegetables. However lack of transport vehicles and storage facilities combined with high market competition for cash crops in the municipal public market (i.e. many farmers and traders from different regions in Luzon export fruits and vegetables in Tanauan) make it difficult for farmers to maintain a high level of income. To support the assessment of the differential vulnerability, a cluster analysis was carried out on the socio-economic and farm attributes of the farmers, which were collected from interviews with 99 farmers in the case study areas. The analysis generated four types of

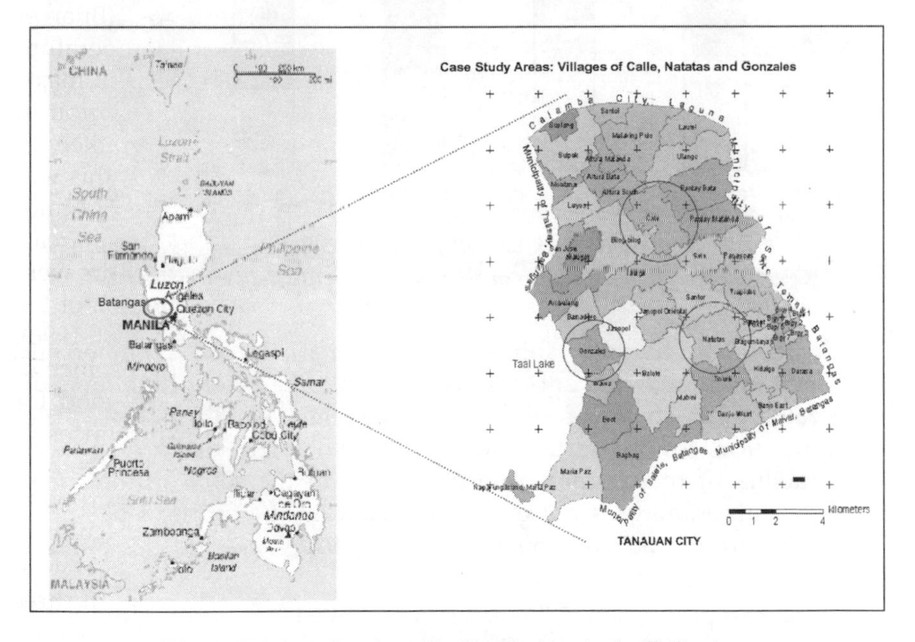

Fig. 6 Batangas Province: Case study Area in the Philippines
Colour image of this figure appears in the colour plate section at the end of the book.

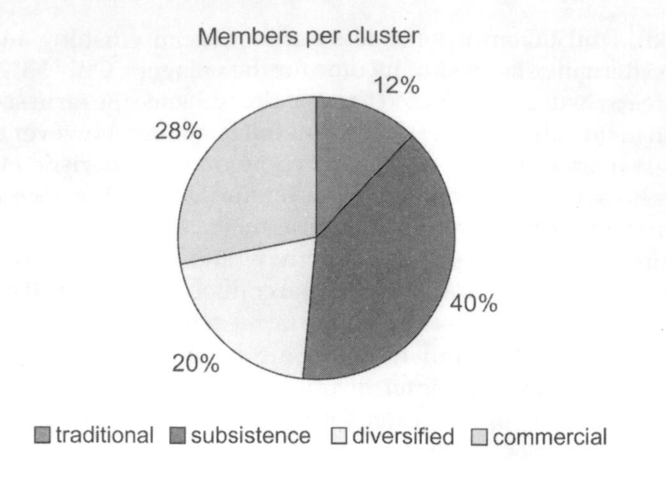

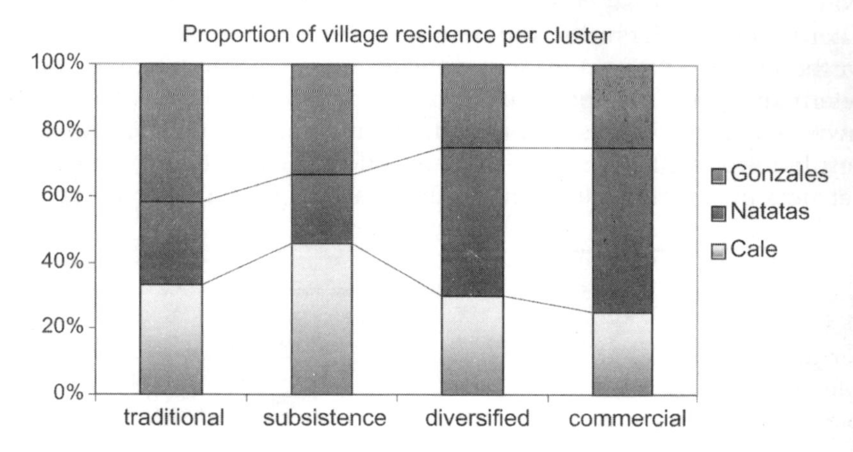

Fig. 7 Distribution of the farmer typologies in the case study areas
Colour image of this figure appears in the colour plate section at the end of the book.

typologies - traditional, subsistence, diversified and commercial. Although traditional and subsistence farmers have a very high level of vulnerability because of their low income, small assets, little diversification, low education and limited knowledge, the degree of vulnerability of these types of farmers are not the same in the three villages. Vulnerability is not only dependent on the farmer's personal attributes, but also on the economic and physical characteristics of their location.

The highest proportion (40%) of the interviewed farmers falls under the subsistence typology (Figure 7). Whilst most subsistence farmers are located in Cale, a large number of the diversified and commercial farmers are in Natatas. Traditional farmers, which account for the smallest number of the farmers, are mostly found in Gonzales. Traditional and subsistence farmers are expected to be very vulnerable in Cale, but not necessarily in Natatas and Gonzales. Whilst Cale village has very good biophysical attributes, the farms are very far from the Tanauan public market and national highway, making farm production and marketing crops more costly. Regardless of the typology, the farmers in Natatas will be able to take advantage of the proximity to Tanauan public market. Moreover, almost all roads in this village are concrete (Barangay Annual Report, 2004), making transportation of products to the public market faster and cheaper. This is particularly important for the diversified and commercial producers of vegetables, because cool warehouses are not very common. Traditional farmers also benefit from their proximity to the public market. They need not rely on traders or middlemen to market their crops. Because Natatas is urbanized, it is easier for the subsistence farmers to find alternative non-farming jobs. Among the three villages, Gonzales is farthest from Tanauan public market. Nevertheless, eco-tourism offers an alternative job to the farmers in Gonzales. Rich families from Tanauan City own lands in Gonzales, and they lease/share the farms to the poorer residents in the village. Most of these farmers fall under the traditional typology. The rich landowners gave the poor farmers part of the land to build their houses (Barangay Annual Report 2004). The good tenant-landowner relationship could, thus help increase the adaptive capacity of the poor farmers. There is greater probability for Natatas and Gonzales to have irrigation systems. Natatas has a river system and relatively plain terrain, which will make construction of irrigation at lower costs. Meanwhile, Gonzales is a neighbouring village of Ambulong and Bañadero, the only two villages with agricultural irrigation system in the municipality. However, according to a municipal agriculturist, construction of irrigation in Cale will not be affordable. The most relevant adaptation support for the farmers particularly in Natatas is the reduction in transport costs and development of irrigation system. Local officials have given priority to irrigation in development plan for Gonzales, but not for Cale and Natatas.

4. LOCAL ADAPTATION THROUGH KYOTO MECHANISMS

The experiences of local communities in different parts of Asia show that many farm households have the cognitive ability, but lack financial capacity and infrastructural support to respond to changes in climate change (i.e. increase in temperature) and variability (e.g. frequency of droughts). They have accumulated valuable knowledge on adaptation to environmental changes that has been passed through generation. Based on the above case studies, the most common measures considered by farmers across Asia are adjusting cropping schedule, changing land use from food to cash crops, diversifying sources of livelihood, identifying alternative sources of water, and strengthening community help. However, the risks associated not only with the rapid change in intensity and variability of climate change impacts, but also the persistent lack of appropriate infrastructure and institutional support to meet the challenges of global market make these measures unsustainable. For example, shifting to cash crops that demand less water will fail to generate high income for the farmers if link to market is weak due to poor transport and marketing systems. Small-scale irrigation systems, such as the water harvesting from rain will not be a relevant adaptation option for many farmers, if they do not get institutional support to build the necessary infrastructure. The major barrier to sustainable adaptation in developing Asian countries is thus—the low level of economic development in many rural communities. Promoting economic and rural development remains an important policy goal in these countries, more than mitigating climate change. Since years national and international development projects and programmes dealing among others, with poverty alleviation and food security aim largely at increasing adaptive capacity of communities in developing countries. Because of the close link between adaptation and development, many experts dealing with the adaptation issues in climate change are those who have gained expertise in human development fields particularly poverty and security. This is of course very appropriate because for climate-policies to be more relevant to developing countries, they must be "development-oriented" (IGES 2005).

The only Kyoto mechanism that addresses development issue is the clean development mechanism (CDM). This mechanism aims to achieve not only the reduction of emissions, but also social and economic development (Oxley and Macmillan 2004). In addition to renewable energy promotion, CDM is considered an appropriate means to integrate climate change in developmental policy (Oxley and Macmillan 2004, IGES 2005). Consequently, the Asian countries including China, India and the Philippines have special preferences for CDM projects with large sustainable development benefits or contribution to adaptation capability. Moreover, the significant interest in CDM is evident from the number of CDM projects (i.e. 16 out of 33 registered projects) hosted by Asian countries in 2005 (Jung et al. 2005). According to Oxley and Macmillan (2004), while the CDM intends to harness market power, its structure is more akin to government development assistance. Depending on national development goals, the approval of CDM projects by the governments will tend to be based more on subjective rather than on objective grounds, with the latter being prioritized in the work of the CDM Executive Board. Among others, the conflicting interests between developing countries and the Board may be also contributing to the low registration rate of CDM projects. For example, out of the 107 projects approved by the Indian government only seven were registered by the CDM Executive Board in 2005 (Jung et al. 2005). Moreover, most of the registered projects addressed emissions reduction with little sustainable development and adaptation benefits for local communities.

Whilst the Asian countries expressed concern about the insignificant contribution of CDM-registered projects on local development and adaptation goals, a significant number of the projects approved by the government to date for registration to the Executive Board are dealing with the promotion of renewable energy, with very little consideration of the adaptation needs of the local communities. Jung et al. (2005) mentioned, for example, that India noted the importance of traditional technologies and indigenous knowledge in meeting adaptation needs. Nevertheless, it emphasized the need to focus on funding for the development and transfer of new technologies for adaptation. In the case study area in India (i.e. Uttarakhand state), all the CDM- projects which more approved by the government focus on development of hydroelectric power, using established technology (CDM country websites). However, hydroelectric power will compete with use of water for agriculture irrigation, and so these projects are expected to further aggravate the water shortage problem in the area. So far, none of the approved CDM projects promote the technology for water harvesting, which suits the local need of the farm households in Uttarakhand state. Similarly, the CDM projects proposed for China and the Philippines promote the use of renewable resources without much regard of the adaptation needs of local communities. The Chinese government prefers the 'top-down' approach for CDM planning and implementation for several reasons, including time saved from centralized monitoring of projects, faster propagation of successful projects, ease of monitoring and evaluation (Pan 2003; Xu 2005 as cited in Jung et al. 2005). However, as the case studies presented above show, the degree of vulnerability of farm households to drought is location-specific due to the differences in economic, social, cultural and physical characteristics of the environment. It will thus need a 'bottom-up' planning and implementation if CDM projects were to address adaptation needs of local communities. "Adaptation strategies built on community-based approaches and local knowledge and based on active involvement of local stakeholders are likely to succeed better than those followed using top-down approaches" (IGES 2005). Not only should CDM projects address local adaptation needs, but their geographical distribution should also reflect the regional vulnerability pattern, not only across but also within the countries. The CDM Executive Board and the responsible government agencies should thus give priority to the most vulnerable people and areas when approving CDM projects.

5. CONCLUSIONS

Asia was the most vulnerable to climatic extremes in the past and its agriculture sector is expected to be significantly affected by climate change in the future. The magnitude of damage due to droughts differs across Asia, and the degree of vulnerability varies across different groups of population. The different case studies in China, India and the Philippines showed, however, that adaptation measures adopted in these countries have some similarities such as changing land use, improving farming practices and farm diversification. But the success of these measures has been constrained by the very low level of economic development in many rural areas. Whilst the Asian countries emphasize the key role of CDM in linking sustainable development and adaptation goals, the CDM projects approved in Asia have very little impacts on adaptation of local communities. Some projects may even run a risk of increasing, rather than decreasing vulnerability of farm households. Long-term objectives of reducing emissions should not be favoured over short-term objectives of reducing vulnerability. Hence, CDM projects must be assessed in terms of its impacts on current vulnerability and adaptation. Moreover, the distribution of the projects should reflect the vulnerability pattern not only across the region, but also within countries. As compared to mitigation, funding to achieve adaptation goals is very small. This will influence the success of Kyoto mechanisms in addressing adaptation needs of a large number of vulnerable people living in agricultural communities with very low level of economic, social and institutional development. While the Asian governments opposed the use of Overseas Development Assistance (ODA) for CDM purposes, ODA funded development projects should be able to complement the adaptation goals in Kyoto Protocol.

Acknowledgements

The case studies in China, India and the Philippines were supported by the Advanced Institute on Vulnerability to Global Environmental Change, a program funded by the David and Lucille Packard Foundation and coordinated by START in partnership with IIASA (http://www.start.org/program/advanced_institutes.html).

References

Acosta-Michlik, L. 2005. Intervulnerability Assessment: An Innovative Framework to Assess Vulnerability to Interacting Impacts of Climate Change and Globalization. START Project Report at http://www.start.org/Program/advanced_institutes_3.html.

Amadore, L., W.C. Bolhofer, R.V. Cruz, R.B. Feir, C.A. Freysinger, S. Guill, K.F. Jalal, A. Iglesias, A. Jose, S. Leatherman, S. Lenhart, S. Mukherjee, J.B. Smith and J. Wisniewski. 1996. Climate change vulnerability and adaptation in Asia and the Pacific: workshop summary. Water, Air, and Soil Pollution 92: 1-12.

Barangay Annual Report. 2004. Reports for the Barangay of Cale, Natatas and Gonzales.

Buan, R.D., A.R. Maglinao, P.P. Evangelista and B.G. Pajuelas. 1996. Vulnerability of rice and corn to climate change in the Philippines. Water, Air, and Soil Pollution 92: 41-51.

CDM country websites: China - http://cdm.ccchina.gov.cn/english/, India - http://cdmindia.nic.in/cdmindia/, Philippines - http://www.cdmdna.emb.gov.ph/

CLUP. 2005. Report for the Comprehensive Land Use Plan and Tax Mapping Project for the City of Tanauan, Batangas, Philippines.

EM-DAT: The OFDA/CRED International Disaster Database, Université Catholique de Louvain, Brussels, Belgium.

Iglesias, A., L. Erda and C. Rosenzweig. 1996. Climate change in Asia: A review of the vulnerability and adaptation of crop production. Water, Air, and Soil Pollution 92: 13-27.

Institute for Global Environmental Strategies (IGES). 2005. Sustainable Asia 2005 and Beyond - In the pursuit of innovative policies. IGES White Paper. IGES, Japan. p. 90.

Intergovernmental Panel on Climate Change (IPCC). 2001. Climate Change 2001: Impacts, Adaptation, and Vulnerability. Cambridge University Press, Cambridge, UK.

Jose, A.M. and N.A. Cruz. 1999. Climate change impacts and responses in the Philippines: water resources. Climate Research 12: 77-84.

Jung, T.Y., A. Srinivasan, K. Tamura, T. Sudo, R. Watanabe, K. Shimada and H. Kimura. 2005. Institute for Global Environmental Strategies Concerns, Interests and Priorities Asian Perspectives on Climate Regime Beyond 2012. IGES, Japan.

Kelkar, U., K.K. Narula, V.P. Sharma and U. Chandna. 2005 Vulnerability and adaption to climate variability and warter stress in Uttaranchal State, India. START Project Report.

Liu, Chunling. 2005. Farmers' Coping Response to Low-flows in Lower Yellow River. START Project Report.

Ministry of Water Resources. 2000. China Water Resources Bulletin (in Chinese).

Narula, K.K. and S. Bhadwal. 2003. Impact of climate change on hydrology for better decision-making at a river basin level in India: a case study. In TERI.

2003. Environmental threats, vulnerability, and adaptation: case studies from India. New Delhi: The Energy and Resources Institute.

O'Brien, K., R. Leichenko, U. Kelkar, H. Venema, G. Aandahl, H. Tompkins, A. Javed, S. Bhadwal, S. Barg, L. Nygaard, and J. West. 2004. Mapping vulnerability to multiple stressors: Climate change and globalization in India. Global Environmental Change 14(4): 303-313.

Oxley A. and S. Macmillan. 2004. The Kyoto Protocol and the APEC economies. Australian APEC Study Centre. p. 10

Parikh, J.K. and K. Parikh. 2002. Climate Change: India's Perceptions, Positions, Policies And Possibilities. OECD Report.

Sati V.P. 2005. Systems of Agriculture Farming in the Uttaranchal Himalaya, India. Journal of Mountain Science Vol 2, No. 1: 76-85.

UNFCCC. 2002. A guide to the climate change convention process. Climate Change Secretariat at http://unfccc.int/resource/process/guideprocess-p.pdf.

Watson R.T., M.C. Zinyowera and R.H. Moss. (eds.). 1997. IPCC Special Report on The Regional Impacts of Climate Change: An Assessment of Vulnerability, Cambridge University Press, Cambridge, UK.

World Bank (WB). 2006. The Little Green Data Book 2006. WB, Washington, D.C., USA.

31

CHAPTER

Climate Change and Land Degradation in China: Challenges for Soil Conservation

Alexia Stokes[1], Yibing Chen[2], Jingjing Huang[2] and Chaowen Lin[2]
[1]INRA, AMAP, TA-A51/PS2,
Boulevard de la Lironde, 34398 Montpellier Cedex 5, France
Tel: +(33 4) 67617525, Fax: +(33 4) 67615668
E-mail: alexia.stokes@cirad.fr
[2]Soil and Fertilizer Institute, Sichuan Academy of Agricultural Sciences,
20 Jingjusi Road, Chengdu, Sichuan 610066, P.R. China
Tel: +(86 28) 84784147, Fax: +(86 28) 84791784
E-mail: ybchen@mail.sc.cninfo.net

INTRODUCTION

China is currently facing serious environmental challenges and is listed amongst the world's most serious contributors to pollution and environmental destruction (Wang 2004, Liu and Diamond 2005). With the highest population in the world (currently 1.3 billion) and the fastest rate of economic development, national resources are being depleted dramatically and not replaced quickly enough. Not only are local products being consumed at a rate too fast for sustainable renewal, but China is also a major importer of tropical forest timber, making it largely responsible for current worldwide tropical forest deforestation (Adams and Castano 2001). China's leaders are aware of these environmental problems, which began with serious deforestation in the 1950s, leading to overgrazing, accelerated topsoil erosion (Fig. 1a), landslides (Fig. 1b, 2) and desertification. Such land degradation has been exacerbated over the last 20 years due to rapid industrialization. In the 12 years since the Kyoto Protocol, huge efforts have been undertaken in what can be called the 'greening' of China. Although the Kyoto Protocol recognizes the importance of controlling and reducing greenhouse gas (GHG) emissions which currently come primarily from industrial and transportation sources, it also recognizes the corresponding opportunities to be gained through better management of carbon (C) reservoirs and enhancement of C sinks (sequestration) in forestry and agriculture (Dumanski 2004). Better management of land use change, soil conservation and the restoration of degraded land will help achieve C sequestration. Nevertheless, mitigation of such strategies is not easy, and requires the cooperation of the central government, local authorities and stakeholders. In a vast country like China which is undergoing huge industrialization and economic development, difficulties may be confounded through financial and technical shortcomings, communication problems and the isolation of impoverished farmers and stakeholders.

LAND DEGRADATION IN CHINA

Soil management challenges for China include achieving food security with minimal risks to the environment. Such a situation will be difficult to achieve, given current problems with land degradation, freshwater deficits, resource-poor farmers and extreme weather events associated with climate change. In order to enhance food production, it would be necessary to utilize land appropriately, by intensifying production on prime agricultural land and restoring degraded lands and ecosystems. The need to produce sufficient quantities of food and the development of an ancillary infrastructure to transport it to the markets has led to the cultivation and modification of hill slopes that often have been neglected due to their poor suitability for these purposes. In a country where two-thirds of the land surface is hilly or mountainous, when changes in land use upset the delicate equilibrium between slope stability and environment, then slope failures can occur and the ensuing degradation may lead to an irrecoverable loss of soil by erosion or even damage to property and loss of life through landslides and flash-floods. If predictions concerning global change prove to be correct (UNEP 2002), more and more extreme weather events will occur, resulting in increased drought, flooding, landslides, avalanches and storms. Water contamination and land degradation through erosion and landslides have been listed as China's top environmental priorities, and concerned 3.67 million km^2 of land in 2002. It has been calculated that 30-100 landslides occur per

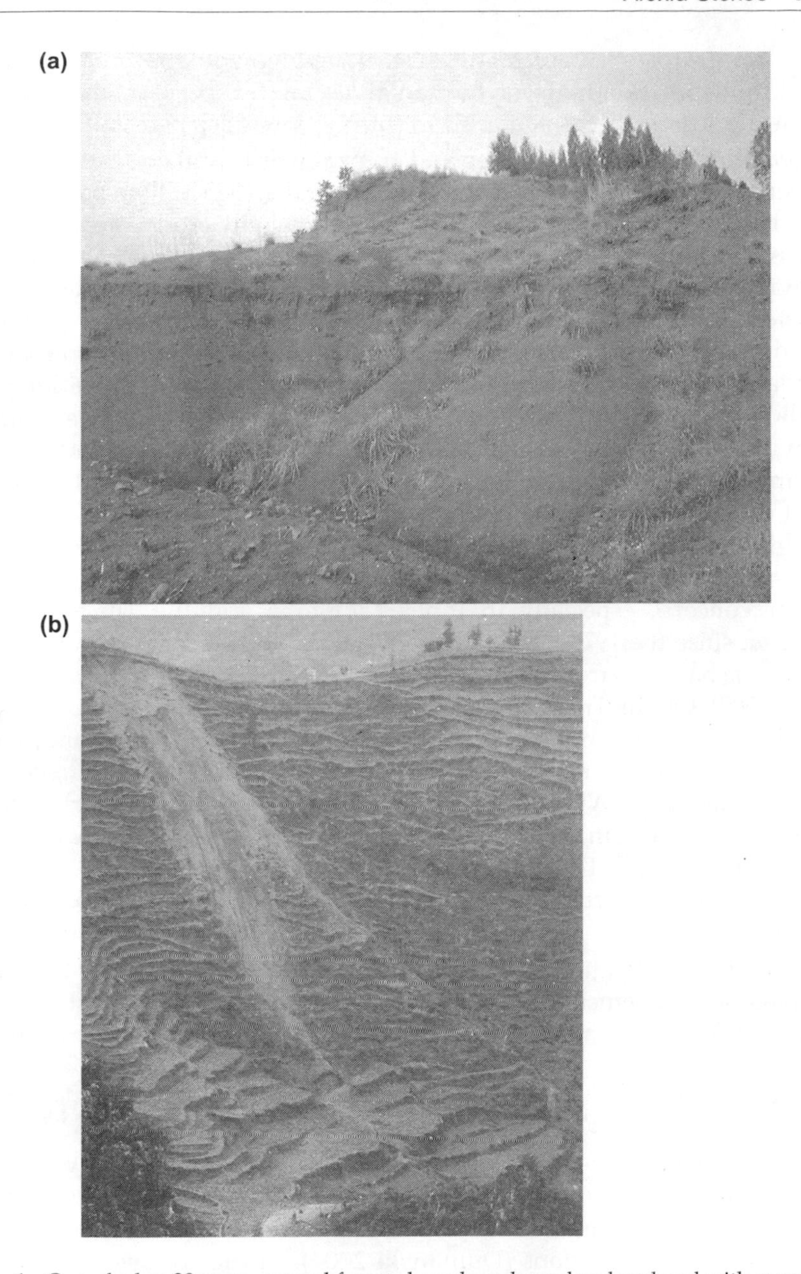

Fig. 1 Over the last 20 years, natural forests have been logged and replaced with crops. In sites which are fragile, such as on steep slopes, the poor soil fixation by unsuitable species can lead to a) soil erosion in the Sichuan province (Photo: Y. Chen) and b) landslides in the Guizhou province; landslides such as this can be seen every few kilometers along major road axes (Photo: T. Fourcaud, CIRAD, France).

Colour image of this figure appears in the colour plate section at the end of the book.

1000 km^2 (http://www.cgs.gov.cn/task/shuigonghuan/zaihai/about/002.htm) with hundreds of human lives lost every year and annual monetary losses of >5.1 billion RMB (http://www.sepa.gov.cn).

Soil erosion by wind, water and freeze-melt is a major cause of soil material removal with an estimation of more than 355 million ha of China affected in 2002 (Zhao et al. 2002). The most heavily eroded areas are the Loess Plateau in north-west China, the Yellow and Yangtze Rivers and the provinces of mountainous southern China. The spatial distribution of erosion is affected by the type of erosion-driving force and geographic region. Wind erosion is most common in the northern provinces of Xinjiang, Gansu, Qinghai, Ningxia, Shanxi and Inner Mongolia (1.9 million km^2 in total) whereas water erosion in prevalent in the southern provinces including Sichuan, Chongqing, Yunnan, Guizhou, Guangdong, Guangxi, Jiangxi, Jiangsu and Fujian (1.65 million km^2 in total). Freeze-melt erosion occurs on the high plateaux and glaciated regions of Tibet-Qinghai, Xinjiang, Gansu and Sichuan (1.27 million km^2 in total). Severe desertification as a result of deforestation and overgrazing has been a major concern, especially in Inner Mongolia and the Tibet-Qinghai plateau, since the 1970s. The mean annual rate of desertification in Inner Mongolia has risen from 1142 km^2 in the 1960s to 2460 km^2 in the 1990s (Li et al. 2003). On the Tibet-Qinghai plateau, desertification and erosion has resulted in 0.345 million km^2 (Yang 2003) of degraded land, and restoration work in this region has been neglected until fairly recently (Yang et al. 2006). With regard to the Loess region, this is a particularly fragile arid area, being 0.62 million km^2 large and with an average soil loss of 10000 t km^2 yr^{-1} (He at al 2006). This soil loss rate can however reach much greater proportions, attaining 24 000 t km^2 yr^{-1} in some areas (Zheng 2006).

Therefore, mitigation strategies are needed urgently in order to better manage soil movement problems. However, such strategies must be cost-effective and sustainable in the long-term, especially for use in rural areas.

CARBON SEQUESTRATION IN DEGRADED LAND

Human-induced soil erosion and desertification, burning of crop residues, grassland degradation, wetland reclamation for agriculture, low water use efficiencies, organic matter and fertility loss and excessive tillage are all sources of C emissions (Dumanski 2004). In soils of degraded land, little C exists, but if planted appropriately, C may be stored in plant leaves, stems and roots, although such a process takes many years to have a significant effect. Soils may also have the potential for greater C storage over a longer timeframe than woody biomass (Lackner 2003) although a very healthy soil microbial ecosystem is needed for C accumulation. In China, the median soil organic carbon (SOC) content is 8.23 kg m^{-1} in the upper 0–1 m of soil with over a third present in the topsoil (defined as the upper 0–0.2 m of soil, Xie et al. 2004). The SOC stock under forests (17.4 Gt) is the highest, accounting for 25% of the total SOC in China despite the fact that forest soil amounts to only 16% of the total territory. Under croplands, SOC is 14.7 Gt, but under deserts is only 3.9 Gt (6% of the total SOC, Xie et al. 2004). Dumanski (2004) estimated that when cultivated, about 20–30% of this topsoil C is released to the atmosphere within the first 20 years in temperate regions, and 50–75% in the tropics. Nevertheless, the potential for C sequestration in China alone is 105–198 Mt C yr^{-1}, provided that land management techniques are suitable. On a worldwide scale, the potential for C sequestration through restoration of degraded lands is about 500–1000 kg C ha^{-1} yr^{-1} (Dumanski 2004). Climatic variables also play a role in C sequestration by soils and in China it was shown that SOC expands with increasing precipitation but decreases with increasing temperature (Xie et al. 2004).

One of the greatest hopes of decision-makers concerned with climate change and GHG, is the use of vegetation to fix C, thus removing carbon dioxide (CO_2), a major GHG, from the atmosphere (Lal 1999). Nevertheless, hopes pinned on C sequestration by forests were dashed to a certain extent by Körner et al. (2005), who showed that mature trees did not accrete more biomass C in stems in response to elevated CO_2. Carbon flux appeared to pass through the system faster when CO_2 levels were higher. It must also be considered that any increases in plant growth which do occur under elevated CO_2 may lead to an increasing input of plant biomass (debris, root biomass and root exudates) in the soil. Therefore, a higher availability of SOC compounds might actually stimulate soil C turnover and hence the emission of C compounds thus resulting in a *reduction* of C storage in soils.

Nevertheless, more research still needs to be carried out on C sequestration by different plant species in a variety of ecosystems. In particular, the estimation of soil C enrichment within different types of soils and soil depths is needed to estimate subsurface C sequestration following reforestation or afforestation activities (Ficklin et al. 2006). Planting of degraded lands should be performed with a view to restoring the damaged soil rather than as a means for sequestering C through the fixation of tree biomass C. Once SOC augments, many other benefits occur including an increase in nutrient cycling, soil biodiversity and fertility as well as an improvement in soil water storage capacity.

HOW CHINA HAS BEEN COMBATTING LAND DEGRADATION SINCE KYOTO

Climate change over the last decade has been held responsible for extreme weather events which are frequent and often catastrophic in China. Due to the massive deforestation which has been carried out over the last 50 years, when extreme rain- and windstorms do occur, soil loss through landslides and erosion is multiplied. Nevertheless, such natural disasters can have useful side effects in that the central Government is obliged to take action. The major 1998 floods along the Yangtze and Yellow Rivers, which affected hundreds of millions of people, resulted in a ban on the logging of natural forests in the upper and middle reaches of watersheds of these rivers. Concerned by the sediment discharge from erosion which has been filling up the Yangtze River since 1949 (Liu and Diamond 2005), the Chinese government instigated a series of engineering measures to mitigate soil loss along the upper reaches of this river. A huge project to restore the vegetation and natural environment was launched in 1988 (Sun et al. 1994, Zhou 1999). This project is planned to last for 30 - 40 years and in the first stage (1988–2004), 7.9 million ha of forest was replanted, thus increasing forest cover from 20 to 40% in the target areas. Statistical data (http://www.riel.whu.edu.cn/show.asp?ID=2439) show that from 1999 to 2004, there were 227 million mu (a mu equals 0.15 ha) being restored to forest land or grassland, amongst which 208 million mu was arable land and 119 million mu were waste land. 50 billion RMB has been invested into this project, of which 12 billion RMB was expended on seed production, 4 billion RMB was used as subsidies to assist farmers and 34 billion RMB was used for providing food for farmers.

China's famous 'Grain for Green' 10-year programme began in 2000 and has already resulted in 79,000 km^2 of cropland being restored to forest or grassland (Xinhua News Agency 2004). This massive conservation programme gives incentives and subsidies to farmers to convert cropland on hill slopes to forestry, orchard or grassland. Farmers can use products derived from these new plantations e.g. fruit and timber, but the system must be sustainable. By transforming potentially unstable slopes prone to over-farming and erosion, benefits will be noticeable not only in the prevention of soil loss, but also river sedimentation and pollution due to the use of pesticides and fertilizers will be seriously reduced. The recent implementation of these soil and water conservation measures has now decreased sediment load in the Yellow River by 25% (He et al. 2006).

To reduce the severe desertification underway in Inner Mongolia, local provincial governments have implemented a series of measures to restore vegetation since the mid 1970s. Grazing has been reduced and desertified grassland has been enclosed using artificial sand arresters (Li et al. 2003). However, the most noticeable measure, which can be seen not only in Inner Mongolia but also in surrounding provinces including Beijing itself, is the famous 'Green Barrier' of poplar trees, whereby millions of poplar trees have been planted to fix soil and reduce atmospheric dust retention. The effects of these artificial forests on soil development and ecological sustainability are only now beginning to be quantified. Nevertheless, significant and positive changes in soil and vegetation parameters have occurred and the Green Barrier is effective to a certain extent in reducing dust deposition (Li et al. 2003).

As the socioeconomic development of the Tibetan Autonomous Region is weak and has lagged behind the rest of China, the financial support which can be devoted to ecological security in this region is limited (Zhao et al. 2006). Only in very recent years, have long-term plans been proposed for a better utilization of pastoral lands and watercourses (Zou et al. 2002, Yang et al. 2006, Zhao et al. 2006). A detailed desertification control plan was worked out in 1997, whereby 3.5 million ha of land will be rehabilitated by 2020 (Zou et al. 2002). Restoration measures include the planting of shelterbelts around farmland, artificial grassland planting, the use of artificial sand-arresters (Fig. 3) and the large-scale planting of sand-adapted economic forest and commercial plants (Zou et al. 2002).

The monitoring of all these restoration programmes will be performed through a national monitoring network. This network will be established under the directive of the National Soil and Water Conservation Law and will incorporate the dynamic monitoring and forecasting of soil erosion in China and the effect of conservation measures. The database to be developed will combine the result of runoff plots (Table 1), watershed management and field surveys of soil and water losses to determine the type and distribution of soil erosion. Such a database can then be used to provide scientific guidance for soil and water conservation (Wang et al. 2005).

Fighting Land Degradation at the Local Scale

Projects instigated by the government, e.g. the Grain for Green Project, mean that farmers have to implement mitigation measures at a local level. Not always having access to local infrastructure, support and guidance, farmers, in particular those who are geographically isolated may make unwise decisions about how to manage their land. As most of the farmland in China is owned by the government and leased over a certain time period, farmers may also take less care in managing their land than if it were theirs, to be passed onto their family. A further problem

Table 1 Examples of soil loss for various soil management conditions and land cover types in China

Area	*Method, slope*	*Cover/treatment*	*Erosion (t ha^{-1} yr^{-1})*	*Reference*
Hongya, Sichuan	Plots, 10°	Alley cropping 1. corn 2. corn + grass	 14.7 0.34	Unpublished, research report of MOA project (2004)
Santai, Sichuan	Plots, 12°	Alley cropping 1. goundnut 2. groundnut + *Eulaliopsis binata*	 3.8 0.17	Unpublished, research report of MOA project (2004)
Jianyang, Sichuan	Plots, 14°	Alley cropping 1. corn 2. corn + pear and day lily	 4.21 1.35	Unpublished, research report of MOA project (2004)
Hongya, Sichuan	Plots, 16°	Alley cropping 1. corn 2. corn + grass	 18.7 2.37	Unpublished, research report of MOA project (2004)
Jiangjin, Chongqing	Plots, 18°	Alley cropping 1. corn 2. corn + Chinese prickly ash	 43.6 12.2	Unpublished, research report of MOA project (2004)
Qingshen, Sichuan	Plots, 21°	Alley cropping 1. corn 2. corn + citrus	 4.6 1.3	Unpublished, research report of MOA project (2004)
Dayao, Yunnan	Plots, 15°	Reforestation, terraces	11.3	Hu et al. 2006
North-east Yunnan	Farm land, 8°	Cultivation along the slope	121.4	Yang et al. 1997
Jianzhen, Fujian	Plots, 20°		106.6	Chen et al. 2006

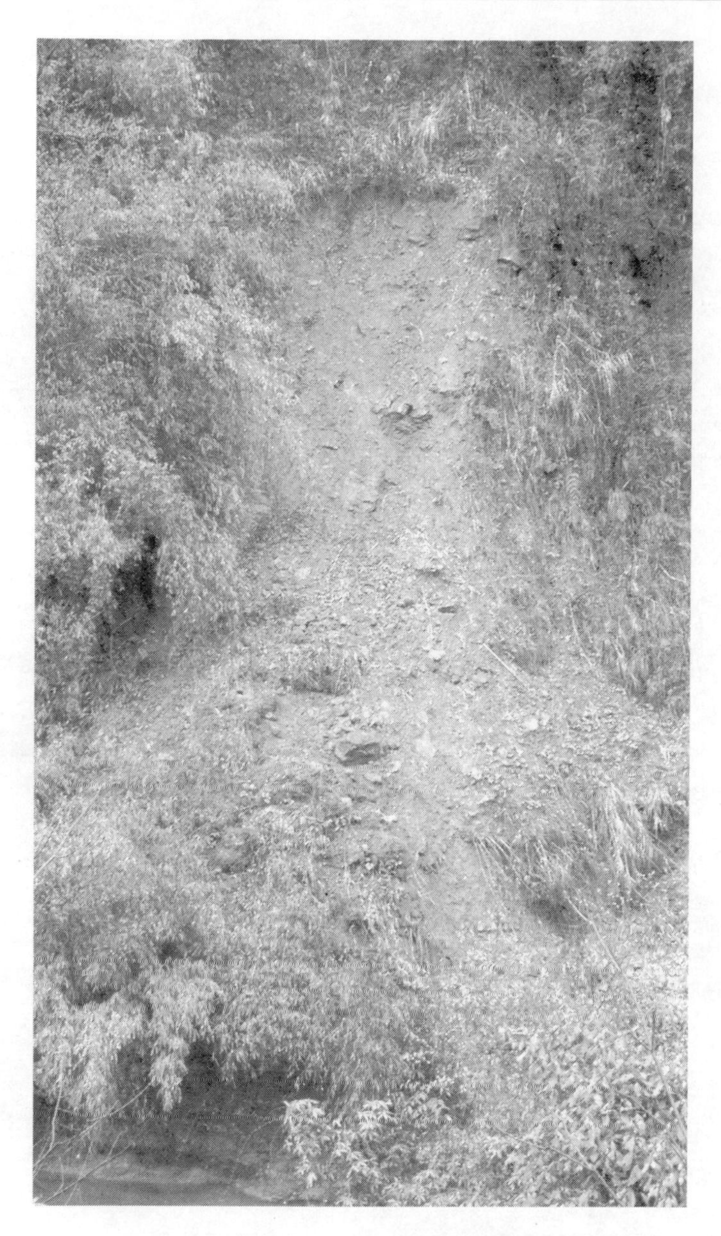

Fig. 2 Some species are not suitable for planting on steep slopes. Landslides are more frequent in plantations of big node bamboo (*Phyllostachys nidularia* Munro) compared to many other vegetation types (Stokes et al. 2007). Nevertheless, new environmental incentives can encourage stakeholders to plant unsuitable species at fragile sites (Photo: L. Jouneau, INRA, France).

Colour image of this figure appears in the colour plate section at the end of the book.

Fig. 3 Sand arresters help prevent the movement of shifting sand onto the railway lines of the new Tibet-Qinghai plateau railway. These simple fences and quadrants of rocks can also be planted with native sand-adapted species to help fix sand and improve substrate quality (Photo: A. Stokes).

Colour image of this figure appears in the colour plate section at the end of the book.

encountered is that traditional sowing, planting and harvesting methods, handed down from one generation to the next, are usually deemed infallible. New farming techniques are not always widely accepted, even when proven to increase yield or reduce topsoil erosion. If traditional methods are used, but not implemented correctly, through lack of training, care or resources, soil loss can be rapidly increased. For example, if hill terraces for the cultivation of crops are poorly constructed or maintained (Fig. 1b), topsoil erosion and slope instability will be exacerbated through water collecting on oversteepened terraces (Sidle et al. 2006). If the terraces collapse, breaches will focus surface runoff leading to gully formation and increased sediment transport down slope (McConchie and Ma 2002). Therefore, when new techniques are suggested to farmers, it is necessary to override the traditional way of thinking, but in the isolated countryside, this lengthy process is usually too difficult to carry out. Notwithstanding these various difficulties, many new methods have been developed to improve soil conservation at the local level.

Reforestation for Soil Conservation

Over the last decade, the Chinese government has put into action different programmes for the restoration and management of eroded and degraded land. However, simply decreeing that trees are to be planted instead of crops is not sufficient to prevent soil mass movement. Unfortunately, as has been recently highlighted, local authorities and lumber companies can profit financially from these new programmes and at the same time can actually contribute to landslides and consequent flash floods. It was shown in the Fujian province of southern China, which has the highest rate of forest coverage in China (62%), that flooding and natural disasters such as landslides, have worsened in the years since Kyoto. The reason behind these problems has been attributed to the replacement of Fujian's natural forests by plantation trees e.g. China fir (*Cunninghamia lanceolata* Lamb.) and high yielding Eucalyptus (*Eucalyptus camaldulensis* Dehnh.). In natural forests, the layers of under storey and dead wood are sufficient to retain rain water (if precipitation events <200 mm), thus controlling surface flow (Feng and Liu 2006). However, in monoculture plantations e.g. Japanese cedar (*Cryptomeria japonica* D. Don), the soil retaining capacity is low, hence soil movement is exacerbated (Genet et al. 2006). Along with the lack of under storey vegetation, erosion rates and sediment yields have been found to increase by over 37 times in coniferous plantations compared to other forest types (Hill and Peart 1998). The plantation of various non-timber species, e.g. fruit trees and bamboo on hill slopes can also be particularly devastating to slope stability, as the contribution of certain species is negligible to soil fixation (Fig. 2, Stokes et al. 2007).

Combined with the problem of planting unsuitable species or mixtures on fragile land, is the corrupt behaviour which can be found in certain local authorities. Although difficult for local people to fight against, outcries have begun in southern China concerning the illegal logging of natural forests by legally registered timber companies (Feng and Liu 2006). Local forest authorities issuing dubious certificates allowing companies to clear natural forest are rarely investigated. Logging companies have also been suspected of burning sites and processing the remaining wood. Yet the planting of fast-growing exotic Eucalyptus and other timber species can be considered as part of the Chinese government's replanting programmes; unfortunately the illegal origin of the bare lands on which the plantations are grown may be overlooked.

Managing Forests for the Long-term Control of Soil Loss

Although it seems evident that natural forest is the best type of vegetation to be used for preventing soil loss through erosion and landslides (Zhou 2004), many forests in China today comprise managed monospecific plantations. Natural processes are important factors governing slope stability in forested areas, but the way in which plantations are managed can severely affect soil conservation. The main interventions carried out by foresters are planting, thinning and felling of trees. The choice of species is governed by geographic and economic factors whereas subsequent management practices are decided by local forestry authorities. In general, mixed stands which include broadleaf species are more efficient in controlling soil loss through erosion or landslides, even when stands are very young (Zhou 2004, Roering et al. 2003). In monospecific conifer stands, it has been assumed that young trees contribute little to slope stability because of their lower root strength (Abe and Ziemer 1991, Sidle et al. 2006). However, young *Cryptomeria japonica* D. Don in the Sichuan is planted at high densities compared to older stands which have been thinned, therefore root biomass density is high and compensates for lower root strength (Genet et al. 2006). Hence, additional root cohesion, i.e. increased soil reinforcement, does not differ between plantations of different ages (when trees >10 yr old). What does influence soil fixation by roots in plantations is the felling of trees, whether it be through thinning or clear-cutting. Once a tree is cut, roots decay over a period of several years and remaining trees have not yet had time to exploit the newly available space (Sidle 1991, Genet et al. 2006). During this period, root cohesion decreases hence landslide risk increases significantly. In rotation forestry, it is of course necessary to cut trees, but care should be taken when to cut e.g. if the monsoon season is from May to August, trees should be felled in September to October, to allow root regrowth of remaining trees— the maximal possible time to occur before the next major rain events. Trees on the most fragile slopes should not be clear-felled, but gradually thinned and replanted. Soil conditions along slopes are usually heterogeneous: if the most vulnerable parts of the slope are identified, such as areas with thin soil or which are frequently waterlogged, trees should be removed with extreme care and should be replanted as quickly as possible.

Farming Practices to Mitigate Soil Conservation

Researchers in southern China have developed several new farming techniques which are also useful for cultivating degraded sloping land. This new methodology has been termed 'agroecology' as agricultural techniques have been combined with ecological principles (Parham 2001). For example, the Sichuan Basin, which is one of the most important foodstuff production bases in China, has suffered severe erosion through extreme rain events in recent years, with 300 million t yr^{-1} of sediment collecting in the Yangtze River. Therefore, local scientists have developed agroecological hedgerows to prevent soil loss. Alley cropping and 'cash crop hedgerows' are now planted perpendicular to the slope (Fig. 4) rather than parallel to the slope, as crops are traditionally planted in this region. Various combinations of different crop species have been tested and the soil loss measured. The most successful hedgerow type was a combination of pear (*Pyrus* spp.) and Chinese day lily (*Hemerocallis citrine* Baroni). After four years, crop yield was reduced by about 10% because the effective planting area was decreased (Fig. 4). However, the net return was two times higher than the traditional planting method, due to the additional crop of Chinese day lily and pear. Yields of 16,500 kg ha^{-1} of pears and 1500 kg ha^{-1} of Chinese day lily flowers are expected. These new hedgerow techniques also reduced soil loss by about 80% (Tu et al. 2006). From discussions with local farmers, feelings are quite positive about these new planting methods and farmers appear willing to adopt such practices. In other areas of China, farmers have also adopted similar practices to control soil erosion and the effect on soil loss is highly significant (Table 1). Successful techniques include the planting of Sabaigrass (*Eulaliopsis binata* Retz.), a perennial, fibre-producing plant which has a growing market in southern China (Fig. 5). Mulberry (*Morus alba* L.) can also be used to stabilize the risers at the edges of paddy fields (Fig. 6).

Fig. 4 New farming methods may help reduce soil loss on steep slopes. In trials in the Sichuan province of China, hedgerows were planted with a combination of pear trees (*Pyrus* spp.) and Chinese day lily (*Hemerocallis citrine* Baroni) flowers. Pears were planted with a within-row spacing of 1 m and day lilies were planted with a spacing of 13.3 × 13.3 cm (two rows in one hedgerow). On sloping land, soil loss was reduced by 80% using this technique (Photo: Y. Chen).

Colour image of this figure appears in the colour plate section at the end of the book.

More data is necessary on how the different farming methods control or at least alleviate erosion problems. A large body of data is now available for Southeast Asia (Sidle et al. 2006), where farming techniques are comparable to those in China. However, collating such data is difficult, methodology is not standard and results are not always easily accessible. Nevertheless, existing data for China (Table 1) show that the most successful farming technique which resulted in the least soil loss was alley cropping a combination of groundnut with *Eulaliopsis binata* (soil loss: 0.17 t ha^{-1} yr^{-1}) Along with the recent input into scientific research in this domain, such studies should be more numerous and widely available in the future.

RESEARCH QUESTIONS FOR THE NEXT 10 YEARS

Although relatively scanty until the last decade, research on land degradation on sloping lands in China is now pulsing along at a major

Fig. 5 After the original forest plantation was destroyed, terraces of Sabaigrass (*Eulaliopsis binata* Retz.) help to stabilize the slope and prevent landslides (Photo: Y. Chen).
Colour image of this figure appears in the colour plate section at the end of the book.

speed. Literature searches will quickly bring to light many new papers dealing with water and wind erosion (e.g. Gao et al. 2002, Zhang et al. 2005, Zhao et al. 2006, Zheng 2006), but less work is currently published on mass movement through shallow landslides on mainland China (Tang et al. 1994, Derbyshire et al. 2000) and how to restore degraded land through various agroecological or ecotechnological methods (e.g. Zhou et al. 2002, Zhang et al. 2004, Li et al. 2006, Yang et al. 2006). What we need to ask ourselves now is; how to plant or manage a potentially unstable or degraded site for a long-term sustainable conservation of soil? Which species should be used, bearing in mind that farmers and stakeholders might need to use this plantation as an income? Where bare land is to be revegetated, should natural regeneration be allowed, or should the soil be planted quickly with young grasses, shrubs or trees? With regard to C sequestration in soils, it is necessary to determine whether an increase in plant growth due to elevated CO_2 in natural conditions can occur and if so under which circumstances? Any such increase may then lead to an increasing input of plant biomass (debris, root biomass and root exudates) in forest and agricultural soils which in turn could cause changes in the C budget of soils. It is thus essential to determine whether higher amounts of

Fig. 6 Mulberry (*Morus alba* L.) is planted at the edges of paddy fields to stabilize the risers. *Colour image of this figure appears in the colour plate section at the end of the book.*

C may be sequestered in soil under these conditions or whether a higher availability of C compounds in soil actually stimulates C turnover and hence the emission of C compounds resulting in a reduction of C storage in soils.

To answer these and more questions, a major national programme is needed which covers not only the scientific study of the problem, but also includes the dissemination of results and teaching of new methodology to local decision-makers, authorities and farmers. Parham (2001) suggests establishing an international research and demonstration site where techniques to improve China's degraded lands can be tested, measured, improved, and demonstrated for farmers, Chinese and foreign researchers and leaders. Decision support systems also need to be freely available to practitioners as well as being easy to use (Jouneau and Stokes 2006). However, for such a programme to be successful in practice, it is vital that local authorities enforce new techniques, but not at the cost of already existing successful systems (e.g. the cutting down of trees to plant new trees). More employees must be hired to enforce the environmental acts and regulations developed by the central government, both now and in the immediate future and offences that damage the ecosystem must be treated seriously (Zhao et al. 2006).

CONCLUSION

Preventing and restoring degraded soils, enhancing soil carbon sequestration to mitigate the greenhouse effect and decreasing contamination of ground water will be priority issues in China in the future. The government aims at greening 600,000 km^2 of eroded areas by 2010 and the control of ALL soil erosion (officially announced at 2 million km^2) is targetted for the mid 21st century (Ministry of Water Resources). To achieve such major goals, soil scientists and engineers need to work hand-in-hand with agricultural engineers, foresters and environmentalists in order to focus innovative and demand-driven research in eco-restoration in the most fragile areas of this vast country. The results of such studies need then to be disseminated correctly to farmers and stakeholders in the field.

References

Abe, K.A. and R.R. Ziemer. 1991. Effect of tree roots on shallow-seated landslides. In: R.M. Rice (ed). Proceedings of the IUFRO Technical Session on Geomorphic Hazards in Managed Forests, August 5-11, 1990, Montreal, Canada. U.S. Dept. of Agriculture, Forest Service, Pacific Southwest Research Station. Gen. Tech. Rep. PSW-GTR-130. pp. 11-20.

Adams, M. and J. Castano. 2001. http://www.itto.or.jp/live/contents/download/tfu/TFU.2001.01.English.pdf (ITTO Newsletter 2001).

Chen, S., Y. Yang, W. Lin and T. Li. 2006. Quantitative research on soil and water loss and counter measures in man-made destroyed red soil region in subtropical region of Fujian Province. J. Soil Water Cons. 20: 6-10.

Derbyshire, E., X. Meng and T.A. Dijkstra. 2000. Landslides in the Thick Loess Terrain of North-West China. Wiley Publishers, USA.

Feng, Y. and Y. Liu. 2006. China's deforestation no longer driven mainly by poverty - Part 2 of 2. http://www.worldwatch.org/node/4495.

Ficklin, R.L., S.R. Mehmood and P.F. Doruska. 2006. Integrating science into public policy: challenges and opportunities for improved forest carbon accounting. J. Agric. Food Environ. Sci. 1 http://www.scientificjournals.org/articles/1026.htm.

Gao, Q., L. Ci and M. Yu. 2002. Modeling wind and water erosion in northern China under climate and land use changes. J. Soil Water Cons. 57: 46-55.

Genet, M., A. Stokes, T. Fourcaud, X. Hu and Y. Lu. 2006. Soil fixation by tree roots: changes in root reinforcement parameters with age in *Cryptomeria japonica* D. Don. plantations. In: H. Marui, T. Marutani, N. Watanabe, H. Kawabe, Y. Gonda, M. Kimura, H. Ochiai, K. Ogawa, G. Fiebiger, kos (eds). Interpraevent – Disaster Mitigation of debris flows, slope failures and landslides. Universal Academy Press, Inc. Tokyo, Japan. pp. 535-542.

He, X., J. Zhou, X. Zhang and K. Tang. 2006. Soil erosion response to climatic change and human activity during the Quaternary on the Loess Plateau, China. Reg. Environ. Change 6: 62-70.

Hill R.D. and M.R. Peart. 1998. Land use, runoff, erosion and their control: a review for southern China. Hydrol. Processes 12: 2029-2042.

Hu, S., Y. Zhou and X. Lu. 2006. Effect of land use change on the soil erosion for the ecological recovery region of DaYao county. YuNan Geographic Environment Research. 18: 40-43.

Jouneau, L. and A. Stokes. 2006. Development of a decision support system for managing unstable terrain: calculating the landslide risk of slopes. In: H. Marui, T. Marutani, N. Watanabe, H. Kawabe, Y.C.G. Heumader, F. Rudolf-Miklau, H. Kienhloz, M. Mikos (eds). Interpraevent - Disaster Mitigation of debris flows, slope failures and landslides. Universal.

Körner, C., R. Asshoff, O. Bignucolo, S. Hättenschwiler, S.G. Keel, S. Peláez-Riedl, S. Steeve Pepin, R.T.W. Siegwolf and G. Zotz. 2005. Carbon flux and growth in mature deciduous forest trees exposed to elevated CO_2. Science 309: 1360-1362.

Lackner, K.S. 2003. A guide to CO_2 sequestration. Science 300: 1677-1678.

Lal, R. 1999. Soil management and restoration for C sequestration to mitigate the accelerated greenhouse effect. Prog. Environ. Science 1: 307-326.

Li, F., H. Zhang, L. Zhao, Y. Shirato and X. Wang. 2003. Pedoecological effects of a sand-fixing poplar (*Populus simonii* Carr.) forest in a desertified sandy land of Inner Mongolia, China. Plant Soil 256: 431-442.

Liu, J.G. and J. Diamond. 2005. China's environment in a globalizing world. Nature 435: 1179-1186.

McConchie, J.A. and H. Ma. 2002. A discussion of the risks and benefits of using rock terracing to limit soil erosion in Guizhou Province. J. Forest. Research (Harbin) 13: 41-47.

Ministry of Water Resources. The water and soil erosion and control measures in China. http://www.lanl.gov/chinawater/documents/niucongheng.pdf.

Parham, W. 2001. Degraded lands: south China's untapped resource. FAS public interest report. J. Fed. American Scientists *54:* http://www.fas.org/china_lands/propose.htm.

Research report of MOA project, Soil and Fertilizer Institute, Sichuan Academy of Agricultural Sciences, Chengdu, China. 2004.

Roering, J.J., K.M. Schmidt, J.D. Stock, W.E. Dietrich and D.R. Montgomery. 2003. Shallow landsliding, root reinforcement, and the spatial distribution of trees in the Oregon Coast Range. Can. Geotech. J. 40: 237-253.

Sidle R.C. 1991. A conceptual model of changes in root cohesion in response to vegetation management. J. Environ. Qual. 20: 43-52.

Sidle, R.C., A.D. Ziegler, J.N. Negishi, A. Rahim Nik, R. Siew and F. Turkelboom. 2006. Erosion processes in steep terrain - Truths, myths, and uncertainties related to forest management in southeast Asia. For. Ecol. Manage. 224: 199-225.

Stokes, A., A. Lucas and L. Jouneau. 2007. Plant biomechanical strategies in response to frequent disturbance: uprooting of Phyllostachys nidularia (Poaceae) growing on landslide prone slopes in Sichuan, China. Am. J. Bot. 94: 1129-1136.

Sun, J.J., Y.Q. Sen, Y.L. Shi, Z.J. Zhang and Q.L. Zhang. 1994. Agricultural Natural Resources and Regional Development of China. Jangsu Press of Science and Technology, Hefei, China.

Tang, B., S. Liu and S. Liu. 1994. Mountain disaster formation in northwest Sichuan. GeoJournal 34: 41–46.

Tu, S., Y. Chen, Q. Zhu, Y. Guo, Z. Zhu and L. Xie. 2006. Integrating cash crop hedgerows and balanced fertilization to control soil and water losses from sloping farmland. Better Crops 90: 36–39.

United Nations Environmental Program (UNEP). 2002. Global Environmental Outlook 3: Past, present and future perspectives.

Wang, Y. 2004. Environmental degradation and environmental threats in China. Env. Monitor. Assess. 90: 161-169.

Wang, L., Y. Zhang, R. Li, P. Cui, X. Yu and Q. Cai. 2005. On key research domain of science and technology for soil and water conservation in China. Sci. Soil Water Cons. 3: 1-6.

Xie, X., B. Sun, H. Zhou and Z. Li. 2004. Soil carbon stocks and their influencing factors under native vegetations in China. Acta Pedologica Sinica 41: 687-699 (in Chinese).

Yang, Z. and Y. Xie. 1997. Analysis of soil and water losses on sloping lands in north-east area of Yunnan. Chinese Agricultural Resoures and Regional Planning. 17-23.

Zhang, B., Y. Yang and H. Zepp. 2004. Effect of vegetation restoration on soil and water erosion and nutrient losses of a severely eroded clayey Plinthudult in southeastern China. Catena 57: 77-90.

Zhang, J.Y., Y. Wang, X. Zhao and T. Zhang. 2005. Grassland recovery by protection from grazing in a semi-arid sandy region of northern China. N. Z. J. Agric. Res. 48: 277-284.

Zhao, X., Z. Zhang, Q. Zhou, B. Liu, W. Tan and C. Wang. 2002. Soil erosion actuality and its synthesis prevention counter measures in China. J. Soil Water Cons. 16: 40-46.

Zhao, C.X., D.W. Zheng, C.J. Stigter, W.Q. He, D.B. Tuo and P.Y. Zhao. 2006. An index guiding temporal planting policies for wind erosion reduction. Arid Land Res. Manag. 20: 233-244.

Zhao, Y.Z., X.Y. Zou, H. Cheng, H.K. Jia, Y.Q. Wu, G.Y. Wang, C.L. Zhang and S.Y. Gao. 2006. Assessing the ecological security of the Tibetan plateau: Methodology and a case study for Lhaze County. J. Environ. Manag 80: 120-131.

Zheng, F.L. 2006. Effect of vegetation changes on soil erosion on the Loess plateau. Pedosphere 16: 420-427.

Zhou, Y. 1999. Effect of Yunnan Pine Forest on Erosion Control. Southwest Jiaotong University Press, China.

Zhou, G. 2004. Relations between typhoon storms, root systems and landslides. In: D.H. Barker, A.J. Watson, S. Sombatpanit, B. Northcutt and A.R. Maglinao (eds). Ground and Water Bioengineering for Erosion Control and Slope Stabilization. Science Publishers Inc. USA, 419 pp.

Zou, X.Y., S. Li, C.L. Zhang, G.R. Dong, Y.X. Dong and P. Yan. 2002. Desertification and control plan in the Tibet Autonomous Region of China. J. Arid Environ. 51: 183-198.

32

CHAPTER

Climate Change: Ten Years After Kyoto – An Australian Perspective

Syed U. Hussainy[1] **and Santosh Kumar**[2]
[1]Institute for Sustainability and Innovation,
Werribee Campus, Victoria University
PO Box 14428, Melbourne City, MC 8001, Australia
[2]School of Computer Science and Mathematics, Victoria University,
PO Box 14428, Melbourne City, MC 8001, Australia,
E-mail: Santosh.Kumar@vu.edu.au
and
Department of Mathematics and Statistics,
University of Melbourne, Parkville, Victoria, Australia

"He, who will not reason, is a bigot;
He, who cannot, is a fool; and
He, who does not, is a slave."
William Drummond, Scottish writer (1585-1649)

1. INTRODUCTION

Climate change is one of the most significant environmentally related problems; the world is facing in recent times. It is also one of the most controversial topics discussed at various levels, be they academics, politicians, businessmen, scientists or the common citizen. Some of the global climate changes do occur naturally and the Earth's natural climate has always been and still is, susceptible to natural changes. It has also been documented that the changes have not been at a constant phase. Over the historical times, the Earth has experienced several periods of warming and cooling. It has also been interpreted from data, that the global temperature over the period has followed a cycle of fairly long term variations in cooling and warming. Grover (2004) provides a comprehensive discussion and a historical perspective on the causes and effect of climate change on Earth due to emission of greenhouse gases (GHG). The United Nations Framework Convention on Climate Change (UNFCCC) defined in 1997 climate change as "a change in climate which can be attributed directly or indirectly to human activity that alters the composition of the global atmosphere and which is in addition to natural climate variability observed over comparable time periods." Although some GHG occur naturally, their remittance and their gross concentration have increased due to anthropogenic activities. A significant portion of the increase has been since the industrial revolution in Europe. It is also estimated that the anthropogenic contribution is only about 5% of the global GHGs, the rest being through the natural processes. This meagre quantity is, however, sufficient to disturb the delicate balance of GHGs in the atmosphere and the relevant changes in its composition. 'Carbon dioxide' and 'methane' play a major role in bringing about climate changes from various GHGs. Thus GHGs are often quoted in terms of CO_2 equivalents. The concentration of carbon dioxide before the industrial revolution was estimated to be about 270 mg/L and currently its concentration is recorded to be above 380 mg/L.

The major source of carbon dioxide emission includes power stations using fossil fuel, biomass burning, motorcars and industrial gases. Rice fields, coalmines, landfills and anaerobic organic waste treatment facilities, on the other hand generally emit methane. Livestock (sheep and cattle) alone produces globally about 67.6 mega tonnes of GHGs per year. A recent publication (Merali 2006) has shown that living plants macrophytes also emit methane, as a normal process. The quantity of methane emitted by the tropical rain forest is estimated to be between 10 and 30% of the global emission, which is 63 to 236 million tons per year. It is also reported that the emission of methane from plants would double with every 10°C rise in temperature and this emission will also increase with the size of foliage. It implies that the trees that are grown to reduce the CO_2, might be emitting another GHG (CH_4) from the same plants. A substance called 'Pectin' present in the plants cell is believed to be responsible for methane formation. The impact of methane on climate change is several times more than of carbon dioxide.

In this chapter an attempt has been made to describe the economic developments in Australia and the emission of GHGs during the process and measures taken, if any, or those proposed by the industry and Government to reduce the impact of the GHGs at the national level. It should also be borne in mind that the leading developed countries like Japan, South Korea, the USA and Australia along with the developing countries like India, China, are not signatures to the Kyoto Protocol. Instead, these countries in recent months have formed a partnership called the 'Asia-Pacific Partnership' (AP6) on clean development and climate to address the current problem on climate change with out any deleterious effect on their economics.

Some of the recent events due to global warming are cited in Section 2. Kyoto Protocol and Australia's response has been described in Section 3. GHG emission in Australia has been presented in Section 4. In Section 5, industrial revolution in China and India and its impact in Australia have been described. Asia-Pacific partnership has been outlined in Section 6. The Kyoto Protocol and the Asia-Pacific partnership have been compared in Section 7. Finally concluding remarks have been given in Section 8.

2. RECENT EVENTS DUE TO GLOBAL WARMING AND CLIMATE CHANGE

Recent bush-fires in Indonesia, Cyclone Larry, a Category 5 storm, in north eastern Australia and wreckage on New Orleans along with calving of ice of the size of small country from Arctic and Antarctic are some of the recent consequences of global warming and climate change. Disasters have always been with us and surely always will be. However, when they come that fast and the emergency is frequent, the trail that is left behind is trying to convey some thing and it is for us to understand that. The Earth is a like a living organism, and its disorders are known storms, floods, fires, massive glacial melts, greenhouse gases into the atmosphere, trapping the heat from sun, rising global temperatures and a subsequent rise in sea level. All these events are happening faster than anticipated. Some of the facts are:

The carbon dioxide concentration in the atmosphere was only a few parts per million, it did contribute to make the Earth a place of comfort for the biota to adapt and propagate. The CO_2 allows sunlight to stream in, but prevents this heat to radiate back into the atmosphere. During the last Ice Age, the concentration of carbon was 180 p.p.m., taking the Earth into a deep freeze state. When glaciers retreated and before the modern era, this CO_2 content raised to 280 p.p.m. and the Earth attained a comfortable level. However, during the past 150 years this CO_2 level has been raised to 381 p.p.m., and we are seeing the consequences. Out of the past 20 hottest years, 19 were from 1980s.

Analysis of data from Canadian and European satellites found that ice around Greenland is melting down twice as fast. Last year alone the ice melted and drained away into the sea in an alarming volume of 220 cu. km as against 90 cu. km in 1990. Note that a cubic kilometre of water is more than the requirement of the entire city of Los Angeles in a year.

The planet's ice cover is decreasing faster and faster and is changing the relationship of the Earth and Sun. Polar ice is reflective that 90% of the sunlight that strikes is reflected back into space. However, ocean water does just the opposite, absorbing 90% of the heat energy received. The more energy it retains, the warmer it gets, and melts more ice faster. This is a feedback loop that accelerates the damage.

A similar feedback loop is arising in melting permafrost, defined as land frozen continuously for two years or longer, in high-altitude regions of Alaska, Canada and Siberia. The soil is warming and decomposing, releasing gases that will turn into methane and CO_2, which in turn leads to more warming and more melting of permafrost.

More draughts are experienced due to global warming. Higher temperature is taking soil moisture out and inhibiting precipitation. These climate changes also promote bushfires and a further increase of CO_2 in the atmosphere.

3. KYOTO PROTOCOL AND AUSTRALIA'S RESPONSE

The Kyoto Protocol presents a variety of policies and measures to deal with the problems currently faced. These include energy efficiency policies, protection of sinks and reservoirs, sustainable forestry practise, sound technologies, encouragement of reforms in the relevant sectors and controlling emissions from the transport sector. The Protocol does not list separate targets for each individual gas, but instead a combined target for all GHGs expressed in CO_2 equivalence. The Pprotocol allows countries to achieve their targets via the use of the "flexible mechanism", which include joint implementation, emission trading, and the 'Clean Development Mechanism'.

The Kyoto targets call on the US to reduce its emission by 7% during 2008-2012, the European Union by 8%, and Japan by 6%, while Australia, Iceland and Norway are allowed to increase their emission by 8%, 10% and 1% respectively (Gupta and Lobsinger 2004). Australia refused to ratify the Kyoto Protocol on Climate Change when 157 nations agreed to do so. Along with Australia, the USA, South Korea, Japan and China also failed to ratify, mainly on economic grounds. The Australian Government objections for not signing the Protocol were:

1. Doubts whether climate change was a scientific reality rather than a false alarm. Even though government now concedes that the threat is real, it still resists that Australia can meet its Kyoto targets without being a signatory.
2. Although the environmental impacts have been canvassed and agreed the other risks include economic and strategic ones. They are:
   2.1 The Federal Government also argued that the Protocol, seemingly unable to signup nations with more than the required 55% of Global GHGs emission was unlikely to come into force, which was subsequently proved to be untrue.
   2.2 Australia abandoned research in treating in carbon emission credits, claiming that this was also unlikely to happen. On the other hand, Russia with 17% GHGs emission supported the Protocol and would speed up ratification. Furthermore with the collapse of Soviet era, the old Soviet industries have left Russia with huge carbon credits.

   The six non-signatory nations (Australia, the USA, South Korea, China, Japan and India) account for 49.5% of Global Gross Domestic Product (GDP), 48% of energy uses and 48% of GHGs emissions (see Table 1). The relationships between the economic annual growth and the energy produced by fossil fuel usage by these nations are clearly evident.

Table 1 Annual growth of GDP in AP6 countries and their GHG emission.

Contribution to global emission			*Average annual growth in GDP*				
Country	*2001 %*	*2050 %*	*Country*	*2001 – 10%*	*2010 30%*	*2030 – 50%*	*2001 – 50%*
Australia	1.6	1.1	Australia	3.4	3.0	2.4	2.8
China	15	26.9	China	7.8	6.4	5.3	6.2
India	5.4	9.4	India	6.0	5.7	5.2	5.6
Japan	4.0	1.6	Japan	1.3	1.2	1.1	1.1
S. Korea	1.6	1.3	S. Korea	4.4	3.5	2.6	3.3
US	22	15.5	US	3.2	2.9	2.3	2.7
Rest of the world	50.5	45.3					

The people of the developed countries like the USA, Australia, Japan, and South Korea have benefited from the market driven growth, whereas the developing countries like China and India intend to join soon. In some business circles China and India have been described as the factory and office of the world respectively. Evidence of global warming driven in part by anthropogenic activity has persuaded nations around the world to adopt the Kyoto Protocol whereas Australia as stated earlier, along with the USA have rejected Kyoto mandatory regimes as too costly, unworkable and inadequate. As per the Protocol, the developing countries are exempted from the binding measures in the treaty's first phase to 2012. During 2002 Australia emitted 540 million tonnes of GHGs, of these the state of New South Wales emitted the highest (29.5%) and Victoria 21.6%. According to the Bureau of Meteorology, the year 2005 was the hottest since comprehensive record keeping began in Australia. Greater than average temperature was experienced in more than 95% of the country for the year, while the global average was 0.48°C above the normal, Australia experienced 1.09°C hotter than the average from 1961–1990. The Bureau assigns this to global warming and there is no other explanation. The year 2005 was also a part of 9 hottest years of the decade providing further powerful evidence. At the present rate of global GHG emission levels, Australia can possibly expect a rise in temperature of between 2 to 5°C over the next 100 years. But Australia's GHG emissions are increasing and there appears to be no check on them.

It was perplexing to note that Australia again refused to sign the Kyoto Protocol when 157 nations agreed to extend it further during 2005. The Government believes that it will still meet its Kyoto target, i.e. not the global goal of 5% cut in 1990 GHG emission by 2012 but a special allowance of an 8% increase. Australia is on track, as the government claims, only when the effect of reducing land clearing is taken into considerations. Actual emissions have soared, particularly in the key areas of power generation and transport. The rhetoric is changing with the weather, but the policy is still a patchwork of hope and promises above the 1990 level.

Kyoto Protocol remains the main global agreement to tackle climate change. The various projects listed under the Protocol are beamed to save about 17 million tonnes of greenhouse pollution a year. Australia is, however, not a signatory for this cooperation and the anticipated progress. Action has already begun around the world in countries like Denmark and Germany in generating solar and wind power to complement the traditional sources. Although the U.S. is not a signatory as stated earlier, to the Protocol; its 188 cities representing 40 million people have committed to meet or beat the Kyoto Protocol target in their community and have urged other States and Federal Governments to do the same. They have also adopted a target of reducing GHG by 80% by 2050. Eleven U.S. states have adopted regulations to cut GHG emission from vehicles by 30%. President George. W. Bush in his speech to the State of Union in January 2006 has appealed to cut the reliance on oil from the Middle-Eastern countries and develop technologies to produce methanol and other energy sources locally. According to the Governor Arnold Schwarzenegger (2006), the state of California is committed to fight global warming by 2010 the level of GHG will be reduced to 2000 level; by 2020 it will be back to 1990 level; and by 2050 to 80% of the 1990 level. He anticipates generating 20% of electricity supply from clean renewable sources by 2010 and a third by 2020.

In Australia encouraging progress is being made at the various state levels. All state-premiers have agreed to explore a state-based nationally consistent emission-trading scheme. New South Wales and Queensland have adopted standards for new domestic buildings to promote energy efficiency in our cities and investigating in renewable powers. Several wind farms have been established through out the state to generate electricity using wind power. The states of New South Wales and western Australia have also chosen to build new lesser polluting gas-filled power stations rather than coal in these states.

4. GREEN HOUSE GAS EMISSION IN AUSTRALIA

Power generation in Australia by and large is from the use of fossil fuel (brown coal). A pollution audit has found that Australia's 24 power stations alone emit 170 million tonnes of GHGs, equivalent to 40 million cars; four times the national car fleet (Fife 2005). A list of the major power stations and the amount of carbon dioxide and equivalent CO_2 car emission is listed in Table 2.

Table 2 Major power stations and their gas emissions

Power station	*State of location*	*Mega tone of CO_2 emitted*	*Equivalent CO_2 car emissions #*
Loy Yang A	Victoria	17.3	4.3
Hazelwood	Victoria	16.3	4.1
Bays water	New South Wales	14.2	3.6
Yallourn W	Victoria	13.4	3.4
Eraring	New South Wales	11.8	3.0
Stanwell	Queensland	10.1	2.5
Gladstone	Queensland	9.8	2.5
Loy Yang B	Victoria	9.7	2.4
Tarong	Queensland	9.4	2.4
Liddell	New South Wales	9.4	2.4

The (#) mark in Table 2 indicates that one mega tonne of CO_2 emitted by a power station in a year is equivalent to CO_2 emission from 250,000 cars a year.

Victoria's major electricity generator, Loy Yang A, is the most polluting power station in the nation (see Table 2). It emits 17.3 million tonnes of CO_2 in the air each year equivalent to the GHGs from 4.3 million cars.

The privatized Victorian generators are inefficient and they are not investing in renewable energy such as wind farm or hydro schemes. Victoria is in a bad state because not only do the generators use brown coal, which is more polluting than black coal, but is totally privatized and this appears to be a disadvantage in making even some moderate improvement to GHGs emission. Victoria's electricity comes from four major generators, Loy Yang A, Hazelwood, Yallourn W and Loy Yong B. Together they produce 57 million tonnes of CO_2 each year. New South Wales generators produced about 60 million tonnes in the year 2002. Australia emitted 540 million tonnes of GHG, of which Victoria produced 21.6%. But Victoria has the highest GHG emission (29.5%) from electricity of other fossil fuels. The Queensland generators Stanwell Corporation and Tarong Energy and the western Australian Government's Western Power, on the other hand, were making remarkable investments in renewable energy. During the recent months some wind farms have been mushrooming in Victoria often with the same objections from the public on cosmetic grounds. Despite some remarkable energy investments, Australia still has heavy reliance on coal-fuelled electricity (47% is from coal, see Table 3).

Table 3 Percentage composition of various energy sources for power production in Asia-Pacific countries and USA. A (*) mark in Table 3 represents production of less than 0.5%.

Country	*Year*	*Coal*	*Oil*	*Gas*	*Nuclear*	*Hydro*	*Non-hydro renewable*
Australia	2001	47	32	20	0	1	*3
	2050	40	36	20	0	1	
China	2001	69	25	3	*	3	*
	2050	52	27	10	9	2	1
India	2001	54	36	7	2	2	*
	2050	37	38	7	16	1	1
Japan	2001	19	49	13	16	2	*
	2050	14	47	13	23	2	1
S. Korea	2001	22	52	10	15	*	*
	2050	17	48	13	22	*	*
US	2001	24	41	24	9	1	*
	2050	20	43	25	10	1	2

Brown coal is one of the cheapest and dirtiest sources of energy. Its competitiveness could come under threat, given the potential cost of reducing emission and calls for taxes and charges on carbon producers and carbon credit trading schemes. There may be cost increases due to the use of clean coal technologies designed to cut GHG emission, but at the same time there will be pressure from stake holders, consumers, and from society generally to reduce costs, such that the real price of goods and services tend to be the same over a period of time.

Coal may overtake oil as the best performing energy investment. That, at least, is the emerging consensus from a diversity of speculators, investors and giant corporations. As coal is the cheapest, most abundant energy source globally, the surge in oil has encouraged people to plan new coal-fuelled power plants and to start using conversion technologies such as coal to diesel.

It is also anticipated that the global energy needs would treble by the end of century, and that being so, the world would require energy from all sources including the renewable and possibly nuclear. Any increase in government taxes and charges would need to be enforced on a global basis. If this is not controlled, pollution would just migrate to another part of the global village. On the other hand if the option of simply 'sweeping under the carpet' is adopted by the rich developed nations, there may be grave implications for reducing poverty in developing countries and could also undermine energy security.

Developing cleaner technology is slow and takes time before it is available on a commercial scale to cut the emission to the desired level. It is also essential to avoid creating false expectations that the problem can be solved so easily by the developed countries. Even though it will not be known until 2012 whether the Kyoto Protocols have had any demonstrable effect on the signatory countries, despite all the optimism, alternate-energy cannot compete with fossil-fuel-feed power stations. The only energy force that could come closer to coal in efficiency is nuclear power with its built-in safety issues.

The Australian Bureau of Agriculture and Resource Economics have called for an action after years of delay. Although the Earth has in the historical past undergone climatic changes, these have taken place over much longer periods, giving humans and ecosystems time to adapt. It is estimated that at least 50% reduction in GHG production by 2050 is needed to ensure a manageable climate change. Even if climate change were not a factor, other issues of environmental sustainability and the utilization of finite resources such as fossil fuel would demand a second look on the patterns of wasteful consumption. Even sensible changes in public behaviour could bring home the 'bacon without pain'.

Australia has one of the highest per capita GHG emissions. This is attributed to the size of country, its low population density, its climate and its heavy reliance on coal for power generation and the support of energy-intensive industries, such as aluminium, which forms the major wealth-generation industry. It is essential that the government of the day must support the aluminium and coal industries in their endeavours to develop lower emission technologies. They are also Australia's biggest export earners, along with natural gas and uranium in recent times, and without economic prosperity no government can support its social and environmental welfare measures so vigorously demanded by the current socio-economic and environmental conditions. To protect the nation's economic future, we need to be a part of the solution to the environmental impact of economic growth of our region currently dominated by China and India.

It is here that the Asia-Pacific partnership could become effective. It could offer Australia not only an opportunity for economic growth and enable it to be a part of the solution to the environmental consequences of the happenings in the region.

Most industrialized countries have relied on fossil fuel for their development. It is unfair and unrealistic to deny developing countries such as India and China, the opportunity to expand their economies as rich countries have done. They can progress only by relying on fossil fuels and increasing their GHG emission. Converting coal into liquid fuel or natural gas becomes economical when oil remains above US $40 a barrel (Martin and Craze 2006). The price of oil has more than doubled since January 2004 and it has not been below US $40 since June 2004. Using more coal is part of the President George W Bush's imitative to make the US less dependent on oil imports.

5. INDUSTRIAL REVOLUTION IN CHINA AND INDIA AND ITS IMPACT IN AUSTRALIA

Currently India and China have been going through an industrial revolution for the past 20 years. Since 1980, their economies has been growing at the rate of 9.5% and 5.5% per year respectively. China and India are also the most populous countries in the world and account for 40% of the world population, where as the rich countries of Europe, North America, Australia and Japan together account only for about 15%. The Chinese are rapidly becoming the chief source of manufactured goods, while the Indians have already captured about half the world's offshore outsourcing businesses. The effect of this is that the two countries' rapidly growing demand for energy and raw materials may outstrip the supply and keep upward pressure on prices. China is already the second largest consumer of energy in the world after the U.S.A., while India has moved to the sixth place (Gittings 2006). Energy demand in both countries is also being boosted by rising income and growing urbanization. Both the countries depend on fossil fuel, both oil and coal; for their energy productions. China imports three-quarter of oil which it needs, while India imports more than 90%. But oil accounts approximately for only a quarter to a third of the two countries total energy consumptions. The rest of the energy is mostly resourced from coal. Both countries meet their coal requirements by importing coal from Australia. According to Gittings (2006) the global ecosystems and the variable resources are simply not sufficient to sustain the current economies of the west and at the same time bring more than two billion people into the global middle class through the same resource intensive industrialization and urbanization.

6. ASIA-PACIFIC PARTNERSHIP ON CLIMATE CHANGE

The Asia-Pacific Partnership on Clean Development and Change (AP6) met in Sydney, Australia during early January 2006. The partnership aims to reducing GHG emission in Australia, China, India, Korea, Japan and the U.S.A., while promoting their economic growths. The partnership also aims to accelerate the development of cleaner, more efficient technologies to meet national pollution reduction, energy security and climate change concensus in a way that promotes economic development.

The pact is believed to work from the bottom up through public-private partnership to build local capacity, improve efficiency and reduce GHG emission from industrial facilities, power plants, mining and buildings. The key difference between the two viewpoints is that the green movement wants government to mandate strict emission control standards, while the U.S.A. and the Australian government maintains it and only through the private sector voluntarily accepting emission reduction in their own self-interest would bring the required change.

Structurally the Asia-Pacific Partnership is unique in that it provides industry with an equal role to that of the government in setting both priorities and objectives for the immediate and long-term challenges of improving the environmental performance of the industry.

The Asia-Pacific Partnership established eight public-private taskforces to encourage the development and spread of technologies that minimize GHG emission. These are:

6.1 Clean Fossil Energy Taskforce - focuses on reducing emissions from coal and gas power stations, and to encourage usage of new technologies that capture and store CO_2. (Chairperson - Australia)

6.2 **Renewable Energy and Distributed Green Task Force** - promotes the spread of renewable energy technologies including hydroelectricity, solar, geo-thermal, wind and tidal power bio-fuels. (Chairperson - Republic of Korea)

6.3 **Power Generation and Transmission Task Force** - examines the ways to improve power plant efficiency and reforms electricity markets. Investigates and promotes demand management technologies that improve efficiency. (Chairperson - The U.S.A.)

6.4 **Steel Task Force** - sets benchmarks for best-practice steel production technologies that minimize power needs. Develops ways to reduce power usage, pollution and GHG emission and increases recycling. (Chairperson - Japan)

6.5 **Aluminium Task Force** - identifies and removes barriers for use of best-practice technologies. (Chairperson - Australia)

6.6 **Cement Task Force** - facilitates demonstration and development of new energy efficient technologies to replace older kiln methods. Provides the world's development with a deeper understanding of best-practice technology. Develops methods of power generation from waste heat recovery. (Chairperson - Australia)

6.7 **Coal Mining Task Force** - promotes best-practice technology for coal mining and facilitates greater reclamation of old mine sites. (Chairperson - The U.S.A.)

6.8 **Building and Appliance Task Force** - supports further uptake of energy-efficient appliances. Promotes best-practice technologies for building design. (Chairperson - Republic of Korea)

Although the group produced no real outcome in their two-day meeting in Sydney, they have clearly put great faith in the working of many industries, including the coal and aluminium. The group will also identify alternative technologies, which either provides new energy sources or cleaner ways to use the existing technologies. They also have the overall objective of making sure that countries have access to energy resources, which they need and they are able to lift people out of poverty and ensure that their economies grow (Franklin 2006).

The U.S. and Australia pledge to develop a multi-million-dollar fund to pursue clean technology designed to reduce GHG emissions. Over the five years the funding is expected to grow to US $ 260 million.

The communiqué also reflected these sentiments with acknowledgements to the growing sale of nuclear power and renewable energy but said fossil fuel underpinned their economies. The communiqué added that it is critical that AP6 work together, demonstrate and develop and implement cleaner and lower emission technologies that allow for the continued economic use of fossil fuel while addressing air-pollution and the reduction in the emission of GHGs.

7. KYOTO PROTOCOL AND THE ASIA-PACIFIC PACT ON CLIMATE CHANGE: A COMPARISON

The Kyoto Protocol defined the international climate change and energy debate by acknowledging that government and industry must work together to reduce GHG emission, which may adversely be affecting the economic growth. Whereas the AP6's Eight Private-Public Task Forces (discussed earlier) were given a year to come up with a plan to push renewable energy forward, accelerate the development of clean-coal technologies and reduce GHG emissions and air pollutants in key industry sectors. According to the Australian Prime Minister, John Howard and the U.S. President, George W. Bush the Partnership's strength is industry-promised participation and investment in the development of low emission technologies that sets the pact part from the Kyoto Protocol of emission reduction target for developed countries.

Although Japan, India, and China subsequently have ratified the Kyoto Protocol; India and China are not required to commit to binding targets, even though they alone produce more than 20% of the global warming GHGs.

The Australian Bureau of Agriculture and Resource Economics reports that the AP6 through its projects and influence, could reduce the global greenhouse growth by about 23% by 2050 but make no net reductions. The Australian Business Council for sustainable energy is of the opinion the future success of AP6 depends on removing barriers to private sector investments in clean energy, like hydro, wind, solar, etc. This appears to be the area of contention surrounding the AP6. Australia's largest renewable energy producers, Hydro Tasmania and Pacific Hydro, want this switch encouraged by specifics, such as charges on carbon contents of fuel or carbon emission, and tax incentives and investments. Australian Federal Government is of the opinion that a price signal may be necessary to push low emission technologies in practice. Bringing the technology up with a push is as important as getting the incentive system right. Although the low emission technology by and large would be available, all that is needed is to make them as cheaper as the promoting fossil fuel-fired power. With financial incentives even companies that are not prepared to spend any money on the environment will reduce emission.

Hodge (2006), based on the Australian Bureau of Agriculture and Rural Economics, reports that the Asia-Pacific climate pact could slash GHG emission in member countries by as much 30% by 2050. The report is based on broad assumptions, e.g. all Australian coal-fired power stations after 2015 would be using carbon capture and storage technology and the use of energy- efficient kind of cars across the member countries. The flow on effect of this around the world would be due to the trade these countries undertake with the rest of the world and the technologies developed here will be filtered out.

Australian Bureau of Agriculture and Rural Economics forecasts on energy consumption, contribution to global emission, average annual growth in GDP and population projection are shown earlier in Tables 2 and 3.

As the two-day Asia-Pacific Partnership on Clean Development and Climate concluded in Sydney in early January 2006, Australia and the U.S. reiterated the six-nation pact was the new model for addressing global warming. 'The new world's pragmatic answer' to Europe created by emission trading and new generation, 'clean' technology. Companies and countries face increasing risk in staying outside the emerging climate change regime.

Therefore, it is necessary to hedge towards cleaner technologies for the use of these fuels. If these were to be replaced by renewable energy sources, such as solar, wind, wave power, it would be financially unaffordable and practically unsustainable in terms of base load energy. Abandoning traditional base load power in favour of renewable energy would result in global economic depression, and the developing countries that are economically deprived may face the full force of the economic reality.

Nuclear Power as a Base-load Energy

Nuclear power energy may be the other option for base-load energy. Australia is a signatory to the nuclear-non-proliferation treaty and takes non-proliferation seriously. It has also strict procedures for uranium sales as well as bilateral agreements to deal with the safe and peaceful use of uranium and disposal of its waste products.

The failure of the U.N. conference in 2005 to strengthen the non-proliferation treaty highlights the danger in relaxing control of the nuclear material trade. The AP6 partnership can play a pivotal role in facilitating the development and use of clean technology especially for the biggest usage economies in the developing world that are not covered by Kyoto-style emission targets. Australia as a major supporter of the clean energy resources (uranium) to Partnership countries and as a potential supporter of clean energy technologies can play a major and a valuable role in achieving global security in both geopolitical, geophysical and climate terms while secure its economic future. Eventually the nearly developed technologies could possibly provide a solution to the greenhouse challenge instead of being part of the problem, China has been described in some quarters as the factory of the world. To meet its energy resources, China is diversifying its resources. It is changing from the traditional to nuclear sources (see Table 2). It is anticipated that by the year 2050, the energy requirements would have grown from 1000 million tons of oil equivalent (Mtoe) to 4800 Mtoe. During this period the forecast is that China will reduce its reliance on coal and increase reliance on nuclear energy source. Australia has above 40% of the world's known uranium resources. Chinese and Australian officials are working towards creating safeguards for the export of uranium to China. They are designed to create a framework to prevent Australian uranium exported to China being used for weapons or to replace domestic uranium supplies used to build nuclear weapons. An agreement could also prevent China from on-selling Australian uranium to other countries that have not signed an export deal with Australia. Uranium export to China will only be considered when adequate safeguards are in place. China has nine nuclear power reactors and is planning a five-fold increase in its nuclear energy by 2020. Demand for uranium in China is booming.

CO_2 Geo-sequestration Project

Australia's first carbon dioxide geo-sequestration project is planned for the Otway Basin in Western Victoria. The Australian $30 million project is expected to begin the trial stage of CO_2 deep underground by the end of 2005. Geo-Sequestration is the geological storage of carbon dioxide being banked on by industry and the federal and state governments. CO_2 will be collected from a power station, pumped several kilometres of pumping into underground storage in a depleted gas reservoir about 2 km beneath the surface. Over a year or two the project anticipates injecting about 100,000 tonnes of carbon dioxide, and studying the safety and effectiveness of the technology. The project is not free from criticism. The Australian Conservation Foundation has referred to geo-sequestration as 'greenhouse dumping' and others worry about the technology's safety and its promotion over strategies to reduce the world's reliance on fossil fuel (Fitzgerald 2006). Resources from a mix of industry and government organizations will fund the project.

Living Plants as a Major Source of Methane

Merali (2006) reports that it is not just farming cows and belching sheep that spew out methane, living plants have been discharging millions of tons of potent GHG into the atmosphere every year. Since industrialization the concentration of methane in the atmosphere has tripled. Hither to it was recognized that the breakdowns of carbonaceous material by anaerobic bacteria in areas such as swamp logs, wetlands, rice fields etc give rise to methane (CH_4). Recent work has revealed that living plants globally produce between 63 and 236 million tonnes of methane per year, while the plant debris contributing another 1 to 7 million tons of methane. Keppler et al. (2005) and Cox (2005) have estimated it to be about 10 to 30% of global methane production. The findings clear up a number of observations that defied explanations. It also creates problems. Growing trees soak up the GHG CO_2, therefore under the Kyoto Protocol countries can promote forestation as an alternative means of reducing the GHG emission. With this new finding, the new problem is CH_4. It is more potent and has a longer half-life than CO_2. While trying to reducing one gas, we may end up in increasing a more potent gas that has a longer life. What is more is the production methane by the plants increases by every 10°C rise in temperature.

A study conducted at the Yale and Columbia universities ranked countries according to 16 indications. The important yardsticks were how the environment affected people's health and the vitality of the ecosystem. Australian credentials on environmental issues are not as good as one generally tends to assume. According to World Economic Forum, Australia's performance is ranked behind those of Malaysia, the Czech Republic, Columbia and Greece (Fife 2006). Despite its wealth, Australia is lagging behind on water consumption, air quality, sustainable energy and biodiversity protection. Out of 133 nations assessed Australia ranked at 20 on environmental performance, and the U.S. was at 28, whereas New Zealand, Australia's next door neighbour is the best performing nation in the world. On water consumption and air quality, Australia's position was 119 and 94 respectively out of 133. On the positive side Australia was ranked third on the environmental health, judge favourably in natural resource category. The researches also commented that the ranking was affected by its high GHG emission, overuse of water and unsustainable agricultural practices. It also ranked the lack of investment in sustainable energy.

8. CONCLUDING STATEMENTS

The partnership recognizes that there is no quick fix to challenges of climate change. A long-term commitment and significant investment are needed to address the sustainable generations and use of energy. The setting of arbitrary targets may not result in practical solutions to global climate change and it was recognized by eight members that cuts in GHG emission can only be achieved through technological co-operation and involvement of large and fast growing economies especially the U.S.A., China and India. The discussion among the AP6 countries also appears to be a useful one on the future low emission of carbon energy and clean technology but unlikely to produce any meaningful result. While everyone at the forum seemingly welcomed the transfer technology (see Table 2). The data from the Bureau of Meteorology is an urgent reminder that time is not on our side. The world is getting hotter at a faster rate. Examples from around the world show the Australian government what it needs to do. The climate partnership should deploy existing venerable energy technologies to deliver large reductions in greenhouse pollution.

With the coal industry and other energy-intensive sectors pushing hard to postpone what may concede inevitable, no one at the AP6 council was willing to advocate a move such as a carbon price. Each of the eight task force of AP6, at first, will be using the opportunities and barriers in each sector in achieving their goal. When once these barriers are recognized, there is a hope that some attempt may be made to over come these hurdles (Franklin 2006).

As our dry summer days get hotter and draught gets worse with climate change, future generations demand management policies have to be driven by reducing the high greenhouse intensity of our power sector globally. There is a unity around the world that there is an urgent need for action on global warming, or as President John F. Kennedy said, "In the final analysis, our most basic common link is that we all inhibit this small planet. We breathe the same air, we all cherish our children's future."

The developed nations of the world in recent times have been spending billions of dollars of 'tax-payers' money trying to find the most elusive 'weapons of mass destruction', so far they have been unsuccessful. The cost of human life cannot be estimated in this event. It has been like a wild goose chase. The real weapons of mass destruction, as we see it, could be the big chimneys as the coal-fired power plants, belching smoke, which include large quantities of smoke that include large amount GHGs, particularly material and many other unburned substances.

With the increasing global temperature and increase in the melting of polar ice due to climate change, it remains to be seen as to how history would judge these nations and the role they have played and how responsible they were in the subsequent drowning of the low-lying areas of the world. It is these weapons of mass destruction that need to be fixed in time, rather than chasing those 'evasive units' yet to be found. The management of GHG and the eventual climate change is a bigger issue than we think and it requires a complete change in attitude and infrastructure. At this juncture let us reiterate the statement by William Drummond (1585–1649), a Scottish writer:

"He who will not reason is a bigot,
He, who can not is a fool, and
He who dare not, is a slave."

References

Cox, P. 2005, "This week" New Scientist; Page 20, 26th March.

Editorial. 2005. The Age, Melbourne, 29th July.

Fife, M. 2006 Nation of Waste Green Index, The Age, Melbourne, 26 January, p. 3.

Fitzgerald, B. 2006. Carbon Dioxide Storage Projects gets Green Light. Western Australian, 9th January, p. 29.

Franklin, M. 2006 Taskforce or the dawn of a new era, Courier Mail, Brisbane, 8th January, p. 29.

Gittings, R. 2006 Headlong to growth overload. The Age, Melbourne, Australia, 8th February, p. 17.

Global Warming - Various Issues, Time (Australia). 2006, April 3, pp. 14-37.

Grover, V. 2004 Climate change five years after Kyoto. Science Publishers Inc., Enfield, N.H., ISBN 1-57808-362-5, pp. 4-51.

Gupta, J. and A. Lobsinger. 2004. Climate negotiations from Rio to Marrakech: An assessment. In: Climate Change Five Years After Kyoto, ISBN 1-57808-362-5, Science Publishers Inc., Enfield, N.H. USA. pp. 69-88.

Hodge, A. 2006. Research Claims in GHG Cut. The Australian, Melbourne, January 15. p. 2.

Keppler, et al., 2005 Nature London. In: New Scientist. Vol. 439, p. 187 2534, January 14th, 2006.

Martin, C. and M. Craze. 2006. Coal may be the new oil, say the big end of the Town, The Age (1/8/06) Business World, p. 6.

Melissa, F. 2005 Victorian Electricity Generator named world pollutor, The Age, April, 10 Melbourne, p. 5.

Merali, Z. 2006. The lungs of the planet belching methane. New Scientist, No. 2534, January 14. pp. 3-13.

Schwarzeneggar, A. 2006. Cool thinking, New Scientist. No. 2535, January, 21. p. 18.

33

CHAPTER

Obstacles to the Adoption of a Holistic Environmental Policy

Alcira Noemí Perlini Montiel
Av. Montes de Oca 1408 5° B (1271)
Buenos Aires, Argentina
Tel/Fax: (54 11) 4303-1527
E-mail: info@etoecologia.org.ar
www.etoecologia.org.ar

Of the Earth's population of 6,500,000,000, 4,000,000,000 are poor. According to the United Nations nearly 852,000,000 people around the world have had serious difficulties in obtaining adequate nourishment in 2005. While most of this Earth's population suffers from the consequences of climate change, however the poor suffer the most, since they have greater difficulty in obtaining water and food, a problem which is increased by the shortage of water and other natural resources. The fundamental preoccupation of those who suffer social exclusion is subsistence. They cannot be concerned with the global situation. At the same time, the world's population as a whole has a general disregard for climate change and its consequences.

Although the rest of humanity has some knowledge of environmental problems, it increasingly acts against environmental survival. These people believe that their own personal environment is essential for their survival. It is, therefore, their main concern. Around the world, however, only a few people respect the environment beyond their own dwellings. Few people recognize that their environment extends beyond the walls of their homes and offices. They do not realize that the outer atmosphere, substantially affects each house's environment and each person who lives within it. Everything that surrounds a person—schools, cities, the workplace, air, water, soil—is part of the environment. The personal environment extends when we enter other ecosystems for vacations, work or migration. As the knowledge of other environments and ecosystems spreads, a person finally may recognize that the whole world is his environment.

In spite of this, only a small group of people is conscious of the Earth's deterioration. They realize that their immediate atmosphere is connected with the rest of the world and that every action—has an effect on the global atmosphere. They understand that climate change can be attributed to each individual's action. Our use of cars, planes, trains, factories and other technological advances also has a bad effect on the environment and herein lies — the dilemma: man values the items he produces more than he does nature.

The space beyond each person's personal environment is little known, badly conserved, and finally, abandoned completely. Nobody makes rational use of available resources. Each one of us undervalues the space beyond our personal environment. This space is mistreated because people believe that it belongs to no one or because it belongs to someone else, and can therefore be exploited. Although the space may be owned, knowledge of that ownership does not stop it from being devastated and stripped of natural resources. Big multinational companies neither respect their own resources nor those of other countries; rather, they degrade and destroy ecosystems. Multinationals have a high international knowledge of all matters concerning climate change, and an immense knowledge about events on the Earth. Inspite of this, no high ranking multinational thinks about climate change, as their responsibility and how to reduce pollution. Both local and multinational companies destroy their own and others' ecosystems while continuing to search for huge, unexploited areas in oceans and continents. Many companies assist international environmental congresses in order to justify their depredation. At congresses, these companies expose their reasons for hiding and sustaining environmentally unsafe practices. It can be said that polluting and depredating companies assist conventions in order to keep abusing ecosystems.

Nonetheless, the greater responsibility lies with those governments that authorize licenses to companies that damage the atmosphere, abuse natural resources and otherwise pollute and damage ecosystems. These governments either forget or despise signed treaties or bilateral, regional, international or United Nations agreements. They do not apply international treaties, even though these treaties were accomplished after long hours of work by the world's certified scientists. The case of the United Nations Framework Convention on Climate Change (UNFCCC) is one example. Many governments do not follow the regulations concerning the Kyoto Protocol (KP), thus allowing overexploitation of their countries' natural resources. Although each state's congress ratifies international agreements, the executive powers reach separate agreements with their respective congresses, enabling them to pass laws that allow them to exploit natural resources without observing international compromises. On one hand, international documents that protect nature by controlling the emission of greenhouse gases are approved. On the other hand, however, the same governmental bodies pass legislations that go against the spirit of the Kyoto Protocol.

In Salta, an Argentinean northern province, the current governor has sold one of the most important natural reservations of the area, the Pizarro Reserve, so it can be deforested for soy cultivation. In Brazil, the Executive Power has sent a proposal to the Parliament that would result in allocating 50% of the Amazon - the great green lung of humanity - for the lumber industry. Both countries have ratified the Kyoto Protocol, and in both countries carbon dioxide (CO_2) emissions have been intensified because of deforestation . According to scientists, 'the great green lung' may, because of uncontrolled tree burning, already emit CO_2.

In another recent case, the Eastern Republic of Uruguay has allowed two cellulose processing plants to be located on the shore of the Uruguay River, inspite of both a bilateral treaty and the Kyoto Protocol. There are many examples of government negligence and indifference towards the protection of the atmosphere, and thus the international community. At the same time, these governments also fail to protect their own citizens, who have given them a mandate to do so.

The Earth is beginning to react to this indolence, lack of consideration and aggressive misuse of natural resources.

The destruction of the environment and the misuse of the Earth's resources have caused greenhouse effect gases to increase on a global level. Both national and international news agencies have described devastation caused by atmospheric phenomenon such as climatic disturbances. Even non-experts have noted that climate change is increasing. When resource extraction is excessive, when nature is ravaged, and when unlimited pollution is produced without concern for the damage it may cause, the planet starts to send distress signals by turning furious.

It happened in New Orleans, thousands of the city's inhabitants, from all social conditions and of all ages, suffered from the Earth's outburst. The world's inhabitants observed, astonished, as the Earth's fury materialized in the destructive power of the Katrina, Rita and Wilma hurricanes in 2005. Similar disasters are occurring all around the world. During the last few decades, experts have been, and continue to, warn us that such extreme weather will only continue. Jeremy Rifkin confirms this when he states that, Katrina is the punishment for increased global warming. He says that it is punishment because of the increase of CO_2 emissions that we, the inhabitants of our world, emit. Then he adds: "They said that we had to pay attention to the Caribbean, that this would be the first place where climate change effects will become clear through the manifestation of really strong, catastrophic hurricanes. And so it did". The high temperatures caused by large emissions of CO_2 from industrial activity, coupled with massive rainforest and forest combustion and deforestation from all around the world, impacts atmospheric gases, causing hurricanes. The continuous round of tornadoes, typhoons and big storms that every one on Earth suffers confirms the existence of irresponsible global emissions.

Nature's attacks result in serious consequences. Materials, natural resources and people are destroyed. The last losses are the most feared. The planet's survival is in peril. As a result of this, the former American vice-president Al Gore has confirmed what has been established by the United Nation Framework Convention on Climate Change, "We have less than ten years to make significant changes", and has qualified this situation as a 'global emergency'.

Atmospheric alterations provoke flooding, avalanche, erosion and sudden torrential rains. An avalanche can make an entire village disappear. This happened in Guatemala in October 2005. The village of Panjab was buried under an avalanche of mud provoked by the 'Stand' hurricane. In February 2006, a village in the Philippines was buried under a mud avalanche and hundreds of people either died or were reported as missing. The inhabitants of these villages did not think about the consequences of climate change. The villagers ignored the fact that increasing rains and floods provoke mud avalanches, which may cause accidents and deaths. This devastating weather is a result of the actions of each of the world's inhabitants – including the inhabitants of the villages that were destroyed.

It is well known that all around the planet high temperatures have been recently recorded, continuing a global tendency that started 25 years ago. However, few measures are taken to combat this. On the contrary, climatic warnings seem to have been silenced. Records reveal that the average temperature of 1998 has been overcome by those registered in the years 2002, 2003 and 2004. Since the mid-1970s, the planet's temperature has risen by 0.6 centigrade. During the 20th century the planet's temperature rose by 0.8 centigrade. Have the different states been acting regarding this data? It seems that they are not . Abnormal temperatures have manifested all over Europe in 2004 and 2005 and throughout the Northern Hemisphere during summer of 2006, when, in south Portugal, the temperature reached 47°C. During the same period, the maximum temperature registered 42°C in Córdoba, Spain. These unforeseen extremes cost the lives of many elderly people. In addition, minimum temperatures increased, and did not drop from 24°C. In July of 2006, Washington D.C. reached 48°C.

Meanwhile, 7000 NASA (The United States National Aeronautics and Space Administration) meteorological stations spread around the Earth provide verified data as NASA's meteorologists register this temperature increase phenomenon. These records have also revealed the consequences of increased temperatures. In 2005 the intense heat resulted in prolonged draughts in many parts of the world. The Gulf of Aden – also known as the Great Horn of East Africa – was under severe stress; a large part of Europe suffered waves of unbearable, suffocating heat and water shortages. Asia, Australia and northern Brazil bore the brunt of weather changes. NASA has also confirmed that the 'strong tendency of global warming' of the Earth has made the Arctic zone warmer than normal. The *Washington Post* states that the greenhouse effect is the reason of the melting of the Arctic as a result of the high temperatures in the Gulf of Mexico and that has provoked the hurricanes that threaten the Caribbean. In all of these cases, however, action against global warming was delayed.

Wind currents are also affected by climate change. The great winds are rapidly modifying coastal geography, reducing the riverside surfaces year after year. In addition to this, the sea level has been increasing as the poles melt, reducing the shores even more. The direct result of the great winds and the increase of sea level is the disappearance of many coasts. Warnings are being made concerning the risks to certain islands. Tortugas island in Hawaii, for example, was almost destroyed and sunk by hurricanes.

The International Organization for the Investigation, Protection and Recovery of Oceans makes it clear that islands are the most vulnerable to climate change because their shores are not able to stand out, like continental shores. The islands' scant elevation above sea level makes them vulnerable to waves and storms, which could result in severe damage or total destruction of parts of those islands. Many islands in the Caribbean, the Pacific, the Indian and the Atlantic oceans have a high chance of disappearing. As the satellites show, the melting of the Antarctic will result in some of the lowest lying islands being submerged sooner than others. The defenseless coastal regions will be rapidly flooded; islands will disappear and the sea will penetrate the large coastal extensions of the continents.

The current erosion of coast lines, such as peninsulas, points and marine capes is related to climate change. If sediment, carried by river water, is not enough to feed the coastal flows, the result is a receding coast line. The coast of Argentine Patagonia presents evidences of erosion along its whole length. The coast line was formed by the last glacial period. The riverside wearing and transformation, shown in the last few years, is the consequence of the rising sea level, which is caused by the melting of the Antarctic Pole.

The astonishing icebergs of the Antarctic, which drift through the Argentine Sea, are created by this melting. They drift very close to Buenos Aires and along the coasts of Santa Cruz and Chubut - so close that - in spite of the intense summer heat - an iceberg arrived in the Buenos Aires port in January, 2006. It progressively lost its frozen mass throughout 5000 km to reach the port with a great size and a longitude of 250 meters. A similar iceberg appeared in the Atlantic Ocean.

The melting of the South Pole is getting worse. The so called 'eternal ice mass' of the Antarctic continent, is no longer eternal and the naked slopes of its mountains are now visible.

For a long time, rain or sleet has been falling on the snowless mountains of the 'White Continent'. Unforeseen fissures, cracks and abysses, previously covered by 'the eternal ice mass' are now appearing. Two Argentine scientists died when they fell through one of those cracks in September 2005. Later that year (October) a Chilean group fell into another fissure of the ground. They, however, were rescued by the personnel of an Argentine base. The Antarctic terrain, which used to be a plain, is becoming unstable because the abnormal rise of temperatures is causing the ice to melt. Nowadays, we know that in 50 years the ice of the Arctic and the Antarctic will disappear, as it decreases at a rate of 8% every year. This is what the scientists have stated during the International Polar Year 2007, as they warned us about a problem that may change the world. The news provided by the Argentine Antarctic Sector confirms this hypothesis: along with a general augmentation in sea levels due the water having a higher temperature and therefore greater volume, the water from the poles will increase sea levels, generating floods, provoking the disappearance of the current coasts and geographically modifying the entire planet. The increase in melting ice is also noticeable in the Andes, an extensive chain of high mountains that runs through South and Central America and includes the Aconcagua, which is 6962 meters high. The glaciers, located in the southern part of the Patagonic Andean Shire, are melting due to the greenhouse effect: in less than three years they have lost more than 5% of their surface.

As snow is rapidly thawing, mountain rivers, in turn, are overflowing. The great amount of water that comes down the mountain overflows the rivers' basin capacity and sweeps through riverside villages on its way down. This has already happened twice in the mountain city of El Bolsón (Rio Negro Province), in January 2004 and in June 2006. The last event was caused by abnormal temperatures of over 15°C in the middle of the winter in an area where temperatures, according to historic registers, have never exceeded 0°C. Tragedies like this have occurred on both sides of the mountain chain. Chile has suffered a devastating storm that seized the southern region of that country in July 2006. The intense rain and floods caused ten deaths, nine missing. More than 29,000 people were affected by this catastrophe on the shores of the Bío Bío River.

Three thousand kilometers away from Buenos Aires, one finds the Province of Tierra del Fuego. Just 1200 km away from Antarctica, it is a beautiful island with mountainous geography in the southern part of the country (between the 25°and 74° western meridians and the 60° parallel south latitude). Winter there is always snowy, but unusual temperatures almost provoked a flood in 2006. Temperatures of more than 16°C were registered during the winter of 2006. These temperatures caused rains and caused the Grande and Turbio Rivers to overflow. The storm and the ferocious outflows destroyed National Route #3—a major highway— and left the cities of Rio Grande and Ushuaia, the southernmost city of the world, isolated. In 2006, in mid-winter, flooding caused hundreds of people to evacuate the southern province of Neuquén. The violent waters of the confluence of the Limay and Neuquen rivers took out more than 15 bridges, from the springs in the north to the capital city. Between 200 and 250 houses belonging to poor families were destroyed. The flooding also affected small farmers in rural zones. Three hundred people had to be evacuated. During the winters of the past decade, heavy snowfall on the borders of Argentina and Chile has caused mountain roads to be closed to traffic— especially truck traffic. During the most recent winter season (July 2006) the roads were closed due to snow at the end of the season, not, as was formerly the case, at the beginning. This climate change greatly affected winter tourism. Astonished residents, who are not used to seeing rain, say: "The climate is changing due to high temperatures".

Moreover, in the north, the 'Salado River', which runs through the city of Santa Fe in Argentina, flooded and razed a big part of the capital of the province of Santa Fe in April of 2003. At the time, the disaster provoked panic and sorrow among the city's inhabitants. This time, the weather gave clear signs that the flooding was imminent: north of the city of Santa

Fe, extreme temperatures and continuous rains saturated the streams and lagoons that ended up in the Salado River. As it was a plain zone, the water moved slowly but forcefully. The situation was well known but nothing was done to prevent it. The cost of this neglect was millions of dollars in losses in one of the richest provinces of the country. Thousands of inhabitants lost everything they had and were left, literally, on the street. All social levels were seriously affected by mud slides. The area's biodiversity was also damaged: today, the ground is smooth land, without the humus characteristic of that agricultural zone. Flood damage, therefore, has increased desertification.This same phenomenon took place in April, 2006 in Tartagal, a city in the northern Argentina province of Salta. The water of river 'Seco' (which, in Spanish, means "dry") flooded, causing a bridge to collapse. The collapse isolated the city's population. These floods occured in Mendoza, Entre Rios, and in other times in the city of Buenos Aires, in Argentina floods have now caused numerous and irreparable damages, just as the floods in Germany did in 2005.

While the poles melt, the mountains remain without snow and the rivers overflow, causing damaging floods, the rest of the cities in the country consume great amounts of fossil fuels. Thermal and hydroelectric energy is used to provide electrical current to the country. The big thermal energy companies supply buildings, the public and factories. All modes of transportation—by land, sea and air—unscrupulously waste petroleum byproducts. Stations that retail fuel have fuel spills. Most of the population travels and vacations, polluting without punishment or consequences. Big petroleum companies sponsor automobiles that emit CO_2, but do not sponsor the intensive planting of trees to clean the atmosphere. Although all vehicles emit CO_2, their drivers, in spite of being alarmed by atmospheric catastrophes, drive without understanding that the petroleum byproducts they use are one of causes of climate change.

Without a doubt, the economic interests of powerful petroleum operations are an impediment to the immediate adoption of alternative energy. The need for change, however, is urgent because global warming is accelerating so quickly. Argentina tends towards modest development of alternative energy sources. For example, in Patagonia, in the provinces of Chubut, Santa Cruz and La Pampas, and in Tandil, in the province of Buenos Aires wind energy is being developed. Solar energy, bio-gas and all forms of clean energy that do not contaminate to the atmosphere need to be used, before we irrevocably lose ample ecosystems to climate change.

At present, information from the United Nations warns us that Argentina is already 74% desert. Desertification will only grow. Argentina, like the other countries of Latin America, is considered one of the most fertile zones of the planet. The overproduction of the fields and atmospheric calamities are reducing the area's fertility. Pesticides and chemical fertilizers are driving the area's flora and fauna to extinction. As pesticides are not selective, they kill all the biodiversity. It is common to see fields full of the dead bodies of the seagulls. Pesticides and chemical fertilizers pollute the air, water and ground and, eventually kill farm workers, who suffer very serious illnesses, such as cancer of the brain and bone cancers, caused by these harmful chemicals. At the same time, the destruction of biodiversity causes putrifaction, which, in turn causes the release of unquantified massive emissions, which remain in the air for a long time. Cattle manure produces methane, which intensifies rural air pollution. Now, however, this advances in soy harvesting have decreased this source of pollution.

The pollution of rivers, where black and grey waters are spilled, produces greenhouse gas emissions. Some rivers in the country have been canalized, and are used as rubbish dumps. This is the case of the Santo Domingo River in Avellaneda. Now it is a domiciliary sewer, from which emanations of methane gas are intoxicating the whole city. Getting rid of waste in this way is cheaper for the government than creating landfills would be. In the long term, however, getting rid of waste in this manner is more expensive, since it has an impact on the health of the people who live near the canal. Pollution leaves these people without an important water source. The stream, unlike sewers that are built for this purpose, is unsuitable for transporting human waste.

The Matanza River, or Riachuelo, is the most representative for having the highest level of pollution. Located in between the city of Buenos Aires and its southern border with the Province of Buenos Aires, the Riachuelo carries liquids and solids through 64 km, like a grand waste collector. Forty-five enterprises throw their liquid and solid waste in its basin without any restrictions. Neighbors and NGOs have been complaining for years, and have even presented a legal appeal to the Argentinean Justice System. The sentence has favored the river, exposing the polluting parties and taking the necessary processes to purify the water. Up until this moment, the only voices heard have been the ones from different political parties giving a verbal solution to the situation. Meanwhile, the Matanza River - Riachuelo is still a grand polluted basin of factory waste and sewage, permanently emanating methane and other deadly gases.

In addition to the pollution of the city of Buenos Aires and its suburbs, the complete lack of trees, the increasing pollution from engines, and the emanations of the Riachuelo, the environment is also affected by the proximity of the Polo Petroquimico, which harbors more than forty petrochemical factories. The CEAMSE a large urban waste landfil site and a great producer of methane was obliged to close its doors thanks to the NGOs gathered in the Assembly of Wilde.

A journalistic report revealed that all of the surface water courses in Argentina and in Latin America are polluted. This investigation was released by one of the most popular broadcasting networks, without the knowledge of the authorities. In the midst of various results, it showed that the autochthonous wildlife from each basin has already disappeared or is about to.

Nowadays, one of the few rivers that is not polluted is the Uruguay River. However, there is a great danger of ruining its water and the air in its vicinity due to polluting gases emanating from the chimneys of two cellulose plants.

The use of chlorine and fluorine, among other highly polluting chemicals, may affect the waters where they are poured and the surrounding environment, due to the fact that the evaporation comes back into the atmosphere as acid rain. The cellulose-making factories Bosnia of Finland and Ence of Spain will be installed on the eastern bank of the Uruguay River, which is shared by Argentina and Uruguay. The exploitation, navigation and use permits of the river were regulated in 1975, as fellow states. However, Uruguay, not taking into account what it has signed, gave unilateral permission to these gigantic industries without previously consulting with Argentina, as required by the Uruguay River Treaty. When the NGOs of Gualeguychú discovered the degree of contamination to which the Uruguay River will be subjected, they strongly protested. Influenced by pleas for the preservation of the environment, Argentinian authorities began to formally protest. Litigation has arrived at the International Court of The Hague. All the ecologists of both countries and a great part of the population - made aware of the case by the press, which has widely publicized it - hope that The High Tribunal will rule that the existing protocols and international treaties must be fulfilled. Those environmentally friendly treaties were signed, not only by Uruguay and Argentina, but by Finland and Spain, who ratified, but have not observed the Kyoto Protocol.

The work of the scientists of the various NGOs was fundamental in changing the position of international organizations. They examined the causes and consequences of the contamination caused by both factories and provided a prediction of what the environmental impact of those factories would be 20 years later. This prediction included effects of contamination on the flora, fauna and health of the population. The cost-benefit for each population, the probabilities that the location or an alternate location only worked if mechanisms of clean development were used. Without these mechanisms six greenhouses gases would be emitted:

carbon dioxide (CO_2), methane (CH_4) and nitrous oxide (NO), and three industrial gases which are fluorides: hydrofluorocarbons (HFC), perfluorocarbons (PFC), and sulfur hexafluoride. It was also predicted that future CO_2 emissions would devastate all the trees of 700,000 hectares of eucalyptus that there are in the Uruguay territory.

In February 2006, the Argentine Foundation of Etoecology - FAE - testified before the International Finance Corporation (a member of the World Bank Group), explaining the reasons why these factories should not be established and pleading for the strict application of all international instruments, including the Kyoto Protocol, signed by the countries involved (Argentina, Uruguay, Finland and Spain). If the signatories do not observe the rules of the documents they sign, those documents are a 'dead letter'.

Contamination of the atmosphere is not the only concern; because of their location, these industries will also seriously pollute both ground water and surface water. One of the greatest water reserves of the planet will be contaminated: the water-bearing Guaraní, shared by Brazil, Paraguay, Uruguay and Argentina is one of the most important water sources of South America.

Another cause of imminent contamination occurs in the moisture zones that exist on the margin of the sea coast. The survival of all of the coastal moisture zones along the Atlantic coast of South America are in serious jeopardy. Those of Argentina are no exception. Ordinary people do not know how to manage these moisture zones. Many of the inhabitants of the coasts of the Argentine Sea and the authorities of many marine municipalities do not know the subject. Poorly handled moisture zones can cause the death of all of the zone's biodiversity. When those life forms die, the rotting vegetal and animal organisms become chronic methane producers. The overuse and misuse of beaches has lead to the loss of one hundred of kilometers of biodiversity. Few people know that a Declaration of Ramsar exists, or that this Declaration, a product of the deliberations and agreements of all the members of the United Nations in Conventions on Moisture Zones, has established guidelines for the care and use of these zones.

Globally, the existing moisture zones in each water current do not get the deserved importance. All of the existing moisture zones in lagoons, estuaries, rivers, streams and lakes contain the original wealth found on the planet. NGOs of the province of La Pampas managed reclamation for the conservation of the humus of the Atuel River basin. The 'Atuel Group' institutions, along with the population of the province of Pampas, started up a program so that the local authorities arranged "the measures conducive to incorporate Atuel River Basin to the list of internationally important moisture zones". In order to enforce this they asked support of other NGOs, people of argentina and other foreign institutions. The aforementioned civil institutions used scientific arguments in a campaign to gain support of companies for these moisture zones. The mission is called 'All by the Atuel' and, by December 2005, it registered more than 20,000 companies. For that reason the authorities agreed to grant the group international rank.

The climatic catastrophes in South America are caused by CO_2 emissions. These emissions have two precise sources: charges of deforestation and urban solid remainders (USR). The deforestation in Argentina and South America is extensive and causes serious damage to all of the area's ecosystems. The serious and complicated consequences [derive in the desertification that to perceive the FAO, (Food and Agriculture Organization of the United Nations)]. They are discharges of all the native forests of the country and the continent. All of them have been overexploited. Through persistent cutting, intentional fires, and ignorance Argentina, like other countries of Latin America, is losing the greatest forest wealth of the planet.

For decades an illogical resolution was proposed. The government of Brazil allowed the logging of the Amazon, the greatest forest of the world. Sanctioned by the Brazilian Congress, multinational companies consumed and depleted the greatest source of biodiversity of the planet. More than 500,000 square kilometers of the 4,750,000 square kilometers were affected. Now, because Brazil passed laws reserving 50% of the designated area for the harvesting of high-quality, valuable wood, the rest will be reserved for the pulp and paper industry.

To the reduction of the Amazon and all native forest we must add the loss of a unique biodiversity, including large and small mammals, a limitless variety of birds, reptiles, insects and microorganisms. All of these life forms are unique and only exist in the Amazon, whose incalculable heterogeneity of virgin and unknown flora that can serve all the humanity. Both agriculture and the pharmaceutical industry are based on the natural information contributed by plants. Massive deforestation reduces the native species of the tropical forests. This happens even though global warming causes temperature increases which favor germination. The human action of deforestation aggravates the greenhouse effect by intentionally causing the loss of species in the tropical zones. The destruction of 500 km^2 of trees of the Amazonian rainforest means that the main gas of greenhouse effect, CO_2, that will collect in the atmosphere and transform into a boomerang of climatic inclemency. Once again, local NGOs and other environmentalists outside of South America oppose this move. It is necessary for the rest of the planet's inhabitants to resist the extermination of the Amazon and all native forest. The economic greed for the Amazon's natural richness will cause it to become the largest continental surface contributing to global warming due to slash and burn agriculture. Recently, due to the difusion of climate catastrophe notices, the governments of the world have been taking measures to reverse the climate change. The government of the Brazil has changed its position and is now, preserving the Amazon from indiscriminate logging.

The systematic and instantaneous discharge of CO_2 that goes along with slash and burn agriculture, along with deforestation throughout the extensive plains of Argentina caused to benefit a single monocrop, soy. The high yield attracts investors, who do not listen to the warnings of environmental experts. The business of soy - with high and steady prices on world-wide markets - is presently very attractive, because it is very profitable. For this and another profitable business the future doesn't matter. They disregard Article 41 of Argentina's constitution, which says: "All citizens have a right to enjoy a healthy atmosphere, balance, and opportunities for human development. Productive activities need to satisfy present needs without jeopardizing the resources we must preserve for future generations. The authorities will protect this right by ensuring the rational use of natural resources and the preservation of the natural and cultural patrimony and biological diversity through information and environmental education".

The present soybean yield, however, may mean more than the written letter of all national and international legislation. All economic gains may be forfeited when increasing desertification prevents seed from sprouting. The vast desertification, which the United Nations has noticed, can change if the letter of the Constitution is respected. Deforestation is practiced because new earth is needed for the farming of soy. Thus when a field becomes sterile, new and fertile territories are carved out from wooded areas. The rising stock market prices increase the propensity for this operation. For that reason it is valued more than the natural patrimony referred to in Argentina's constitution and will continue until all of Latin America is unproductive. Latin America will then become the new Sahara. People who believe that only soybeans can give economic benefits do not realize that native trees can also give economic and environmental benefits.

Tree planting programs may be successfully implemented if they are conceived like CDM (Clean Development Mechanism) programs. During the 1930s, U.S. president Franklin D. Roosevelt took similar measures. At the beginning of the Depression, Roosevelt, attempting to bring his country out of poverty, instituted the 'New Deal', a system of public works which included planting trees throughout the extensive territory of the United States. To deforest such areas is to defy the continuance of human life on the surface of the planet. As trees that contribute to our well-being, we must achieve an ecological balance between human activity and each ecosystem. This is because each species of tree brings with it a particular biodiversity, from microorganisms up to reptiles, birds and mammals that live in each species of tree. Trees also limit erosion in the hydrographic river basins and influence the variations in climate.

In Latin America the indigenous tribes and the rural communities that traditionally take care of the forests know how to handle them. These people know about the forests' multiple uses and, without depredating them, can harvest diverse products needed for consumption while leaving the forests intact. These communities use, not only wood, but food, fuel, animal feed, fibers and fertilizers. The indigenous people make rational use of the forest resources. But, as multinational companies in the region advance, they displace the indigenous population and, with them, the conservation and permanence of the arboreal zones.

According to experts, Argentina, at the beginning of the 20th century, had more than 700,000,000 hectares of native trees. Presently, it has 33,000,000 hectares. A report reveals that the country loses 250,000 hectares per year. Thousands of forests in northern Argentina, in the provinces of Salta and Chaco, are affected by indiscriminate cuttings. The so-called 'impenetrable' forest does not exist any more because, in a few hours, electrical equipment can demolish trees that took several hundred years to grow. In the deforested region of the Dry Chaco, 70% of the native forests were eliminated for the benefit of the agricultural production.

Last year, on the Buenos Aires campus, Catholic University of Argentina (UCA), carried out a study on the deforestation in northern Argentina, Bolivia and Paraguay. In this vast region scientists can observe – by the means of satellites – red points. The zone seems to be afflicted with the measles. The red spots are actually bonfires burning the remnants of trees – branches, leaves – to clear fields in order to plant soybean. Thousands of square kilometers of trees fall before the axes, the saws and butters to give rise to monoculture plantations. Yungas are looted by whoever clears the land. This includes territory in the provinces of Salta, Jujuy and Tucumán. NGOs claim that the Reserve of Pizzarro, in Salta, which was illegally sold by the governor of that province, is a typical case. Far from there, in the Province of Entre Ríos – in Mesopotamia – all the forest of the Cleaver of Montiel is quickly being lost for the same reasons. On all sides, deforestation results in increased CO_2 emissions and desertification. In the forests of Missiones, in spite of the prohibitions, the branches of growing trees are surreptitiously cut down and are transported by river. One of the most biodiverse zones of the South American continent is in danger of gradually disappearing. At the confluence of the Parana and Iguazú rivers, the world famous Iguazú Falls—a voluminous waterfall 17 km wide, with a declivity of 198 ms—has diminished in volume by 80%. While this anomalous event is associated with dams that Brazil has constructed on the Paraná River, it is also the product of the massive cutting of trees, which in turn affects the region's rainfall. Diminished water levels adversely affect the area's rivers since rainfall in region, which oscillates between 1500 and 1800 mm annually, is rapidly diminishing. In the north of the country, the once 'impenetrable' forest of the Argentine Chaco extended to the Paraguayan Chaco. There, 500-year-old hard wood trees (lapachos, quebrachos) are quickly transformed into wood 'chips' and furniture. Thus cleared, the forest no longer appears inaccessible. The culture's 'fever' for the soybean's economic yields brings with it the contamination of earth and water by agro-toxins as pesticides drain from the furrows in the fields. Intensive soybean cultivation puts the health of the population at serious risk.

The forests of the territory of Andean Patagonia are also being ruined. What does not yield to the soybean yields to logging for furniture, or the building of cottages. Native species, unique and unstudied by science, are endangered. The 'ñair' - one of the trees that consumes the most CO_2 - as well as the 'radal', and 'cohigüe' area endangered. The cleared native trees are being replaced by 'ponderosa' pines, a fast-growing, exotic European species that ruins and the arboreal quality of the region and disfigures and deforms the original landscape of the native ecosystem.

Native trees are being exterminated wholesale, providing a glimpse of the apocalypse. In the most austral province of the planet, Tierra del Fuego the forests of 'lengas' are today almost nonexistent. It has become evident that their loss has resulted in catastrophic effects. A North American real estate company Trillium Company (which, at other times, operates under the names of Lenga Patagonia, Forest Trillium, Bayside, Forest Sap, etc.) was apparently authorized by the government of Tierra del Fuego Province to log a thousand of hectares of native forests in two years as part of the Río Grande project. With the consent of the authorities of the Province, the company destroyed most of the trees in the forests on the island. The company deemed reforestation an annoyance, and the authorities were apparently too preoccupied to make them plant trees. In addition, an exotic species animal, the beaver, was illegally introduced to the island and reproduced massively. The beavers devastated the remaining 'lengas'. The great winds and intense rains furthered the area's ecological devastation. Recently, the Trillium Company bought 75,000 ha of forest to continue harvesting.

It is evident that the catastrophic effects such as floods and climate changes must - among other things - be caused by clear cutting, which is practiced all over South America. The most dramatic corollary, however, is the evacuation of thousand of farmers. Displaced by floods, they move to cities and increase the urban marginal population.

The lack of trees is also noticeable in large cities. Green spaces in urban areas can serve to mitigate the effect of gases and preserve the mechanisms that originate entropic activity. Buenos Aires and Greater Buenos Aires are the paradigm of cities with much air pollution and few trees. One of the best ways to calculate how much CO_2 the urban forest absorbs is to calculate city's arboreal cover—that is, the number of trees per capita in a given city. This is calculated by comparing the number of inhabitants of a city with the number of trees within the same area. In Buenos Aires the arboreal cover is only 10%. Of this, the most contaminated, the famous district of La Boca, accounts for 1.02% of cover. Buenos Aires is far from being able to absorb the CO_2 that is suspended above the city. In comparison, in Montreal which has an arboreal cover of 45% trees, every tree is considered an addition to the citizens' quality of life.

Another great influence on the greenhouse effect is the disposal of urban solid waste (USR). The integral treatment of USR is a problem in Argentina and in all the countries of the region. Individual disrespect for the environment, together with the excessive consumption, leads to gigantic concentrations of waste. A lack of environmental education is one of the reasons that so much waste is produced and dumped. Along the side of Argentinean highways, and other highways throughout the continent, it is possible to see plastics, papers, bottles and all kinds of waste and recyclables.

Open-air garbage dumps are rapidly increasing in the suburbs of Buenos Aires and other cities and towns of Argentina and Latin America. In every Latin American cava, ragpickers try to benefit from existing dumps. Children who were born on dumpsites or areas surrounded with trash are often manipulated by adults, who force them to pick through trash illegally. Planning for integrated waste management should include planning for the people who live with the rubbish, so that they, as workers, can find dignified jobs. Furthermore, integrated waste management would also mean better policies to manage waste disposal thus reducing methane release in the atmosphere.

In order to appropriately treat USR, some programs, designed to turn waste into methane for fuel, have been launched in different cities throughout the country. Olavarría was the first city to achieve the construction of a biogas plant, taking advantage of the possibility of generating electricity. Other cities, such as Avellaneda, which intend to develop the sanitary landfil CEAMSE, and Mercedes are also taking part in this project. It is important to note that, most of these products will not generate energy—they will only burn the gas they produce, emitting GHGs anyway.

In some smaller cities - those with less than 300,000 inhabitants - the complete USR treatment is performed. The cities of Rauch, Laprida, Trenque Lauquen and Baradero have exemplary plants. In them, all of the residential waste is recycled, refined and transformed. Waste is separated in organic and inorganic waste. Compost is made out of organic waste mainly using the vermicomposting method. Solid inorganic waste is separated into: glasses, plastics, cardboards and papers, metals. This town's population has been educated so that waste is separated and put in different bags. As a result, people also think about how to reduce their waste. This procedure helps to reduce methane emissions. However, processing plants still do not have a big presence in Argentina.

Another good example of reducing GHGs—is the foresting of areas with autochthones trees, which is presently done by Argentine Foundation of Etoecology—AFE. Efforts to decrease greenhouse gas emissions by the AFE's reforestation of specific zones with trees native to the Area has been successful. If all of the businesses were harmonized and urged to cooperate on social and environmental issues; if adequate information was published by the right judicial institutions (NGOs - as stated in Chapter IV of the Kyoto Protocol); and if all of the people received environmental education, our planet would be a better place to live. Rather than implementing old economic recipes, that do not allow for sustainable growth, governments need to engage in consciousness efforts to raise environmental education so that people know about sustainable growth and how it can be achieved.

In most of the places, it is the NGOs who take the lead role in case of environmental disasters and in cases where social issues need to be addressed. NGOs work with community to raise funds, to education, and to preserve and clean the environment.

The task of publicity and education is also carried out by several NGOs in Latin America. In 2005, the Argentine Foundation of Etoecology (FAE) - one of the 740 United Nations Observers for climate change - has joined with a chorus of associations in 14 cities of Argentina to spread the message of the climate change. At that time, many demonstrations were held and petitions were signed, all complaining about the current state of natural resources and the population. At the same time, the groups demanded action on environmental protection and climate normalization. This program was carried out as a side event of the COP (United Nations Conference of the Parties on Climate Change) 11 in Montreal.

But despite the encouragement coming from NGOs in Latin America and elsewhere, governments make only mild attempts to change damaging environmental practices. Unusual to the business world, some companies have joined an initiative, UN Global Compact, to deal with issues of human rights, labour, the environment and anti-corruption. About 220 campanies from argentina have also signed this framework. In this region, only a few companies sponsor independent NGOs programs aimed at defending and conserving the environment. People and Organizations are becoming aware of the economic support available from international organizations and philanthropists because authorities use the internet to encourage international cooperation. It is too optimistic and naive to think that climate change, which is an environmental emergency that affects the whole planet, is a problem that can be solved by civil society alone. Although society can fix some guidelines, solving climate change is not its main purpose. Despite this, there are hopes that the planet's civil society will unite in order to respond to demands beyond its own frontiers.

Like the subject of climate change, the NGOs are expanding their efforts. With hope, they are working to reduce greenhouse gases and stabilize climate change. However, hope is not the only thing needed to deal with a complex situation even as climate change, a problem which is global in scope. All citizens - governing and governed, civil society and its organizations, individual countries and their neighbors - must work jointly. However, there is not much time. First, it is necessary for the press to truthfully inform the population. The authenticity and urgency of the message will be accepted if all human beings respect the need to build a sustainable planet. Education is the best tool to obtain this goal.

As can be seen by the examples discussed above - and there are many, many more - Argentina and all the countries of the region that have ratified the Kyoto Protocol do very little to achieve the goals or targets set by the protocol. On the contrary, there is now more environmental deterioration and there are now more centers of CO_2 emission. The countries of the region have interpreted the Kyoto Protocol like an economic instrument, not like a document of environmental cleanup meant to improve the quality of life on the planet. For this reason, Latin America does not have an environmental policy for sustainable development.

The abovementioned circumstances in Argentina, show that the country lacks a holistic environmental program. If climate change continues, the number of poor people in Argentina, other regions, and the planet will increase. If desertification increases, food will become scarce; if water is not more potable, thirst will be a world-wide problem. Argentina, South America and all fertile countries must continue to preserve their resources, otherwise 852,000,000 impoverished people will not be the only ones to experience hunger. Hundreds of millions more will experience it as well.

34

CHAPTER

Impact of Global Warming on Antarctica and Its Flow on Effect on Australian Environment

Syed U. Hussainy[1] and Santosh Kumar[2]
[1]Institute for Sustainability and Innovation
Victoria University (Werribee Campus)
PO Box 14428, Melbourne City, MC 8001, Australia
[2]Department of Mathematics and Statistics,
University of Melbourne, Parkville, Victoria, 3010, Australia
and
School of Computer Science and Mathematics
Victoria University, Footscray Park Campus,
PO Box 14428, Melbourne City, MC 8001, Australia
E-mail: Santosh.Kumar@vu.edu.au

INTRODUCTION

The greenhouse scenario postulates a worldwide increase in the sea level due to eustatic changes. Until recently the conventional belief was that the sea level was rising at a rate of 1.0–1.5 mm y^{-1} (Barnett 1983). The 1985 Villach Conference proposed a rise of 20–140 cm in the next century attributable mainly to thermal expansion of ocean water (WMO 1986). A sea level rise of this magnitude has destructive implications on the world's coastline. Beach erosion would be accelerated, lowlying areas would be permanently flooded or subjected to more frequent inundation during storm events, and the base-line for the water table would be raised.

The global application of this accelerated eustatic rate poses a problem for two reasons. First an eustatic rate currently is difficult to measure and by no means is it uniform. Secondly the recorded data of worldwide eustatic is sparse, but sea level is exhibiting great temporal and spatial variability. Besides the spatial variation in long-term sea-level trends, sea level at a station is characterized by temporal variability operating at time spans of days to years. There are fluctuations caused by the same factors but experience a wide range in diverse locations, such as the Bay of Bengal, west coast of Mexico, Northeastern Siberia, and Australia. Thirdly, the persistent winter-annual fluctuations in sea level associated with El Niño/Southern Oscillation (ENSO) events in the Pacific region. Sea-level increases in the Pacific region may not be indicating an eustatic rise, but simply reflecting the regional consequences of more frequent ENSO events. Finally there are long-term fluctuations reflecting climate and oceanographic changes.

In this chapter an attempt has been made on the impact of global warming on the melting of Antarctic ice and the subsequent rise on sea level and its impacts on the Australian coastline environment.

ANTARCTIC ENVIRONMENT AND HUMAN INVASION

The Australian Antarctic Territory covers 42% of the continent, 5.6 million square kilometres of land, almost all of it covered by ice with a rugged 7,000 kilometres of coastline and a huge off shore economic zone. Due to the climate changes and an increase in the anthropogenic activities on the continent, plants and animal species alien to Antarctic are invading the virgin land. Non-native species are already establishing themselves in what is one of the world's last wilderness (Edward 2006).

The Larsemann Hills in Antarctica seems to be the last place on earth one would think of :an international conflict. These hills are bare rock hummocks surrounded by ice on the shore of Prydz Bay in Eastern Antarctica. Now India has decided to build a base there sparking the first dispute to publicly break the diplomatic calm of the Antarctic Treaty System in years. It told the other treaty nations that the sacred Godavari River would have flowed there millions of years ago, when the Antarctica and India were a part of the Gondwana land. It was not as if the Godavari flowed to that promontory only, the geology matches for hundreds of kilometres around (Darby 2006). India still has to produce a comprehensive environmental evaluation and there will be, no doubt, an on going dialogue of the project's future. Russia, on the other hand, is proceeding with the plan to drill a giant lake (Lake Vostok, average temperature 3°C) locked under Antarctic ice. The lake covers almost 13,000 sq. kilometres and has not been exposed to the outside world for more than 400,000 years, although the lake might have been isolated for millions of years. Due to the long isolation, it is believed that the water inside may contain a new form of life and unique geochemical processes. Russians have so far drilled through 3,650 metres of ice to come within 130 metres of the lake's surface far inland in the Australian Antarctic Territory. Lake Vostok is also the seventh largest fresh water lake on earth. There are evidences indicating that it has geothermal-heated areas and could support life including fish (Darby 2006). With the possibility of CO_2 concentration doubling or even tripling in the next 100 to 250 years. The Antarctic landscape may lead to a similar landscape that was 40 trillion years ago (Darby 2006).

GREENHOUSE GASES AND ITS IMPACT ON THE ANTARCTICA

The greenhouse-effect can be characterized by both local and global, occurring on a time scale of decades to centuries with some changes being inevitable. Changes in average air and sea surface temperature, and global mean sea levels have occurred which are consistent with the greenhouse-effect. Dunbar (2006) postulates that there would be an increase in the rates of ice melt, in the next 100 years experienced by warmer temperature in Antarctica.

Carbon dioxide, a common atmospheric gas, with the properties of being transparent to solar radiation, but somewhat opaque to terrestrial infrared radiation (heat). This along with the other gases with similar properties, e.g. methane (CH_4), nitrous oxide (N_2O) and the chlorofluorocarbons (CFC_5), tend to increase the Earth's surface temperature in comparison to its otherwise would be temperature. This suggests that any event that tends to change the concentration of the greenhouse gases in the atmosphere would change the global climate.

It is an established fact that there is an increase in CO_2 in the global atmosphere. Pearman (1988) observed an increase of about 1.5 ppm y^{-1} (0.4% y^{-1}), over a period of 15 years. It is estimated that each year about 5GT of carbon is released into the atmosphere due to combustion of fossil fuel. This carbon, as CO_2, is exchanged with the biosphere through the photosynthetic and respiratory processes and is exchanged with the of oceans. The rate of increase of CO_2 in the atmosphere, therefore, depends on the rate of exchange of CO_2 between the reservoirs and their capacity to accumulate them. Predictions of the future CO_2 concentrations are usually performed using global carbon cycle models (Enting and Pearson 1987,

Seigenthaler and Oesetiger 1981). Methane, on the other hand, has been increasing in the atmosphere at about 1% per year (Khalil and Rusmussen 1983). There are a number of sources of atmospheric CH_4, many of them are related to anthropogenic activities. The CH_4 increase impact can also be possibly related to decreasing levels of hydroxyl radical and increasing concentration of carbon monoxide (Khalil and Rusmussen 1985). But as the global warming causes the permafrost to melt, lakes worldwide could emit even more methane, reinforcing climate changes. Until now the input from the lakes was not realized as an important source. Over the coming years methane flux from the lakes is likely to increase as bacteria convert carbon into methane (Walter 2006). The radioactive flux divergence at the Earth's surface due to increase of CO_2 by 2030 is expected to be 0.9-3.2 $w.m^{-2}$ with a best estimate of the warming due to CO_2 being 0.7°C. The combined effect of all other gases will be 1.3–3.9 wm^{-2} or a best estimate of the warming of 0.7°C which is as much as CO_2 alone. This means that CO_2 alone would be responsible for about half the total greenhouse warming. A double of CO_2 alone would theoretically lead to an average equilibrium surface warming of about 2 to 4°C with greater warming occurring in winter and high altitudes. The reinforcing effect (positive feedback) of warmer temperatures leading to less snow and ice cover and thus to more sunlight being absorbed by the surface.

CLIMATE CHANGE AND ITS IMPACT ON AUSTRALIAN ENVIRONMENT

Due to climate change, it is postulated that in land areas the temperature increase are expected to be about 2°C in Australia and up to 3 or 4°C further south. Near the coast these increases could be moderate by the lagging behind of sea-surface temperatures, and it is possible that winter and overnight minimum temperatures will warm more than summer and daily maximum temperatures. Rainfall would also vary with season and location. It would seem likely that the summer rainfall regime will intensify and push further south. This has already started to happen when the average spring, summer and autumn rainfall in same months in central NWS from the Hunter Valley, west through Dubbo has increased by 30-40% since this early century. The increased surface temperatures make it possible for the air to hold a greater absolute amount of water vapour so that in situations where the air is continuously lofted to create intense rainfall, greater maximum rainfall rates are possible. This is a critical factor in the intensity and frequency of flush floods and in the design of dam spills ways intended to cope with maximum runoff in water catchments.

Global average sea level is expected to rise by 20 to 140 cm with a global surface warming of around 3.5°C (Botins et al. 1986). Future sea level rise will not be geographically uniform, as it will be affected by the changes in ocean circulation and wind stress, as well as, continuing movement of land surfaces due to tectonic activity and local subsidence and changes in the distribution of atmospheric pressure. Besides thermal expansion of water, sea level will also be affected by the melting of mount glaciers and by changes in melting and accumulation of the Antarctic and Greenland ice sheet. The unstable is the West Antarctic ice sheet, which in itself would contribute 5–7 m rise in sea level. The snow line would on an average rise about 100 m for every 1°C warming. However, local variations related storm frequency and precipitation might be significant.

The concept of a global sea level is based on analysis of tide gauge records from around the world's coastline. Tide gauge record show changes of sea level relative to the land—the resultant of upward or downward movement of the sea. In Australia only three tides gauge records were recorded as reliable over a period of more than 30 years (Bird 1988). These were New Castle rise of 2.0 mm y^{-1}, Sydney (Fort Dennison) during 1897–1983 an annual rise of 0.7 mm y^{-1} and Port Adelaide (1882–1976) an annual rate of 2.5 mm y^{-1}.

A general indication of sea-level rise will be an increase in the rate and extent of coast-line erosion. A sea-level rise brings wave action to a higher level; erosion is likely to accelerate on cliffs, beaches, marshlands and deltas. The effect will be particularly severe on coastline subject to recurrent storm surges, which will penetrate further than do at present.

EXPECTED SEA LEVEL RISE FROM CLIMATIC WARMING

Projections have been made over the years based on evaluations of possible changes and rates of changes on the ice-sheet in the Antarctic particularly the West Antarctic ice-sheet (Thomas 1985, Lingle 1985). It is noteworthy that the floating ice shelves and parts of the ice-sheets below sea level do not contribute to changes in sea level. It is the part of ice sheet above the sea level over the major ice streams, which can provide the potential sea-level rise. It is estimated that the Antarctic sheet has a net accumulation over the ground, ice of about 2×10^3 km^3 y^{-1} over its area of approximately 13.6×10^6 km^2 (Budd and Smith 1985).

One of the anticipated consequences of climate change is increased precipitation over Antarctica. Any increase in precipitation over the Antarctic would have a cumulative effect on sea level due to long residence time of the snow on the continent before reaching the sea. These would be greater than 1000 years of most of the grounded continental ice except the coastal one. A doubling of CO_2 concentration would lead to 30% increase in Antarctic precipitation (Budd 1988).

GLOBAL WARMING AND ICE SHEET MELTING RATES

The presence of large ice shelves in coastal embayment may be responsible for keeping the strain rate of the thick ice near the grounding line low. The disappearance of ice shelves from increased melting could allow the strain rates to increase resulting in high coastal flow rates.

In the case of substantial warming (about 4°C), the melt rates could increase greatly. Such high melt rates could remove the bulk of the large ice shelves in about 50 years (Budd 1988). With 50 years being required for the 4°C warming accompanied by gradually increasing melt rates, the anticipated time for the removal of ice shelves will be about 100 years. As the ice shelves in the large embayment are thin, the grounding lines retreat and high strain rates develop near the grounding line. These higher strain rates occurring as the ice streams become afloat to allow the ice stream to flow rapidly (Budd 1988). The major concern in regard to the prospect for rapid flow of the Antarctic ice sheets has been for those zones with deep ice which could develop lower basalt stress and slide rapidly thus reaching speeds of surging glaciers, some tens of km y^{-1}. The current ice-stream velocities are a few hundred meters per year and model results of Budd (1988) indicate that increases in several km y^{-1} are likely, but not tens of km y^{-1}. As a consequence, it would take 100 years or longer to drain the ice from the deep ice regions of West Antarctica.

The region of ground ice in the Antarctica, which is most vulnerable to change resulting from the global warming in the deep ice in the West Antarctica between Ross Ice Shelf and the Pine Island Bay in the Amundsen Sea. According to Budd (1988), the large contribution to sea level rise from increased Antarctic ice-flow rates is expected to set in until some 50 to 100 years later. The net effect of Antarctic contribution to sea level rise expected would be in 100 of years. Although the projected long-term sea level increases of several metres are a long way in the future, these changes should be viewed as serious because they may be practically irreversible.

The various factors that favour the initiation of acceleration of beach erosion are widespread and more so assuming that the near shore waters deepen, so that wave erosion becomes stronger and more frequent. As the sea level rises, the proportion of retreating sandy coastline will increase as beaches that are at present stable, alternating or prograding pass into an erosional regime (Budd 1988). The prevalence of beach erosion is due to several factors (Table 1) that operate in varying associations around the world coastline. A relative sea-level rise is only one of these.

Table 1 Factors that favour the initiation or acceleration of beach erosion.

1. Diminution of fluvial sand supply to the coast as a result of the reduced runoff or sediments yield from a river catchment (e.g. because of reduced rainfall, or dam construction leading to sand entrapment in reservoir, or successful soil conservation works).
2. Reduction in sand supply from eroding cliff or shore outcrops (e.g. because of diminished runoff, a decline in the strength and frequency of wave attack, or the building of sea walls to half cliff recession).
3. Reduction of sand supplies to the shore where dunes that had been moving from inland are stabilized, either by natural vegetation colonization or by conservation works, or where the sand supply from this source has run out.
4. Diminution of sand supply washed in by waves and currents from the adjacent sea floor, either because the sand supply has run out or because the transverse profile has attained a form, which no longer permits such shoreward drifting.
5. Reduction in sand supply from alongshore sources as the result of interception (e.g. by a constructed breakwater).
6. Increased losses of sand from the beach to the backshore and hinterland areas by landward drifting of dunes, notably where backshore dunes have lost their retaining vegetation cover and drifted inland, lowering the terrain immediately behind the beach and thus reducing the volume of sand to be removed to achieve coastline recession.
7. Removal of sand from the beach by quarrying, and losses of sand from intensively used recreational beaches.
8. Increased wave energy reaching the shore because of the deepening of near shore water (e.g. where a shoal has drifted away, where seagrass vegetation has disappeared, or where dredging has taken place).
9. Increased wave attack due to climatic change yielding a higher frequency, duration, or severity of storms in coastal waters.
10. Diminution in the volume of beach material as the result of weathering, solution or attrition of beach sand grains, leading to a lowering of the beach face and a consequent increase in wave attack on the backshore.
11. A rise in the water table within the beach, due to increased rainfall or local drainage modification, rendering the beach sand wet and more readily eroded.
12. Increased losses of sand alongshore as a result of a change in the angle of incidence of waves (e.g. as the result of reflection growth or removal of a shoal or reef, or breakwater construction).
13. Intensification of wave attack as the result of lowering of the beach face on an adjacent sector (e.g. as the result of reflection scour inducted by sea wall construction).
14. Migration of beach lobes or forelands as the result of longshore drifting – progradation as these features arrive at a point on the beach is followed by erosion as they move away down drift.
15. Submerged and increased wave attack as a result of a sea level rise relative to the land.

AUSTRALIAN COASTS AND ITS POTENTIAL TO SEA LEVEL INUNDATION

Australia is surrounded by one of the world's longest national coastlines. It contain over 30,000 km of main land coast (including Tasmania), plus an equal magnitude of coast contained in the surrounding island and in the estuaries and coastal lagoons, lakes and bays. The total coastline is of the order of 70,000 km (Galloway 1982). The projected sea-level rise would impact the entire coast. The degree of impact would, however, vary according to the rate and magnitude of sea-level rise and the morphology and location of each coastal sector. The nature and significance of the rise would include both the physical impacts of inundation and shoreline readjustment together with ecological impacts as all littoral communities readjust to the physical changes.

Permanent Inundation

All sections of the Australian coast would experience permanent vertical inundation of a magnitude equal to a sea-level rise. The degree of horizontal inundation would, however, vary considerably depending on the magnitude of rise relative to the cross-shore gradient combined with secondary morphological adjustment to the rise. Areas most prone to permanent inundation can be classed by the cross-shore gradient. In general terms the lower the gradient the greater the horizontal inundation. Australia has open coast which contains 22,500 km of open sedimentary shore line, consisting approximately 10,000 km of low gradient intertidal and supra-tidal mud flats and sand flats and chenier-beach ridge plains with an area of over 60,000 km^2.

A second category of low gradient shoreline not classified as open coast is contained in estuaries, coastal lagoons, lakes and bays. All of Australia's estuaries/wetlands will be severely affected by the sea level rise due to the generally low shore line gradient (Shot 1988). This impact will also affect many coastal lakes seasonally or permanently blocked to the sea whose water level, however, is a function of sea level.

Permanent inundation will, therefore, have its greatest horizontal impact across the tidal flats, chenier-beach ridge plains of northern Australia and South Australian gulfs, and in the coastal wetlands that occur widely in Australia, though particularly along the north, east, south east and south-west coasts.

Periodic Inundation

Several studies on periodic inundation have been conducted around Australia and these include Geary and Riffin (1985). Periodic inundation is defined as sea-level inundation above mean sea level due to astronomical (tides) and atmospheric forcing (waves, set-up, storm surges, flooding etc.). Periodic inundation currently occurs on all coasts. The magnitude and frequency ranges from the highly predictable oscillations through to incident, infra-gravity and shelf waves and other longer periodic sea oscillations together with episodic storm surges and coastal flooding. Periodic inundation is always recognized as a major hazard around Australian coasts. In all areas exposed to coastal flooding a rise in sea level would act to raise the flood levels in those areas by a similar magnitude. This would be further aggravated when coastal flooding coincides with strong on shore winds such as the ones generated by tropical cyclones and east coast lows producing high waves (up rush and set-up) and storm surges leading to extreme elevations in sea levels. The full impact of sea-level rise around Australia will depend on both the juxtaposition of the rise on pre-existing cross shore gradient, coupled with the secondary effect of periodic inundation riding on the top of the rise. Whereas the impact of a permanent vertical rise in sea level is highly predictable, the spatial and temporal impact of associated periodic inundation range from highly predictable, e.g. tides to moderately predictable, e.g. wave up rushed to more unpredictable, e.g. storms surge and flooding.

EARLY INDICATIONS OF SEA LEVEL RISE

As cited earlier, there are problems associated in the determination of global trends on sea level fluctuations by direct method measurements (Bird 1988). It may, therefore, be useful to explore the use of ecological indicators of marine submergence as a source of additional evidence of sea level change, especially in low-lying areas most likely to be affected. Intertidal communities such as salt-marshes, mangrove swamps, coral reefs and the invertebrate fauna of rocky shore and artificial structures in the intertidal zone have all been suggested as possible indicators of sea level change. Most occur on coastlines which have developed during the period of relative sea level still standing following the Holocene marine transgression and could show changes in response to predicted sea level rise.

Many of the communities exhibit some form of zonation largely in response to the severe environmental gradient imposed by alternating tidal inundation and sub-aerial exposure. In order to be useful as ecological indicators, communities of intertidal organisms should exhibit a clear pattern, ideally with distinct boundaries, which reflect the relative tolerance of individual species or association to inundation or exposure.

The communities must be sufficiently responsive to exhibit some detectable change in their structure or extent as a result of sea level rise. Individual organism should be sessile or persistent and limited in their distribution largely by their tolerance to inundation or exposure. Sessile organisms are not able respond by moving positions with respect to relative sea level and so provide a good indication of a change in ecological conditions. They need to be sufficiently persistent to provide a record overtime in the form of either a change in population age structure or as remains. In addition supportive evidence obtained from a series of zones within a site and from several locations is desirable prior to attributing any changes to a rise in sea level, in order to eliminate factors such as individual species response or localized disturbances.

Monitoring of changes in the structure and composition of intertidal fauna and flora may possibly provide an earlier warning on the sea level rise than could be obtained from the sea analysis of global sea level trends.

THE POTENTIAL IMPACT DUE TO PERMANENT AND PERIODIC INUNDATION AROUND AUSTRALIA

The potential impact of sea level change and subsequent inundation either periodic or permanent around the Australian coast is summarized as follows:

Some features of the Australian coast, which are of particular relevance to the effects of climate change and prone to flooding include:

- The coastal flood plains of the east coast streams, which are regarded as the most flood-prone areas of Australia.
- The Great Barrier Reef coast of central and northern Queensland.
- The central and northern Queensland and the Gulf of Carpenteria, tropical cyclones are major cause of flooding.
- The zone of tropical cyclone hazard extends from Geraldton on the west coast to Brisbane to Gold coast on the east coast. However, cyclones have been recorded as far south as Perth in the west and Coffs Harbour in the east (Leivesley 1984).
- South of latitude 25° S the coast receives high-energy swell from the Southern Ocean because of its position Adelaide is also at risk.
- Much of the dune-backed shoreline of South East and mineral sand provinces is recording and this is very little accretion.

- Most of the species that make up the nation's seafood catch depend on coastal estuaries and tidal marshes during some or all of this life cycle. These areas also form a pollution control device that cleanses inland waters on their way to sea. Anthropocentrically speaking, the major function and uses of the Australian coastal zone are to provide sites for.
- Residential, recreational and commercial use.
- Ports and sea transport.
- Commercial and recreational fishing.
- Tourism.
- Conservation of natural environment, and
- Industrial use, e.g. salt production, mining and agriculture.

Of all parts of Australia, the coastal zone is likely to be particularly affected by climate change and the melting of Antarctic ice. The physical consequences of sea level rise *per se* can be broadly classified into three categories: Shoreline retreat, temporary flooding and salt water intrusion (Titus and Barth 1984). Effects on natural (unmanaged) systems are difficult to predict, but these may include abrupt transition to new systems domains, e.g. intertidal mangroves and shore bird habitat; upstream translations of salt tolerant ecotones; gradual inundation of wetlands on law energy coast etc. In addition to shoreline retreat due to higher sea level *per se* erosion of beached coastlines under storm conditions. Salt water intrusion via both breached barrier formation (e.g. the lower Murray, South Australia) and high sea level (e.g. Koowerup Swamp in western Port Bay, Victoria) will affect water supplies to coastal agriculture in a limited number of locations. Perth (Western Australia) and New Castle (New South Wales) are the only large coastal centres dependant on ground water, but are unlikely to be affected in this way (Cooks et al. 1988). Changes in lifestyle for most Australians as a result of coastal zone changes are likely to be more those of degree rather than kind. Coastal resort towns, existing largely for their beach recreation and/ or seaside holiday opportunity will be under special threat. The higher priority regions which are susceptible to sea level rises include:

- Metropolitan Brisbane and Sydney,
- Major urban area: New South Wales North Coast, Townsville, Cairns and Coast, and
- Minor urban: Darwin and Geraldton

The impact of sea level rise is also left on developments, particulate marine structure and set-back distances:

- Coastal recessions on the open coast and in estuaries,
- Flooding of low-lying coastal plains.

- Changes in the salinity require of estuaries and water tables.
- Alterations of site-specific coastal processes include changes to:
    1. Gross sediments transport.
    2. Net sediments transport.
    3. Wave climate (including wave direction).
    4. Coastal alignment.
    5. Movement of sediments into estuaries.
    6. Losses of overshore sediments sinks.
    7. Over topping and roll back of dune system.

The impact of permanent and periodic sea level inundation around the Australian coastal line is listed on a state-by-state basis in relation to the geology of the area.

Tasmania

Tasmania is the island state on the south of the main land Australia. Except on the northwest coast, Tasmania has waves dominated by beaches and high dunes or rocky coast that dominated the open coast. Open coast beaches and fire dunes would experience accelerated erosion from wave uprush, set-up and moderate amplitude shelf waves. The high dunes and rocky coast restrict horizontal inundation to numerous estuaries and river mouth particularly on north, east and south coasts.

Victoria

The state of Victoria occupies the southeast sections of the main-level. High wave energy beaches and beaches and dunes would restrict permanent inundation of extensive coastal estuaries and lakes, particularly in eastern Victoria. In open areas wave uprush, set-ups and shelf-waves to 0.5 m amplitude accelerate beach and foreshore erosion.

New South Wales

Permanent inundation would have greatest impact in estuaries and coastal lagoons and lakes particularly on wetlands. Storm surges and shelf waves are minimal along the New South Wales coast. However, wave uprush, and wave set-up coupled with coastal flooding will continue to pose most risk to the coast (Geary and Griffins 1985, Short 1988). Erosion of beaches and foredunes would accelerate, particularly in association with the east coast cyclones.

South Australia

The South Australian coast contains well protected high-energy beaches, dunes and extensive low gradient gulf shorelines. The gulf tidal flats and beach ridge plans would get flooded by permanent inundation raising the water level in the large Coorong estuary. The entire coast would continue to be affected by shelf waves coupled with extreme waves uprush and set-up on the open coast. This would eventually result in foredune erosion and increasing frontal dune instability.

South Western Australia

The high wave energy coast dominated by steep and rocky gradient would resist permanent inundation of the numerous estuaries and coastal lakes around the southwest. The open beach system would be impacted by shelf waves particularly when coupled with waves uprush and set-up leading to beach erosion and dune scarping. Protection from waves would increase north of Geographe Bay as the lower shelf gradient minimizes wave attack and shelf waves decrease in amplitude.

Queensland

Cross-shore gradients are generally higher in the north, with variable ones along the central and in the southern coast. There are several wetlands, river mouths, deltas and estuaries distributed all along the entire coast. Permanent inundation would vary considerably depending on cross-shore gradients, but would be generally minimum on wave dominated open coast and maximum in more protected tidal and chenier beach ridges, plains and wetlands. Tropical cyclones and storm surges pose the greatest risk in north and central Queensland. Extreme waves uprush and wave set-up will accelerate beach erosion and dune scarping on the open southern Queensland coast.

Gulf of Carpenteria

The gulf is predominated by low gradient tidal flats, beaches and chenier ridge plains, particularly in the south together with river mouths and estuaries around the coast prone to extensive permanent inundation and low shelf gradient produce a high risk of storm surge inundation, particularly in the meso-tidal southern gulf. The coast is also prone to monsoonal (summer) flooding compounded by episodic tropical cyclone rainfall.

North Western Australia and the Northern Territory

This coastal region of Australia is characterized by increasing tidal range, decreasing cross-shore gradients and increasing tropical storm surges would result in increasing permanent and periodic inundation. The high tide ranges would, however, minimize the storm surges impact. Many west coast tidal flats and beaches ridge plains are backed by steeper gradient beaches and foredunes which would minimize spring tidal inundation. Numerous river mouths and estuaries would experience wetland inundation.

EFFECT OF SEA LEVEL RISE ON THE GREAT BARRIER REEF

Coral reefs have been built up to about low tide level by organic growth and sedimentation, and where the surfaces are exposed at low tide, dead coral and algae dominate them. The growth, morphology and functioning of coral reef are intimately related to sea level and sea level changes. The majority of modern reef have foundations, which date back to at least the late Tertiary and have evolved discontinuously during an era of rapidly fluctuating sea levels (Hopley and Kingsey 1988). Coral and reef may grow at 100 m below sea level, effective reef growth is restricted to approximately the upper 40 m of the euphotic zone. Modern reefs have not simply grown upwards with the post-glacial transgression but have developed through recognition of previous foundations by organisms including corals which have migrated across the shelf in their free-floating larval stage within the water column.

Both geological and metabolic data suggest that a continuous sea level rise at a rate greater than 10-12 mm y^{-1} will be greater than can be matched by any reef at surface and will result in the progressive drowning of the reef. Even at inundation rates of 10 mm y^{-1}, a mat dry (reefal) accretion rate is unlikely as this presumes that the reef surface being submerged will be able to achieve immediately a carbonate productivity at or close to the maximum possible.

The overall effect of the projected greenhouse rise in sea level in the short term is expected to be a rejuvenation of the reef flat areas of the Great Barrier Reef, particularly in coral growth. Senile reef tops dominated by sediments with revert to mature active form in the short to medium terms (50 to 150 years). If inundation continues, immature reef tops with limited zonation will result, whole reef tops being dominated by living corals. Inundation of at least 3 m will be necessary to produce this stage.

A rise in sea level of approximately 0.8 m in the 0.8 m in the next 50 years or even 1.6 m in the next century should not be regarded as traumatic experience for most of the Great Barrier Reef. The 50 years projection of sea level may be further put into context by suggesting that the rise is of a similar magnitude to the 2-weekly change from neap to spring high tides although critical level such as mean low water springs (higher level open water coral growth) and the mean sea level (highest level of extensive moated coral growth) will be elevated, presuming no alteration to tidal range takes place.

In the short term (50 to 100 years), the effect of a rapid rise in sea level on the Great Barrier Reef, on balance may be beneficial rather than detrimental. Rejuvenation on the longer depauperate reef flat of the province by renewed coral growth will certainly make them aesthetically more pleasing.

Increasing water temperature may eventually disadvantage coral reef biota. Many other aspects of a greenhouse modified environment may have a more subtle effect, though most would be long term rather immediate. Increasing cyclone incidence and severity may retard decolonization of submerged reef roots. Increased rainfall and greater cloudiness may have a detrimental effect, particularly, on the inshore fringing reef through the present suggest climate change may be relatively trivial for coral reefs. Changes to run-off, upwelling and oceanic circulation may affect the delicate nutrient status of shelf waters, but again the responses are most likely long term rather than short term. The anthropocentric developments on the adjustment mainland and highland, including the animal feedlots, and increasing tourist pressure on the reef itself may be of greater significance rather than the modified greenhouse climate (Hopley and Kinsey 1988).

The scientific information discussed above demonstrates the impact of climate change on Antarctica and its flow-on effect on Australia and the surrounding environment. This should assist us in developing strategies and implement them to address the issues related to climate change. The strategies should focus on the sustainable work practices to reduce the impact on climate change for the short term as well as for long term. The responsibility and accountability rest with us to act now. We cannot afford to procrastinate. Thus it may be appropriate to remind us, the views of Thomas A. Kempis (1380-1471) who said, **"Verily when the day of judgement comes, we shall not be asked what we have read, but what we have done?"**

References

Barnett, T.P. 1983. Recent changes in sea level and their possible causes. Climate Changes 5: 15-38.

Bird, E.C.F. 1988. Physiographic inundation of a ses level rise: In: G.I. Peasman (ed.). Greenhouse Planning for Climate Changes, CSIRO, Australia, ISBN: 06434048030. pp. 60-73.

Budd, W.F. 1988. The expected sea level rise from climate warming in the Antarctic. In: G.I. Peasman (ed.). Greenhouse Planning for Climate Changes, CSIRO, Australia, ISBN 06434048030. pp. 74-82.

Budd, W.F. and I.N. Smith. 1985. The state of balance of the Antarctic ice sheet, Glaciers - Ice sheet and sea level: Effect of CO_2 induced Climate Changes, US Department of Energy Report: DOE/EV/60235-1. pp. 172-177.

Cocks, K.D., A.J. Gilmour and N.H. Wood. 1988. Regional impact of rising sea-levels in coastal Australia. In: G.I. Peasman (ed.). Greenhouse Planning for Climate Changes, CSIRO, Australia, ISBN: 06434048030.

Darby, A. 2006a. Russia ignores plea on drilling Antarctic Lake. The Age, Melbourne, 13th July 2006, p. 8.

Darby, A. 2006b. Black Ice, Antarctica Focus. The Age, Melbourne, July 11, 2006, p. 11.

Dunbar, R. 2006. Frozen land to bloom. Herald Sun, Melbourne, 13th July 2006.

Edwards, R. 2006. Alien species at the gates of Antarctic. New Scientist No. 2588, pp. 16.

Enting, I.G. and G.I. Pearman. 1987. Description of a one-dimensional carbon cycle model calibrated using techniques of constrained in version. Tellus, 39B, 459-476.

Galloway, R.N. 1982 Australian Islands. Cartography. 12: 233-237.

Geary, M.G. and A.G. Griffin. 1985. Significance of oceanographic effect on coastal flooding, N.S.W. Australian Conference on Coastal Ocean Engineering, Christ Church, The Institute of Engineers, Australia. pp. 509-520.

Gornitz, V., L. Lebedeff and J. Hansen. 1982. Global sea level trends in the last Century. Science 215: 1611-1614.

Hopley, D. and D.W. Kinsey. 1988. The effect of a rapid short term sea-level rise on the Great Barrier Reef. In: G.I. Peasman (ed.). Greenhouse Planning for Climate Change, CSIRO, Australia, ISBN: 0643048030.

Khalil, M.A.K. and R.A. Rasmussen. 1985. Sources, Sinks and Seasonal Cycles of Atmospheric Methane. J. Geophys. Res. 88: 5131-5144.

Khalil, M.A.K. and R.A. Rasmussen. 1985. Causes of increasing methane depletions of hydroxyl radicals and the rise of emissions. Atmos. Environ. 19: 397-407.

Leivesley, S. 1984 Natural disaster in Australia. Disaster 18: 82-88.

Leingle, C.S. 1985. A model of a polar ice stream and future sea level rise due to possible drastic retreat of west Antarctic ice sheet. In: Glaciers: Ice Sheets and Sea Level: Effect of CO_2 Induced Climate Change US Department of Energy Report, DOE/EV/60235-1. pp. 317-330.

Pearman, G.I. 1988. Greenhouse gases: Evidence for atmospheric changes and anthropogenic causes. In: G.I. Peasman (ed.). Green House Planning for Climate Change, CSIRO, Australia, ISBN: 064 304 8634. pp. 3-22.

Seigenthaler, U. and H. Oeschgh. 1987. Biospheric C emission during the past 200 years reconstructed by de-convolution of ice core data. Tellus, 39B: 140-154.

Titus, J.G. and M.C. Barth. 1984. An overview of the causes and effect of sea-level rise. In: M.C. Barth and Tilus (eds). Greenhouse Effect and Sea Level Rise; A Challenge for This Generation, Van Norstrand Reinhold, New York. pp. 1-56.

W.M.O. 1986. Statement from UNEP/WMO/ICSU conference, Assessment of the role of CO_2 and Other Greenhouse Gases in Climate Variations and Associated Impacts. World Meterological Organization, WMO, Bull. 35: 129-134.

Walter. 2006. Siberia's pool burp out nasty surprise. Nature Vol. 443, pp. 71. In: New Scientist, No. 2568, 9th September. p. 20.

Section VII

Gender and Climate Change

35

CHAPTER

Solidarity in the Greenhouse: Gender Equality and Climate Change

Ulrike Röhr[1] and Minu Hemmati[2]
[1]Genanet – focal point gender justice and sustainability
LIFE e.V., Dircksenstr. 47, D-10178 Berlin, Germany
E-mail: roehr@life-online.de
[2]Ansbacher Str. 45, 10777 Berlin, Germany
E-mail: minu@minuhemmati.net

INTRODUCTION

The United Nations is formally committed to gender mainstreaming within all policies and programmes. However, gender equality is not yet realized in any society, in any part of the world. Men and women have different roles, responsibilities and decision-making power, leading to disadvantages for women. It is therefore not surprising to note that although gender plays a role in relation to climate change, yet the topic has not been explored much, and many people still find it difficult to understand in what way gender might be a factor in climate change, or how it should be addressed.

Gender aspects are rarely addressed in climate change policy. This applies – with few exemptions – to the national as well as the international level. Various reasons account for this neglect: gender aspects in climate change are often not self-evident, and there is few data, research, or case studies clarifying and exemplifying the linkages between gender justice and climate change. This is a gap that needs to be addressed – but it is also an invitation and an opportunity for those doing research in climate change. Gender is not only relevant in its own right but also constitutes an opportunity to introduce a focus on social aspects into the climate change agenda.

There are a number of issues that point to the crucial role of gender when understanding the causes of climate change, aiming to mitigate it, and working towards successful adaptation to inevitable climate change. We will address the following issues in the this chapter:

- Women and men – in their respective social roles – are differently affected by the *effects of climate change;*
- Women and men – in their respective social roles – are differently affected by *climate protection instruments and measures;*
- Women and men differ with regard to their respective *perceptions of and reactions to climate change;*
- Women's and men's contribution to climate change differs, especially with regard to their respective *CO_2 emissions;*
- As the male perspective is dominating, climate protection measures often fail to take into account the *needs* of large parts of populations (e.g. infrastructure, energy supply); and
- The *participation of women in decision-making* regarding climate policy and its implementation in instruments and measures is very low.

Women and men are not homogenous groups but include people who vary with regard to age, ethnic groups, education, income and differences related to those, such as influence, attitudes, contribution to climate change and being affected by it. This applies to developed as well as developing countries. Principally, however, the situation of women in the global South differs significantly from the situation in the global North. While women in the South are more affected by climate change, women in the North play a significant role as consumers but have little influence on decision-making relating to emissions reductions. Hence, we will first describe developing and developed countries separately and in the conclusion regarding a future climate 'regime', we will re-integrate both perspectives.[1]

First, however, we will look at the gender sensitivity of climate-related negotiations, the participation of women in international climate related-policy making processes and the influence they may have (had).

---

[1]The present chapter is aiming to provide an overview and cannot review findings in detail. In order to review original research, readers may refer to sources listed at the end.

GENDER AND CLIMATE CHANGE POLICY – DOES IT MAKE A DIFFERENCE?

International negotiations about global climate protection over the last 10 years have been slow, delivering meagre results. The debate started 20 years ago with a target of achieving 20% in CO_2 emission reductions, and ended up in the Kyoto Protocol with mere 5% - and even this has been questioned time and again. We know from some European countries that women are more supportive of their governments' climate protection policies than men, and would also be more supportive of more ambitious reduction goals, basically expecting their countries and the EU to take a leadership role. The international climate negotiations are in fact in dire need of such support.

In general, Skutsch (no year) identifies two rationales which may prompt the explicit inclusion of gender considerations in climate change policy development process:

1. the idea that inclusion of gender considerations may increase the *efficiency* of the climate change process; and
2. the idea that if gender considerations are not included, progress towards gender *equity* may be threatened.

   A third one could be added:

3. the idea that inclusion of gender considerations may increase the *quality* of the climate policy process.

Putting social impacts of climate change commitments and targets, mitigation and adaptation policies onto the agenda - would broaden the debate and change it into a discussion framed by the principles of sustainability. This will also provide entry points for gender considerations. A broadening of the debate may have the following positive impacts on climate protection:

1. The climate change debate is known as a very narrow one, focusing mostly on economic impacts, efficiency or technological problems. However, policies and measures, targetting human behaviour more holistically, should also be taken into account. This may help not only to integrate gender perspectives in the negotiations, but also to increase the likelihood of implementation of commitments and targets. This, in turn, would also improve the recognition and acceptance of the international policy process by the general public.
2. Looking at requirements and impacts from a variety of perspectives will lead to improved measures and mechanisms: It will lead to solutions that are not only reflecting the interest of the

most powerful societal groups, but support the integration of perspectives of less influential groups whose voices are rarely heard at international conferences.

3. The broadening of perspectives will attract representatives from women's organizations to take part in the policy process and influence the debate. We know very well from sustainability policy processes (e.g. UN CSD) that it indeed makes a difference if women representatives and gender experts stand up for women's rights and gender considerations - in terms of the process development itself as well as the content and outcomes of negotiations. This applies even more if gender expertize is coupled with in-depth knowledge of the issues under consideration, which is why networks such as ENERGIA, GWA or WOCAN [2] have been founded and are operating successfully to deliver gender mainstreaming in international agreements.

In addition, taking into account gender perspectives may avoid possible negative impacts of climate change measures and mechanisms on gender equality, e.g.:

- Market-based instruments/mechanisms can affect women in another way than men because of differences in income levels as well as in access to markets and services; they would need to be carefully designed in order to create gender equitable effects.
- If the domestic sector is expected to contribute to CO_2 emission reductions, this may have adverse impacts on gender equality as gender-specific roles and division of labour are affected. At the same time, private households are the ones with the least influence and representation of their interests in the context of climate negotiations.
- Considering that technological solutions are not always the solutions preferred by women: 'faster-bigger-further' is a rather masculine principle, which one may also find in the climate change policy process. Women tend to believe that technical solutions, such as further development of biofuels or carbon capture and storage, are not sufficient to meet the requirements of developing a low carbon economy.
- When flexible mechanisms, such as in CDM projects, do not explicitly take into account women's energy needs, especially for

[2]ENERGIA: International Network on Gender and Sustainable Energy. GWA: The Gender and Water Alliance. WOCAN: Women Organizing for Change in Agriculture and Natural Resource Management.

doing housework and/or for income-generating measures, then women may not benefit from them. In order for energy, produced within CDM projects, to be accessible and affordable, local women must be involved in their development.

In conclusion, climate protection instruments and measures can exacerbate existing inequalities if they do not take full account of gender differences. However, when developed with a gender lens, such instruments and measures can indeed contribute to increasing gender equality.

HISTORY OF WOMEN'S PARTICIPATION IN UN CLIMATE CHANGE NEGOTIATIONS

The UNFCCC was adopted 1992 in Rio de Janeiro during the UN Conference on Environment and Development ('Earth Summit'). All other outcomes of the conference, like Agenda 21, the Rio Declaration, or the Conventions on Biodiversity and on Desertification and Drought, include a strong focus on women's concerns and recommendations. Only the UNFCCC is lacking a gender perspective. One might have reasonably expected that gender would be brought forward for consideration at subsequent UNFCCC Conferences of the Parties (COPs), particularly in light of the agreement's overall lack of specificity around targets and rules for mitigating climate change. Indeed, some efforts were made in this direction, but they quickly fell to the wayside at later negotiations. Until today, no gender analyses have been conducted in relation to the instruments and articles of the UNFCCC and the Kyoto Protocol. We can assume that this gap is linked to the lack of participation by gender experts in the negotiations: women are not one of the 'constituencies' included as observers in the UNFCCC process, hence, experts for equal opportunities/women's rights are not likely to be represented and gender issues are not addressed.

Women's Activities at the UNFCCC Conferences of the Parties (COPs)

COP1: An international women's forum 'Solidarity in the Greenhouse' ran parallel to the first COP (Berlin 1995). It attracted 200 women from 25 countries who came together to discuss their views on climate protection. A list of demands was developed for consideration by the Parties, and a letter written to the chair of the conference (Mrs. Merkel, the former Minister for Environment and Chancellor of Germany since 2006).

In hindsight, this encouraging start can be understood as the result of the drive and euphoria that flowed from the Earth Summit. The international women's movement acted on the assumption that, from Rio onwards, Agenda 21 and all other UN decisions to integrate women/ gender perspectives into policies would be incorporated into every future process, at least at the UN level. But history has shown this to be too optimistic: from the Parties and the UNFCCC Secretariat to the scientific community, climate protection has been and continues to be presented as a gender neutral issue.

COP6: After the remarkable beginning in Berlin, it took five years until women/gender perspectives appeared again in the conference programme. This happened at COP6 in The Hague, when a side-event 'The Power of Feminine Values in Climate Change' was held. Banished to the back corner of the exhibition hall outside of the conference centre, there was little opportunity to draw attention to the issues. Notable at this COP, were the many statements published in the daily newsletters of the NGO community bemoaning the low participation of women - even though COP6 actually had the highest share of women yet (see figure below). The articles highlighted the important role of women in the negotiations as they were serving as key bridge builders between opposing parties.

COP7: The first (and only) official mentioning of women is contained in the text of a COP7 Marrakech resolution: Decision FCCC/CP/2001/13/ add.4 (2001) calls for more nominations of women to UNFCCC and Kyoto Protocol bodies. It also tasks the Secretariat with determining the gender composition of these bodies and with bringing their results to the attention of the Parties.

COP8: At COP8 in New Delhi a workshop was organized by ENERGIA, in cooperation with United Nations Development Program (UNDP): "Is the Gender Dimension of the Climate Debate Forgotten? Engendering the Climate Debate: Vulnerability, Adaptation, Mitigation and Financial Mechanisms" (Parikh and Denton 2002). This workshop received a lot of attention, but focused exclusively on the situation in developing countries.

COP9: Two women's organizations (ENERGIA and LIFE) invited those interested in gender and climate change to attend a meeting at COP9 in Milan. Thirty people came to discuss strategies on how to increase cooperation and improve lobbying efforts for the stronger integration of gender perspectives into the negotiations and the implementation of outcomes. There was also a side event 'Promoting Gender Equality, Providing Energy Solutions, Preventing Climate Change' organized by the Swedish environment minister and her colleagues from the Network of Women Ministers of the Environment.

COP10: Building on activities at COP9, two side events were organized at COP10 in Buenos Aires by the emerging 'gender and climate change network', one focusing on adaptation in the South, the other on mitigation and women in industrialized countries. During the conference, the network released the statement 'Mainstreaming Gender into the Climate Change Regime'.

COP11–COP/MOP1: A shift in women's activities was achieved in Montreal. In preparation for the conference, a strategy paper was drafted by LIFE and WECF, identifying possible entry points for gender aspects in the climate change debate. Women engaged in three complementary activities:

- Raising awareness and disseminating information via an exhibition booth 'gender–justice–climate', two 'Climate Talk' events and a statement in plenary;
- Building women's capacity and joint strategizing on gender mainstreaming in climate policy via women's caucus meetings; and
- Developing a future research agenda and initiating a gender and climate change research network via convening a research workshop.

These activities helped kick-start a new era in women's involvement and gender issues in the UNFCCC process. After almost ten years of discontinuous and uncoordinated participation by women's organizations, the path from COP1 has finally been picked up again.

Quantitative Participation of Women in the UNFCCC Negotiations and Their Impacts

In general, besides the parties there are five different groups ('constituencies') of observers at UNFCCC: industry representatives, environmental organizations, municipal/regional networks and local governments, indigenous peoples, and the research community. Delia Villagrasa, who for many years directed the non-governmental organizations Climate Action Network-Europe and e5 (European Business Council for a Sustainable Energy Future), reflects on the role of women in the negotiations: "Women were able to play a strong and generally positive role for climate protection based on their networking and interpersonal skills, and their ability to think and plan for the long term, even though they were generally underrepresented in the decision-making positions in their respective communities" (Villagrasa 2002).

Constituencies and Delegations

Governmental delegations are composed of senior staff from research, industry, and associations, in addition to state ministerial representatives. Most often the host country's delegation is remarkably large, also including many representatives from the non-governmental sector. The following graph shows the progression of women's representation in governmental delegations at UNFCCC COPs:

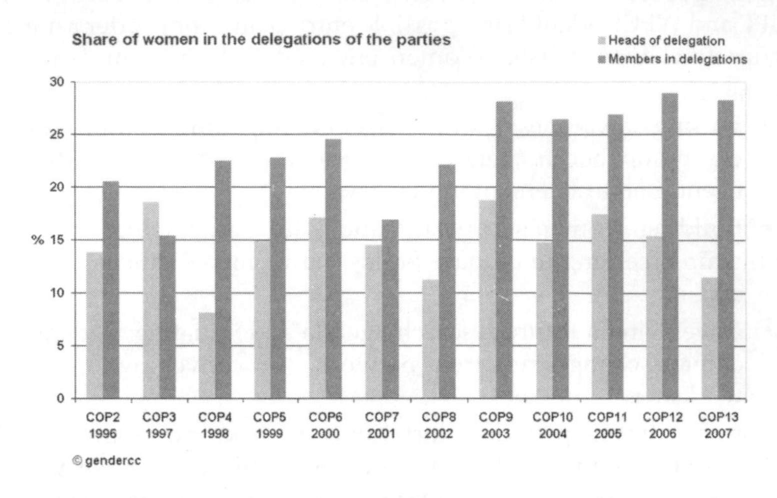

At the highest level - heads of delegations - women are substantially less represented. Yet, according to Villagrasa (2002), women still played the most important roles in shaping the Kyoto Protocol. For example, the German and Swiss (female) lead negotiators are described as "both acting in ways which differentiated them from their male colleagues in a crucial manner: they actively and often went out of their 'bunker', interacting strongly with other delegations beyond formal sessions. In particular, they were proactive in linking with delegations from developing countries, and furthering their integration into decision-making processes" (Villagrasa 2002). As developing countries delegations were often only one- or two-person strong, they were not able to follow the negotiations in numerous parallel meetings. Although "they had to represent their countries' interests in the negotiations, their personal integrity and openness earned the respect and trust necessary to 'build bridges' between nations. (...) This type of female interaction helped to forge links and mutual understanding (...) allowing to build the alliance necessary to achieve the adoption of the Kyoto Protocol " (ibid). Jennifer Morgan, former coordinator of the Climate Action Network US and then director of WWF's international climate program, described the situation at COP1 as follows: "Although women are in the vast minority in this male-dominated UN-structure, they are (...) the individuals who stand out and say – let's cut the rhetoric, break the gap between the negotiations inside the building and events in the real world and let's move forward" (Women for Peace and Ecology 1996).

The constituency of business and industry (BINGO) representatives is still an almost exclusively male club. The BINGO group has the smallest percentage of women as compared to all other constituencies, especially among their decision-making lobbyists. BINGOs also form the group with the least geographical diversity: the vast majority of representatives are from the USA and less than 5% are from developing countries.

While, on average, men dominate environmental NGOs (ENGOs) as well, some of the biggest ENGOs are or were in the past led by women. This is certainly the case within the global network of climate organizations, the Climate Action Network (CAN). Within CAN, for a long time 'women power' has prevailed: some of the largest and most active regional nodes were led by women. Women have ensured that NGOs were working together, so (...) that strong messages went out to the world" (Villagrasa 2002).

Women are not a separate constituency in the UNFCCC. At the 10 COPs examined here, there were a total of only 23 representatives explicitly representing women's organizations, half of them as members of the larger NGO delegation and the others as small women's delegations unto themselves. There are four women's organizations accredited to the UNFCCC as observer organizations, at most one or two of them are taking part in the annual conferences with their own delegations. The unusually low level of participation of women's organizations at UN Climate Conferences may be due to the different structures of participation (no Major Groups like at UN CSD meetings), but also to the lack of recognition of women and gender aspects in the negotiations.

GENDER RELATIONS AS AN ISSUE IN CLIMATE CHANGE NEGOTIATIONS

As mentioned above, individual women in leading positions were largely responsible for giving shape to the negotiations at critical junctures. But what does that mean with regards to the impacts, if any, of women's representation in COP negotiations? Apart from changing the structures or culture of negotiations, does gender-balanced participation actually result in gender-sensible outcomes?

The total absence of gender aspects in the climate change debate is striking, both in negotiations and in the development of mechanisms, instruments and measures. Skutsch (2002) attributes inattention to gender aspects to a demand for generalized intended outcomes. She shows, for example, that in shaping the Kyoto Protocol negotiators needed to focus on universal issues in order to create necessary consensus, and speculates that gender issues might have diverted attention. According to Skutsch, gender aspects simply did not have a place in the crisis atmosphere, in which the whole debate around the Kyoto Protocol took place.

There could be a link between the profiling of gender/women's issues at the COPs and the importance and state-of-play of issues in the negotiations. When negotiations are bogged down, or when they are prolonged and boring, space may open for 'gender' or 'women's participation' topics. Female lobbyists and negotiators have asserted that gender issues are being used to fill gaps rather than as substantive agenda points in the negotiation process (e.g. Wichterich 1992), and similar observations are reported by indigenous peoples representatives.[3]

The community of environmental NGOs, having a history of raising all sorts of climate-related topics at the negotiations, has hardly mentioned the need to integrate gender aspects into the debate. This is surprising considering that 'equity' is such an important part of discussions within the environment and development community. However, as is evident from NGO publications, equity is primarily understood as global or North-South-justice, with hardly any attention given to justice within nations or gender justice. The so-called Bali Principles of Climate Justice are equally disappointing in this respect. Although developed by representatives of people's movements, social and environmental organizations, gender justice is totally absent, and women's aspects are only very weakly represented.

Neither the Parties, nor the Observers, nor the UNFCCC Secretariat have undertaken efforts to integrate gender aspects into the negotiations. The same applies to the Intergovernmental Panel on Climate Change (IPCC), the most important and influential scientific body on climate change and a necessary key player in any effort to mainstream gender into the negotiations. Unfortunately, the IPCC continues to lack gender sensitivity in its review of research and published reports. IPCCs 20th meeting report (2003), published in preparation for the 4th Assessment Report on the effects of climate change, mitigation and adaptation, is a case in point. The Panel concluded that greater attention should be paid to

[3]G. Alber, Director of the Climate Alliance of European Cities with Indigenous Rainforest People; personal communication.

cross-cutting issues. Seven subjects[4] were identified as essential for consideration in all workgroups and themes. Contrary to the UN Commission for Sustainable Development (CSD), which also identified cross-cutting issues for its future work, gender was not included. As the accepted authority on climate change, the IPCC 'position' can be viewed as a key foundation of the climate change debate, in which climate change is treated as a gender-neutral issue. The forthcoming 4th Assessment Report (AR4) brought at least some recognition to these issues: Chapter 17 in part II highlights gender aspects of vulnerability and adaptive capacity (Adger, Agrawala, Mirza, Conde, O'Brien, Pulhin, Pulwarty, Smit and Takahashi 2007). This might be a first step in the right direction.

GENDER AND CLIMATE CHANGE IN DEVELOPING AND DEVELOPED COUNTRIES

Climate protection cannot be easily defined within clear limits, quite the contrary, the climate and its protection is affected by and affects a range of issues, such as: energy; mobility and transport; agriculture and forestry; water; biodiversity; disasters and extreme weather; land use and urban planning; building and housing; health; and so on. Each of these issues is being addressed within their various disciplines with a more or less strong focus on climate protection, and their influence on climate policy varies as much as the impacts climate change is having and will have on developments in these areas.

The availability of gender sensitive data varies considerably, depending which issues and which regions of the world we look at, and it seems that gender sensitive data are particularly rare with regard to issues that are particularly relevant for climate protection. Before we go into the details of available findings with regard to developing and developed countries, the following provides a summary overview of topics studied so far:

- In industrialized countries, most available gender sensitive data relate to questions of mobility and transport. For many countries, for example, we know that women and men have different access to different kinds of cars and that they use them for different purposes. Gender roles in society define travel purposes and this is reflected in different needs regarding public transport, and consequently regarding urban planning. However, women's

[4]Cross-cutting issues are: 1. Uncertainty and risk, 2. Integration of adaptation and mitigation, 3. Issues related to Article 2 UNFCCC and key vulnerabilities, 4. Sustainable development, 5. Regional integration, 6. Water, 7. Technology.

participation in planning and decision-making is minimal and hence, transport planning is largely reflecting men's perspectives and needs. Although it would be interesting to see if increased participation of women in transport and urban planning would, via promoting public/low emission transport, lead to more climate sensitive results, the possible linkages between the described patterns and climate protection have not been studied yet.

- Energy production and demand are particularly closely linked to climate protection. However, gender disaggregated data are very hard to come by, and conceptualizations of gender relations and their impact on energy use are equally rare. We do know, however, that women are much more critical towards high risk options such as nuclear energy than men. Gender differences with regard to perceiving and dealing with risks are important in this context.
- Energy from a women's perspective in developing countries, however, is well studied. We have data about the lack of women's access to energy; the need for affordable energy supporting women's income generating activities; the high number of victims of indoor air pollution, particularly women and children; the physical burden of collecting firewood and the impacts on women's time; and so on. Furthermore, there is an excellent global network, ENERGIA, whose expertise in the area of gender/women and energy is well known and well respected. ENERGIA produces reports and positions from nearly all (developing) regions in the world but is only marginally active in the climate policy processes.
- From industrialized countries, as compared to developing countries, we know more about women's and men's perceptions of climate change and its risks, the impacts of climatic changes on their daily lives (e.g. hot summers), which climate protection measures they prefer and how they have changed their own behaviour or would be prepared to do so.
- Research on impacts of climate change is focusing on developing rather than developed countries. With regard to vulnerability and risk management, long-term experiences are available and there are networks that deal with the effects on women and women's experience and knowledge.

These available data demonstrate *gender differences*. On the one hand, taking such an approach carries the risk of perpetuating gender roles. On the other hand, such gender differences tend to even out with increasing legal and economic gender equality. However, this does not imply increased *gender justice*: gender specific roles, gendered segregation of the labour market, division of labour at the household level and power differences remain even if quantifiable gender differences are minimized.

Hence, while data about gender differences can provide information about men's and women's interests with regard to climate protection policies, they are not sufficient in terms of gender justice. Only a detailed analysis of the societal, economic and cultural factors that determine gender differences, as well as the mechanisms that keep perpetuating and reproducing them, can provide insights into the actual effects of climate protection instruments and measures on women and men and on gender relations.

Gender and Climate Change in the South

Gender aspects of climate change in developing countries are strongly linked a) to access to energy, b) to impacts of climate change on daily life, and c) vulnerability to climate change and extreme weather events/ disasters. In each of these areas the participation of women in designing measures and in decision-making is crucial, but mostly lacking.

(Lack of) Access to Energy: Women's Burdens – Women's Contributions to CO_2-emissions?

Main sources of CO_2 emissions include energy production from fossil fuels and from biomass. A large percentage of households in the developing world, especially in rural areas, is lacking access to sustainable, clean and affordable energy systems. This lack and the effects of climate change are interlinked, and the impacts on women's lives reflect the gender inequalities and inequities prevailing in the social, economic and political arenas (United Nations 2006).

More than two billion people in developing countries, particularly in rural areas, use traditional fuels and lack basic modern energy services. It is mostly women and girls who are responsible for the daily household energy production in rural areas (e.g. for cooking, heating, preservation of agricultural products). They are spending many hours day by day by gathering fuelwood, crop residues, and animal dung, walking long distances and carrying heavy loads. The collection of firewood impacts their health, and also their security in remote areas. The time spent for producing household energy is lacking for education and income generating activities. Indoor air pollution from unvented bio-fuel cooking stoves is directly linked to respiratory diseases and to 1.6 million people, most of them women and children, die of these diseases every year (WHO 2005 and 2006).

Thus, the lack of access to affordable energy services is a serious barrier to sustainable livelihoods and emergence from poverty. Seventy percent of the 1.3 billion people in developing countries living on less than one dollar a day are women: energy poverty is a problem that disproportionately affects women (United Nations 2006).

While available data about the amount of CO_2 emissions caused by unsustainable use of biomass for household energy purposes in developing countries vary, depending on regions and mode of calculation, there is certainly an urgent need to improve the situation and provide sustainable, clean energy systems for household energy. Improving efficiency of cooking stoves alone can reduce the use of traditional fuels by 10 to 40 per cent (karekezi et al. 2004). This will help to reduce CO_2 emissions as well as improve women's economic situations by reducing heavy and time-consuming workloads.

There is a lot of literature - research, case studies, training manuals - available regarding women and energy in most of the regions of the developing world[5]. We urgently recommend to use these resources and take their results into account while planning any energy-related climate change projects, e.g. in CDM projects.

Impacts of Climate Change on Women's Lives

As predicted by the Intergovernmental Panel on Climate Change (IPCC), "climate change impacts will be differently distributed among different regions, generations, age classes, income group, occupations and genders" (IPCC 2001). The IPCC also notes that impacts of climate change will hamper development and harm human living conditions and lifestyles. The impacts will fall disproportionately upon developing countries and the poor within all countries, and thereby exacerbate inequities in health status and access to adequate food, clean water, and other resources."

Most of the key areas of the negative consequences of climate change are strongly connected to gender equality issues, like decreased availability of water in both quantity and quality or decreased agricultural productivity with increased risks of famine. Women are responsible for 70-80 per cent of household food production in Sub Saharan Africa, 65 per cent in Asia, and 45 per cent in Latin America and the Caribbean. As women usually are involved in a very labour-intensive, low-emission subsistence agriculture, while men are more often found in mechanized agriculture, women's livelihood strategies and food security are seen to be disproportionately affected by impacts of climate change. There is a need

[5]See e.g. www.energia.org or the other networks listed at the end of this article

to research into this assumption. Climate change is likely to exacerbate existing shortages of water. In general, it is women responsible for fetching water, while men are responsible for irrigation. Often the norm is that irrigation water should only serve men's businesses. Thus, women's workloads increase when they have to cover long distances, or it may further weaken their economic resource base. Additionally, droughts and flooding can be detrimental to women who keep livestock as a source of income and for security (Lambrou and Piana 2005; Women's Statement at COP10/2004).

Moreover, because of gender differences in property rights, access to information and in cultural, social and economic roles, the effects of Climate Change are likely to affect men and women differently: Following the cyclone and flood of 1991 in Bangladesh the death rate was almost five times as high for women as for men. Warning information was transmitted by men to men in public spaces, but rarely communicated to the rest of the family and as many women are not allowed to leave the house without a male relative, they therefore perished waiting for their relatives to return home and take them to a safe place (Aguilar 2004). Another clear illustration of the different vulnerabilities women and men face is offered by the fact that more men died than women during Hurricane Mitch. It has been suggested that this was due to existing gender norms in which ideas about masculinity encouraged risky 'heroic' action in a disaster (Nelson et al. 2002).

The effects of climate change on gender inequality are not limited to immediate impacts and changing behaviours but also lead to subsequent changes in gender relations. Shortfall of resources like water and fuelwood or care-giving in post-disaster-situations may increase women's workloads. Spending more time on traditional reproductive tasks re-enforces traditional work roles and works against a change in which women might begin to play other roles or take up non-traditional activities.

While women are known as experts in post-disaster management, their involvement in disaster mitigation and response efforts is often lacking. "Although women's social, economic and political position in many societies makes them more vulnerable to natural hazards, they are not helpless victims. Women are important agents for change and need to be further strengthened as such, recognising and mobilising their skills and capacities as social force and channelling it to enhance efforts to protect their safety and that of their communities and dependants is a major task in any disaster reduction strategy" (Briceño 2002).

Gender in Adaptation Measures

As discussed above, climate change impacts women's lives differently than men's. Consequently, adaptation policies and measures need to be gender sensitive. Equal involvement of men and women in adaptation planning is important not only to ensure that the measures developed actually benefit those who are supposed to implement them, but also to ensure that all relevant knowledge is integrated into policy and projects.

In principle, gender could be relatively easily integrated into, for example, stakeholder analyses, livelihoods analyses and multi-criteria decision tools if the users are aware of the need and choose to do this. Several recommendations have already been made to increase awareness of the importance of gender and the knowledge of possible approaches and to include a gender perspective in adaptation studies, projects and policy-making. Above all, there is a need for empirical evidence demonstrating (a) the gender differences in climate impacts and adaptive capacities, and (b) the positive effects of using gender analysis on the choice of investment in particular adaptation projects. In addition, existing and future toolkits related to vulnerability analysis as well as adaptation project implementation need to be reviewed by gender experts to answer (a) how gender awareness among users can be stimulated, and (b) whether gender should be integrated in the existing tools or whether new tools are needed.

Do Women Benefit from Mechanisms of the Kyoto Protocol?

To date, the question of what the flexible mechanisms of the Kyoto Protocol mean in gender terms has hardly been addressed in the UNFCCC negotiations, nor have the technology transfer or the adaptation fund been subject to gender analysis. Women/gender and energy organizations from the South have voiced some clear demands regarding the participation of women in CDM projects. In addition, there are concerns regarding the project size: profitable returns on investments and large-scale organizational efforts are most often associated with large projects that do not normally take into account women's rather low energy needs. Exclusive funding of large projects would be a major mistake, considering their impact, or lack thereof, on women and on poverty reduction (Skutsch and Wamukonya 2001, Roehr 2004).

Skutsch (2004) draws the following conclusion from analysis of CDM and especially LULUCF: "It may well be argued that solving women's energy needs is not the point of the CDM. Its primary object is, basically, reducing carbon dioxide levels in the atmosphere as cheaply as possible, while contributing to sustainable development in developing countries.

However, women's energy needs may be considered one aspect of sustainable development, and given that energy is central to the carbon dioxide question, and that biomass fuels for cooking are central to the energy problem in developing countries, the lack of focus on this is unacceptable."

Gender and Climate Change in the North

Although there is a lack of data and research regarding gender and climate change in developing countries, the gaps are even more dramatic with regard to industrialized countries. However, there is sufficient evidence to suggest that climate change and policy and measures to mitigate climate change will affect women and men differently in industrialized countries. The following summarizes some of what we already know:

Do Women have a Smaller Ecological Footprint (Lower Carbon Emissions) than Men?

Relating to the different roles they fulfill in society, there seem to be significant differences in energy use between women and men. Existing data, and gender roles, suggest that men produce more emissions, and more 'selfish' ones – that is, related to maintaining and exercising their social status, whereas women produce emissions when caring and catering for other people (children, elderly, sick). CO_2 emissions produced by the burning of fossil fuels are the principal cause of climate change. A very close link thus emerges with the topic of 'energy', but arguments about the over-consumption of energy in industrialized countries shall not be repeated here. Relevant questions include: who produces CO_2 emissions and through what activities? How is the reduction of such emissions affected by underlying social, political and planning conditions? What role do gender (relations) play in this context? Assessing the CO_2 emissions of women and men can best be done at the private households level. Apportioning emissions to the activities of women and men appears difficult at first, but a start can at least be made by examining the gender-specific division of labour and the very different consumption and leisure habits of men and women. For example, all emissions connected with mobility have a clear gender component. In Europe, women travel by car less frequently and over shorter distances, they use smaller (energy-saving) cars and they fly considerably less frequently than men (Eurostat 2005, Hamilton et al. 2005, Linden 1999). This applies to journeys connected with employment, looking after the family, and leisure activities.

Education and income level also play a role - admittedly a contradictory one - where CO_2 emissions are concerned. On the one hand, the higher the income, the higher the CO_2 emissions (because the size of houses, the amount of electrical equipment and the cars used by higher income earners). On the other hand, low income leads to a situation where older, less energy-efficient appliances are used, housing is built in a non-energy-saving way and the purchasing of electricity and heat produced from renewables is seen as being too expensive, all of which can result in higher overall CO_2 emissions. Education also has an effect as changes in behaviour and corresponding consumer decisions cannot be made without knowledge of the options for reducing CO_2 emissions.

Does it help to know about gender differences in CO_2 emissions? It does not help the climate if we use these data to blame some people or portray others and the good, climate friendly folks. However, these data help us to analyze reasons and amounts of energy use, and enable us to develop concrete reduction strategies, and communicate them more effectively. Doing this with gender in mind, we will develop more effective strategies, which will be more accepted by consumers.

Yet participation of women in planning and decision-making regarding climate protection programmes is equally important as the information quoted above. The level of women's participation in this area is very low even in industrialized countries, and this is linked above all to the largely technical nature and traditional male dominance of the areas of work concerned (energy, transport, town planning; Climate Alliance 2005). Similarly, it is generally men who benefit more from the new jobs created in these areas, whether in renewable energies or in emissions trading. Where measures and instruments are drawn up predominantly from the viewpoint of just one gender, a gender-specific view of the problems is likely to emerge. This gives rise to the following questions:

- What is the socio-economic situation forming the backdrop to these measures?
- Are caring work and its requirements recognized and taken into account? and
- How is this reflected in the general situation (e.g. financial aid, information, supportive measures)?

Does Climate Change Impact Women's Lives Differently than Men's?

The effects of climate change that are already evident today (e.g. extreme weather conditions) can impact differently on women and men. For example, the 20,000 people who died in France following the extreme heat-wave in Europe in 2003 included significantly more women than men (15-20% in all age groups; WHO 2004). Poor provision for the elderly, in particular, is deemed to be the reason. Studies in Germany by the Potsdam Institute for Climate Impact Research show that women are more severely affected by extreme weather conditions in coping with everyday life. The double burden of working to provide for the family and outside the home plays a role here, as does caring for the health of family members (higher ozone values, burns from UV radiation, adjusting eating habits, etc. (PIK 2000).

As in developing countries, it is primarily the poor who suffer in the aftermath of natural disasters, as evident, for example, in hurricanes in the USA or floods in Europe. Allover the world, these are mostly women, and in developed countries mostly elderly women and single mothers. Sound analyses of gender and disaster research have accumulated in recent years, which describe the effects of natural disasters on women as well as their crucial role in disaster management. There are astonishingly few differences between women in developing countries and in industrialized countries in this regard, as was shown just recently in the case of Hurricane Katrina in New Orleans. Not only do women constitute the majority of victims of floods, but are also often confronted with sexual violence in the aftermath of disasters. Yet, "though not this simple, it is often said that men rebuild buildings while women reweave the social fabric of community life" (Enarson 2005).

How do Women and Men Perceive Climate Change?

In general, women and men perceive risks differently, and this also applies to climate change. In Germany, more than half of the women, as compared to only 40 per cent of the men, rate climate change brought about by global warming as 'extremely' or 'very dangerous'. Women are also more strongly convinced that global warming will be unavoidable in the next 20-50 years, and they doubt that politicians will be capable of dealing with the consequences. By contrast, they believe very firmly that each individual can contribute toward protecting the climate through his/her individual actions (Greenpeace 2004, Federal Ministry for the Environment Germany 2004, Roehr et al. 2004). Nonetheless, two-thirds of the women are in favour of Germany taking a leadership role in climate policy. Policy planning, however, does not in any way reflect these perceptions.

Do Women and Men Prefer the Same Measures for Climate Change Mitigation?

In the case of mobility, substantial work has already been done to analyze the gender dimension of policies and measures. Existing transport systems in many countries have been designed with a specific view to middle-aged full-time working men, neglecting women's higher dependency on public transport and their specific needs when they look after children and elderly. Integrating a gender perspective would make transport systems both more user-friendly and more climate-friendly. Regarding energy saving and climate change mitigation measures, gender analysis is still outstanding.

But there are some more general linkage points which also apply to energy and climate change measures:

Women are more willing to alter environmentally harmful behaviour. They do not rely as much on science and technology to solve environmental problems, to the exclusion of lifestyle changes. As a result, they place a higher value on the influence exerted by each and every individual on preventing climate change.

Climate change mitigation strategies are more and more driven by technological developments and economic arguments. Changes of behaviour and lifestyles are hardly addressed any more. This also implies that the areas where women want to and are able to contribute to climate change mitigation are not valued any more.

While technological advances - such as energy efficiency or production of biofuels - receive significant support and are communicated effectively to the general public, social and behavioural aspects are virtually ignored. Energy-efficiency technologies are not concerned with the ultimate usefulness of the appliance being developed, or if it is environmentally desirable to even produce it. Rather, it is only focusing on the amount of energy being saved. Biofuels are a similar case. Here, there are close linkages with questions of biodiversity, health (pesticides, allergies) and nutrition (land use competition), which are neglected in favour of possible CO_2 emission reductions. Hardly anybody seems to question the need of 300 KW cars at all - regardless of it running on bio-diesel or most efficiently - or if 30 KW would not suffice. This means that there is no discussion about structural change, but only substitution within the same model and invitation for business as usual.

Typical, traditional male fascination with technology is particularly evident regarding the question of nuclear power: women strongly reject replacing one risk - climate change - with another - nuclear power.

Women also reject more strongly than men the economic instruments such as increasing prices for energy or introducing eco-taxation (PIK). This is not surprising as women have significantly less disposable income than men: Within the EU, the income gap between women and men is 16% (average for all 25 Member Countries), ranging from 4% (Malta) to Germany, Estonia and the Czech Republic with 23-24% (European Commission 2004). For the same reason, women have less options to avoid higher cost. If such gender specific aspects are not considered when developing climate protection measures, such measures will not be as effective as they could be.

HOW SHOULD GENDER ISSUES BE ADDRESSED IN THE POST-KYOTO REGIME?

Gender mainstreaming is a key prerequisite for a successful post-Kyoto regime. Gender *must* form an integral part of the whole process and must be integrated into the suggestions towards a future climate regime. From other policy processes we know very well that only if gender aspects are integrated in the documents, there will be a chance to refer to them and hold governments accountable to their commitments.

The climate change debate in general, and the development of future commitments under the Kyoto Protocol, needs to be set firmly into the context of *Sustainable Development* and its inclusion of social/equity aspects in environmental issues. The Principles of the Rio Declaration should serve as the overall framework for developing fair and effective policies for mitigation and adaptation. Therefore, the preparations for the 2nd commitment period under the Kyoto Protocol need to draw upon experiences, indicators, and other tools developed in international sustainable development processes as well as regional and national sustainable development strategies.

Actions should include the setting up of a process or mechanism that ensures that all suggested commitments, and mechanisms to help meet them are checked for their environmental, social and economic impacts. When preparing contributions to the post-2012 process, inputs should be requested from relevant international bodies such as the UN Division for the Advancement of Women, UNIFEM, and women's organizations and networks as well as gender experts should be invited to comment on draft documents.

Gender-sensitivity in the Post-2012 Process

The international climate change negotiation process - as well as climate policies at regional, national and local levels - must adopt the principles of gender equity at all stages: from research, to analysis, and the design and implementation of mitigation and adaptation strategies.

Actions to be undertaken include ensuring gender mainstreaming in all mitigation and adaptation policies, drawing on experiences with gender mainstreaming in environmental policy, e.g. regarding gender impacts assessment tools, affirmative action policies, etc. There is a need to invest in research to obtain more comprehensive data on gender aspects of climate change, relating both to mitigation and adaptation (e.g. in what ways women and men are vulnerable to climate change? What are the strengths and skills of women and men that we need to build on?). Finally, gender analysis should be included in all preparations of commitments and mechanisms to help meet commitments, addressing questions such as: what do climate policies mean for women and men? Are there differences? And how can such differences be addressed to ensure gender and climate justice?

Women's Participation in the Post-2012 Process

In order to ensure gender mainstreaming in the post-2012 discussions, it is important to draw upon the expertize of women and gender experts. Therefore, governments should aim to ensure the involvement of women and gender experts when they prepare their contributions for the international process, and ensure women's participation at international meetings.

Governments should draw on the expertize of international institutions, such as the UN Division for the Advancement of Women (DAW; serving CSW); UNIFEM; the CEDAW Committee, INSTRAW, FAO, UNDP, UNEP, and others, through inviting and supporting their active participation in the UNFCCC process. They should also draw on the expertize at the national level, through including gender departments and experts within national environment/energy ministries and agencies, and inviting other relevant cabinet ministries (e.g. Women's/Gender Affairs) to actively participate in the work on the UNFCCC process, and consider inviting representatives of women's organizations and gender experts to join national delegations to international meetings. Furthermore, governments - particularly donor governments - should provide funding for supporting the contributions and participation of women and gender experts in the international climate change process. Finally, the UNFCCC Secretariat, the Chair and Bureau of international meetings relating to climate change should be advised to ensure that women can actively participate, e.g. through inviting statements in plenary, providing a meeting room for the Women's Climate Caucus, and providing a booth in the exhibition area.

OUTLOOK

During recent UNFCCC Conferences of the Parties, women have picked up the thread from the first COP and further developed their work towards integrating a gender perspective into climate policy-making. An initial breakthrough was achieved at COP11-COP/MOP1 in Montreal, where the first-ever statement of behalf of women was delivered to the closing plenary. A research workshop, also held on the margins of COP11, helped to develop open questions and priority issues for future research. As a first step, a review of existing research is being conducted at the time of publication by the two authors, in collaboration with FAO. The review will deliver a systematic overview, categorization and assessment of existing data and knowledge. This will hopefully create a basis for future research as well as for integrating existing knowledge into current climate debates, and will serve to raise awareness among those who are working to create a strong and effective post-Kyoto regime.

We will also put out efforts into strengthening the involvement of the international women's movement in the climate process, thus also strengthening the demand for gender mainstreaming. The topic of climate change is not well established within the international women's movement, and there is a need for information about the issues as well as the political processes. As has been demonstrated in recent years, it is particularly important to increase the participation of women who have expertize both on gender and on climate.

The general advocacy for women's rights and gender equity is crucial, but so is the ability to argue in detail, why and how gender should and can be integrated into climate-related policy-making. Such expertize is hardly available within the process right now, and the participation of gender and climate experts - from the governmental or other constituencies' side - will help formulate policies that are avoiding to put women at a disadvantage but rather be gender neutral or help to further gender equity.

In terms of gender mainstreaming of climate-related policy and implementation, there is a range of established analytical and practical tools that can and should be used, such as gender analysis, gender impact assessment, gender budgeting, and the knowledge and practical experience gained in participatory community development that is particularly relevant for adaptation.

We will also continue to build alliances with other constituencies, who share our concerns about social and economic justice and human rights issues in the context of climate change, such as trade unions and indigenous peoples.

Climate change is the crucial challenge today, a fact that is becoming more apparent every year. The increasing number and intensity of natural disasters is but one indicator, which even climate sceptics cannot deny anymore. Time is passing and matters are becoming ever more urgent. Ambitious attempts at climate protection are still rare, which makes the task ever more difficult. In this process, one point is becoming more and more clear: we will not be able to master this challenge without justice – and there is no true justice without gender justice.

Acknowledgements

We thank Gotelind Alber and Lars Friberg who kindly commented on earlier drafts of this chapter. We also thank all colleagues who have contributed to raise the awareness of gender and climate change issues through researching and discussing individual aspects of these issues. We are unable to list all of them but want to highlight in particular the contributions from ENERGIA, the Gender & Disaster Network, and Yianna Lambrou at FAO.

References

Adger, W.N., S. Agrawala, M.M.Q. Mirza, C. Conde, K. O'Brien, J. Pulhin, R. Pulwarty, B. Smit and K. Takahashi. 2007. Assessment of adaptation practices, options, constraints and capacity. Climate Change 2007: Impacts, Adaptation and Vulnerability. Contribution of Working Group II to the Fourth Assessment Report of the Intergovernmental Panel on Climate Change, M.L. Parry, O.F. Canziani, J.P. Palutikof, P.J. van der Linden and C.E. Hanson, Eds., Cambridge University Press, Cambridge, UK, 717-743.

Aguilar, L. 2004. Climate Change and Disaster Mitigation. Available on-line: www.iucn.org/congress/women/Climate.pdf

Briceño, S./UN-ISDR 2002. Gender Mainstreaming in Disaster Reduction. Presentation at the Commission on the Status of Women, 6 March 2002, New York, USA.

Enarson, E. 2005. Women and Girls Last? Averting the Second Post-Katrina Disaster. SSRC project on Understanding Katrina (http://understanding katrina.ssrc.org/Enarson).

Eurostat (eds). 2005. Short distance passenger mobility in Europe. Statistics in focus. Luxembourg.

European Commission. 2004. Report on equality between women and men. COM(2004) 115 final. Brussels, Belgium.

Federal Ministry for the Environment Germany. 2004. Umweltbewusstsein in Deutschland 2004 [Environmental awareness in Germany 2004]. Berlin, Germany.

Greenpeace 2004. So grün ist Deutschland [How green is Germany]. Greenpeace Magazine 5/04

Hamilton, K., L. Jenkins, F. Hodgson and J. Turner. 2005. Promoting gender equality in transport. Working Paper Series No. 34 Equal Opportunities Commission, Manchester, UK.

[IPCC] International Panel on Climate Change. 2001. Climate Change 2001. Impacts, Adaptation and Vulnerability. Contributions of the Working Group

III to the Third Assessment Report of the Intergovernmental Panel on Climate Change. Cambridge, UK.

Karekezi, S., K. Lata and S.T. Coelho. 2004. Traditional Biomasse Energy: Improving its Use and Moving to Modern Energy Use. Thematic Background Paper 11. Secretariat of the International Conference for Renewable Energies (ed). Bonn, Germany.

Lambrou, Y. and G. Piana. 2005. Gender: the missing component in the response to climate change. Food and Agriculture Organization of the UN. Rome, Italy.

Linden, A., A. Carlson-Kanyama and A. Thelander. 1999. Gender Differences in Environmental Impacts from Patterns of Transportation. In: Society and Natural Resources, Vol. 12, issue 4, pp 335-369.

Nelson, V., K. Meadows, T. Cannon, J. Morton and A. Martin. 2002. Uncertain predictions, invisible impacts, and the need to mainstream gender in climate change adaptation. In: Gender and Development - Climate Change. Oxfam Journal, Vol. 10, Number 2: 30-39.

Parikh, J. and F. Denton. 2002. Gender and Climate Change: Vulnerability, Adaptation, Mitigation and Financial Mechanisms. Proceedings and Thematic Paper prepared for COP8, 2002.

[PIK] Potsdam Institute for Climate Impact Research. 2000. Weather Impacts on Natural, Social and Economic Systems. PIK-Report Nr. 59. Potsdam, Germany.

Roehr, U. 2004. Gender relations in international climate change negotiations. Published in German. In: U. Roehr, I. Schultz, G. Seltman and I. Stieß (eds). Klimapolitik und Gender. Einer Sondierung möglicher Gender Impacts des europäischen Emissionshandelssystems. Frankfurt a.M., Germany.

Skutsch, M. and N. Wamukonya. 2001. Is there a gender angle to climate change negotiations? Position paper.

Skutsch, Margaret. 2002. Protocols, treaties, and action: the 'climate change process' viewed through gender spectacles. In: Gender and Development - Climate Change. Oxfam Journal, Vol. 10. Number 2, pp. 30-39.

Skutsch, M. (no year). Revisted paper Protocols, treaties and action: the 'climate change process' through gender spectacle (see above).

Skutsch, M. 2004. CDM and LULUCF: what's in for women? A note for the gender and climate change network. Enschede, The Netherlands.

[United Nations] United Nations Economic and Social Council. 2006. Discussion papers submitted by major groups: Contribution by women. E/CN.17/2006/1.

Villagrasa, D. 2002. Kyoto Protocol negotiations: reflections on the role of women. In: Gender and Development - Climate Change. Oxfam Journal, Vol. 2, pp. 40-44.

Women for Peace and Ecology (eds). 1996: Solidarity in the Greenhouse. International Women's Forum to the UN Climate Summit, April 1-2, 1995. Documentation. Berlin, Germany. p. 19.

[WHO] WHO European Centre for Environment and Health. 2004. Extreme weather and climate events and public health responses. Report on a WHO meeting in Bratislava, Slovakia, 09-10 February. p. 21.

[WHO] World Health Organization. 2005. Indoor Air Pollution. Fact Sheet No 292. Geneva, Switzerland.

[WHO] World Health Organization. 2006. Fuel for Life. Household Energy and Health. Geneva, Switzerland.

Wichterich, C. 1992. Die Erde bemuttern. Frauen und Ökologie nach dem Erdgipfel in Rio [To mother the earth. Women and ecology after the Earthsummit in Rio]. Köln, Germany.

Relevant Networks

gendercc - women for climate justice
www.gendercc.net
genanet - focal point gender, environment, sustainability
www.genanet.de/unfccc.html?&L=1 and www.genanet.de/klimaschutz.html?&L=1
International network on gender and sustainable energy
www.energia.org
Gender and disaster network
www.gdnonline.org
Mailing list on gender and climate change
gender_cc-subscribe@yahoogroups.com

Section VIII

Safe Landing: Protecting the Climate for Future Generations and Health

36

CHAPTER

Climate, Health and the Changing Canadian North

Christopher Furgal[1,2], Pierre Gosselin[3,4,5] and Nicolas Vézeau[3]

[1]Trent University, Environment & Resource Studies Program, 1600 West Bank Drive, Peterborough, ON Canada K9J7B8

[2]Unité de recherche en santé publique, CHUQ-CHUL Research Center

[3]Université Laval, Québec, Canada G1K7 P4

[4]Institut national de santé publique du Québec

[5]Ouranos

"The living person and the land are actually tied together because without one the other doesn't survive and vice versa. You have to protect the land in order to receive from the land. If you start mistreating the land, then it won't support you... The land is so important to us to survive and live on; that's why we treat it as part of ourselves".

Mariano Aupilarjuk, Inuit poet

INTRODUCTION

There is strong evidence that the Canadian Arctic, like other circumpolar regions, is experiencing changes in its climate (e.g. McBean 2005, Bonsal and Prowse, 2006; Ouranos 2005; Huntington and Fox, 2005). Over the past 30 to 50 years the western and central Canadian Arctic have warmed, especially during winter months, by approximately 2-3°C (Weller, 2005). During the same period, the eastern Canadian Arctic cooled but has since followed this warming trend as well. Local aboriginal people have reported significant warming throughout the Arctic in recent decades, further strengthening these scientific findings (e.g. Huntington and Fox, 2005; Nickels et al., 2006). According to both scientists and local residents these changes are resulting in significant impacts for the Arctic ecosystem. Observations of such things as decreases in the extent and thickness of winter sea ice throughout the Canadian Arctic, melting and destabilization of permafrost, increased coastal erosion in low lying areas, and shifts in the distribution and migratory behaviour of some Arctic wildlife species exist. The implications of these changes in the Arctic environment, where a significant number of people still rely on the ecosystem for aspects of their physical, socio-cultural, mental and economic well-being are far-reaching. For these reasons, Canadian Arctic, and other circumpolar populations have been identified as some of the potentially most vulnerable to the impacts of climate change (Furgal and Seguin, 2006). The information presented in this chapter in support of this argument is drawn from material developed for two recent Canadian national climate impacts and adaptation assessments, as well as recent scientific literature on the topic.

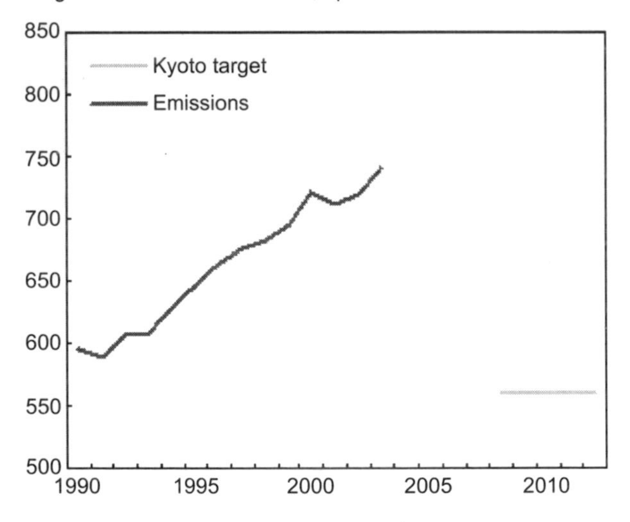

Fig. 1 Emission levels of greenhouse gases in Canada since 1990 (from Statistics Canada, 2005a)

THE CURRENT POLITICS OF CLIMATE CHANGE IN CANADA

Adopted in 1997, the Kyoto Protocol aimed at reducing Canadian GHG emissions by 6% from 1990 levels. Strategically, this reduction would be achieved by the years 2008 to 2012. To this day, according to Statistics Canada (2005a), Canadian emissions have increased approximately 24%. Although 59% of Canadians feel that Kyoto is important for the country and that Canada should not withdraw from the agreement (Doskooch, 2006), recent changes in the policy position of the Canadian Government are likely to influence the country's role in Kyoto and the approach to reducing GHG emissions.

Mr Ryan Sparrow, Press secretary to the federal Minister of the Environment, indicated that the "Conservatives do not feel bound to the former liberal government's Kyoto undertakings"(Doskooch, 2006). As of August 2006, the Federal government web portal dedicated to the subject of climate change read "The Government of Canada Climate Change site is currently unavailable" (Government of Canada, 2006a). There has since been no further indication on its reopening. Furthermore, a 2006 Conservative platform document stipulated that their strategy resides in a "made-in-Canada plan for reducing greenhouse gases and ensuring clean air, water, land and energy for Canadians"which was widely criticized when released in the fall of the same year (Environment Canada, 2006). However, the poor management performance of previous Liberal governments in addressing Kyoto commitments has also been severely criticized by the Commissioner of the Environment and Sustainable Development, a bureau of the Office of the Auditor General of Canada (Government of Canada, 2006b) and thus no one ruling political party in the country has effectively led Canada towards meeting these goals since making the commitment to do so.

The current government has reported that as the environment also falls under provincial jurisdiction according to the Canadian constitution, that the provincial governments could fund and run their own plans. To date a few provinces, such as Québec, Manitoba and Newfoundland and Labrador have said they are prepared to work independently in meeting their Kyoto targets if the federal government defaults on their deal (Doskooch, 2006). In fact, Québec has tabled and commenced implementation of its 2006-2012 Action Plan (Gouvernement du Québec, 2006). Currently, Ontario is reported to still expect its financial agreement with the previous government to be honoured. As a result of this lack of effective vision and action, no coordinated approach exists today for addressing the issue at the national scale.

Research on Climate and Health in the Canadian North

Research on climate change and health impacts among northern Canadian residents is in its infancy (Hassi et al., 2005; Berner, 2005; Furgal and Seguin, 2006). Most recently it has given particular attention to aboriginal populations. It has worked to include aboriginal knowledge and local observations of environmental change along with scientific assessments of the impacts associated with these and other forms of change in remote and rapidly changing communities (e.g. Berner and Furgal, 2005; Huntington and Fox, 2005).

A common definition of Canada's North adopted here includes the three territorial administrative regions north of 60° N (Yukon, Northwest Territories and Nunavut) as well as the regions of Nunavik, north of 55° in the province of Québec and the Inuit settlement region of Nunatsiavut within the province of Newfoundland and Labrador. The latter two regions comprise communities with large aboriginal populations and share many biogeographical characteristics with the territorial Arctic. The majority of the 'northern' population lives in the three territories with comparatively smaller but significant populations living in Nunavik and Nunatsiavut (Table 1). Together, this region covers approximately 60% of Canada's landmass. (Furgal and Seguin, 2006)

Table 1 Current (2005) and projected (2031) populations (thousands) for Canadian northern regions.

	2005 population	**Mean annual growth rate for scenario 3* (range of scenarios 1-6)**	**Projected population in 2031 scenario 3* (range scenarios 1-6)**
Canada	32,270.5	7.3 (4.5–10.0)	39,024.4 (36,261.2–41,810.0)
Nunatsiavut (Labrador)	23.9	**	**
Nunavik	9.6	**	**
Nunavut	30.0	4.0 (1.2–6.6)	33.3 (30.0–35.6)
Northwest Territories	43.0	9.1 (5.8–11.4)	54.4 (49.9–57.7)
Yukon	31.0	3.6 (0.7–5.5)	34.0 (31.5–35.7)

*Scenario 3 assumes medium growth and medium migration rates with medium fertility, life expectancy, immigration and inter-provincial migration as outlined in 'Population projections for Canada, Provinces and Territories, 2005-2031', Catalogue no. 91-520-XIE, Statistics Canada, 2006.

** Data not available at the regional level.

Source: Population projections for Canada, Provinces and Territories, 2005-2031', Catalogue no. 91-520-XIE, Statistics Canada, 2006a.

Nearly two-thirds of Canadian northern communities are coastal and the large majority are small (100-500 residents) and isolated in nature (e.g. only 3 centres of more than 5,000 people exist). However the large centres account for significant proportions of regional populations (e.g. Yukon 58.7%; Table 2). In some regions the large majority (67%, Nunavut) live in communities of less than 1000 people.

Table 2 Population characteristics of Canadian northern regions.

Indicator	*Canada*	*Yukon*	*NWT*	*Nunavut*	*Nunavik*	*Nunatsiavut (Labrador)**
Population density (per sqr. km)	3.33	0.06	0.03	0.01	0.02	0.11
% Urban of total population	79.6	58.7	58.3	32.4	0.0	68.3
% Aboriginal population	3.4	22.9	50.5	85.2	91.3	34.1

*Health Labrador Corporation health region which includes most mainland Labrador communities including Happy Valley Goose Bay and the south coast

1. Data source: Statistics Canada, 2001 Census (20% sample)
2. Urban areas are those continuously built-up areas having a population concentration of 1,000 or more and a population density of 400 or more per square kilometre based on the previous census; rural areas have concentrations or densities below these thresholds.
3. Aboriginal people are those persons who reported identifying with at least one Aboriginal group (e.g., North American Indian, Métis or Inuit)

and/or those who reported being a Treaty Indian or a Registered Indian as defined by the Indian Act and/or those who reported being a Treaty

Indian or a Registered Indian as defined by the Indian Act and/or those who were members of an Indian Band or First Nation.

Most growth since the establishment of communities throughout the North has occurred primarily in these three main urban centres (Whitehorse, Yellowknife and Iqaluit) and the population density remains low outside of these locations (AHDR, 2005). More recently, much of the growth has been attributed to an increase in the non-aboriginal population associated with resource development and public administration which is the largest secondary sector activity in the circumpolar north (AHDR, 2005; Chapin et al., 2005). Over the next 25 years the greatest growth is projected in the Northwest Territories whose population is projected to increase beyond 50,000 residents (Table 1). On average, the northern population is considerably younger than the national average with Nunavut and Nunavik having significant segments of their population under the age of 15. Meanwhile a significantly smaller percentage of residents are over the age of 65 in the North than in the rest of the country (Statistics Canada, 2006a).

Just over half of northern residents are aboriginal and represent diverse cultural and language groups from the 14 Yukon First Nations in the west to Nunatsiavut in the east, some of which have been in these regions for thousands of years (Table 2). However significant numbers of non-aboriginal people are also resident in these regions too. The majority of small communities are predominantly aboriginal in composition and are places where various aspects of traditional lifestyles are still very strong components of day to day life.

Due to their relatively poorer socio-economic status and their close interconnection with the environment northern aboriginal populations are more highly exposed to climate related environmental hazards than many people in other regions of the country. It follows then that disruptions in the local environment have direct and indirect effects on the health status of more vulnerable segments of the populations in these regions. For example, Inuit community members from Kangiqsujuaq, Nunavik have reported that "the unpredictability of the weather causes further difficulties with travel [on the land], as it is no longer possible to tell if future weather will be good or bad" (Community of Kangiqsujuaq et al. 2005). The adaptations implemented by communities to date to cope with climate change reveal critical information that may be useful to their southern neighbours or other northern regions. In this sense, the experience that is being lived in the Canadian Arctic today may teach valuable lessons about adapting to climate change and protecting public health in the rest of Canada and elsewhere around the world facing similar changes today or in the near future.

CURRENT HEALTH STATUS OF CANADIAN NORTHERN POPULATIONS

In the holistic vision of health adopted by the World Health Organization (1967), which is very similar to that supported by many indigenous groups throughout the world, the well-being of individuals and communities is tied to that of their environment.

Health Summary

A summary of selected health status and mortality indicators are presented in Tables 3 and 4. In general, northern populations report much lower rates of death from such things as cardiovascular disease than the national average (Table 4; Statistics Canada, 2001). However these are crude rates (i.e. unadjusted for the age structure of the population) which can be misleading as the northern population is much younger than in other parts of Canada. For instance, a recent publication of age-adjusted rates in Nunavik (INSPQ, 2006) has shown significantly higher mortality rates for cardiovascular diseases than in the rest of Québec. Moreover, northern populations report higher than national rates of mortality from such things as lung cancer, and unintentional injuries associated with motor vehicle accidents and drownings. The latter is likely associated, in

Table 3 Selected health status indicators for Canadian northern regions (2001).

Indicator	*Canada*	*Yukon*	*NWT*	*Nunavut*	*Nunavik*	*Nunatsiavut (Labrador)**
Public health spending per capita ($)	2,535	4,063	5,862	7,049		
Life expectancy at birth (males, 2002)	75.4	73.9	73.2	67.2	63.3	73.6
Life expectancy at birth (females, 2002)	81.2	80.3	79.6	69.6	70.2	78.7
Life expectancy at age 65 (males, 2002)	17.1	15.6	14.5	16.3	**	**
Life expectancy at age 65 (females, 2002)	20.6	19.5	19.2	11.4	**	**
Infant mortality rate (per 1,000 live births, 500 grams or more, 2001)	4.4	8.7	4.9	15.6	17.8	**
Low birth weight rate (% of births less than 2,500 grams)	5.5	4.7	4.7	7.6	**	**
Potential years of life lost due to unintentional injury (deaths per 100,000)	628	1,066	1,878	2,128	**	**
Self reported health (% aged 12 and over reporting very good or excellent health)	59.6	54	54	51	51	64
Physical activity (% aged 12 and over reporting physically active or moderately active)	42.6	57.9	38.4	42.9	**	48.7

*Identified as the region covered by the Health Labrador Corporation (Labrador), which includes all of mainland Labrador including the north coast communities of what is now the Inuit land claim area of Nunatsiavut.

**Not available at the regional level.

1. Data source: Statistics Canada, Canadian Community Health Survey, 2000/01
2. Population aged 12 and over reporting level of physical activity, based on their responses to questions about the frequency, duration and intensity of their participation in leisure-time physical activity.

part, with the high level of dependence on various modes of transport (e.g. skidoo, four wheel all terrain vehicle, boat, etc) for land based activities that are strong part of livelihoods and traditional activities in these areas (i.e. hunting, fishing and collection activities). Greater than 70% of northern aboriginal adults report harvesting natural resources via hunting and fishing, and more than 96% do so for subsistence purposes (Statistics Canada, 2001). Finally, significantly higher rates of mortality are reported from intentional self-harm (suicide) than other regions of the country and this is particularly the case in Nunavut (Table 4; Statistics Canada, 2001).

This pattern of social stress is reflected in the results for the perceived level of social network support by individuals across the North (Table 5; Statistics Canada, 2001).

Table 4 Selected mortality indicators (per 100,000 deaths) for Canadian northern regions*.

Indicator	*Canada*	*Yukon*	*NWT*	*Nunavut*	*Nunavik*	*Nunatsiavut (Labrador)*
Major cardiovascular diseases	233.2	111.3	118.5	78.9	**	**
Acute myocardial infarction	58.9	6.5	35.5	10.3	**	**
Deaths due to heart attacks	52.1	37.1	28	3.7	**	**
Lung cancer	48.2	73.2	61	209.5	**	**
Accidents, unintentional injuries	28.6	65.5	59.2	30.9	**	**
Transport accidents (motor vehicle, other land transport, water, air and unspecified)	9.9	19.6	16.6	27.5	**	**
Accidental drowning	0.8	9.8	7.1	0.0	**	**
Intentional self-harm (suicide)	11.9	19.6	23.7	106.4	**	**

*Crude mortality rates/100,000 in 2003

**Not available at the regional level.

Source: Statistics Canada, 2006b. Mortality, Summary List of Causes, 2003. Health Statistics Division. Catalogue no. 84F0209XIE

The Patterns of mortality differ among communities. For example in the NWT, cancer was the leading cause of death (32%) for Yellowknife residents between 2000 and 2002, followed by cardiovascular disease (20%), injury (16%), digestive diseases (8%) and respiratory ailments (7%). In the regional centres of Hay River, Fort Smith and Inuvik proportionately more deaths occurred during the same period from cardiovascular diseases (29%), followed by cancer (21%), injuries (17%) and respiratory disease (11%). In the smaller communities in the NWT during the same period, injuries accounted for 22% of all deaths, along with cancer (22%), followed by cardiovascular diseases (20%) and then respiratory diseases (13%; GNWT, 2005). The injury mortality and hospitalization rates among Inuit and Dene in the NWT is more than twice as high as that for other residents. Finally, residents in the regional centres and smaller community outside of Yellowknife were more likely to be hospitalized or die because of an injury than were residents of the city of Yellowknife (GNWT, 2004). In terms of other key health behaviours, there are higher rates of smoking, consumption of alcohol and obesity among northerners yet they report feeling less stress than the average Canadian (Statistics Canada 2002). Approximately 80% of all Canadians responding

Table 5 Selected social and economic indicators for Canadian northern regions

Indicator	*Canada*	*Yukon*	*NWT*	*Nunavut*	*Nunavik*	*Nunatsiavut (Labrador)**
High Social Support (1,2)	–	78.0	74.5	58.1	–	85.8
Sense of belonging to local community (very strong or somewhat strong) (3)	62.3	69.3	72.3	80.9	72.0	87.6**
Proportion of Census families that are lone female parent families (3)	15.7	19.8	21.0	25.7	35.5	15.5
Personal average income $ (in the yr 2000) (3)	29 769	31 917	35 012	26 924	23 215	28 478
Government transfer income as proportion of total % (2000) (3)	11.6	8.6	7.3	12.9	17.0	10.2
% long-term unemployed (labour force aged 15 and over) (3,4)	3.7	6.0	4.8	11.2	8.7	9.3
% of population aged 25-29 that are high school graduates (3)	85.3	85.4	77.5	64.7	52.7	83.9

*Identified as the region covered by the Health Labrador Corporation (Labrador), which includes all of mainland Labrador including the north coast communities of what is now the Inuit land claim area of Nunatsiavut.

**Identified as the region covered by the Labrador-Grenfell Health region which includes all of mainland Labrador.

1. Data source: Statistics Canada, Canadian Community Health Survey, 2000/01.
2. Level of perceived social support reported by population aged 12 and over, based on their responses to eight questions about having someone to confide in, someone they can count on in a crisis, someone they can count on for advice, and someone with whom they can share worries and concerns.
3. Data source: Statistics Canada, 2001 Census (20% sample)
4. Labour force aged 15 and over who did not have a job any time during the current or previous year.

to the 2001 Census had contact with a medical doctor in the year previous to the survey with some variation between regions (Statistics Canada, 2001).

Food Security

Due to social, cultural (Freeman, 1996), economic (Chabot, 2004) and nutritional (Blanchet et al., 2000) factors, food security of northern aboriginal groups is significantly supported by the inclusion of traditional food sources. More than just subsistence, traditional hunting, gathering and fishing activities are central for aboriginal physical, economical, spiritual, and cultural health and well-being (VanOostdam et al., 2005). The consumption of foods such as marine mammals, fish, birds and berries are important sources of some nutrients among these populations. For example, high concentrations of omega-3 fatty acids found in high concentrations in fish and marine mammal tissues consumed are beneficial for foetal development (Lucas et al., 2004), provide protection against arteriosclerosis and ischemic heart disease (Bjerregaard et al., 1997) and some cancers. Also, beluga muktuk (whale skin and fat) eaten by Inuit throughout the North is a dominant source of selenium, an antioxidant and known anticarcinogen that may also help protect individuals from mercury toxicity (Blanchet et al., 2000).

Even though northern aboriginal groups place particular value on country foods, there is an increasing integration of western cultural diet items in Arctic regions despite the high costs of many market foods. A comparison of a list of standard market items ('food basket') and their average prices throughout the country shows that prices can be as much as three times higher in the North for the same products (Statistics Canada, 2005b). As well, via their consumption of Arctic flora and fauna northern aboriginal people are also more highly exposed to some environmental contaminants originating primarily in more industrialized regions of the world. The atmospheric transport and deposition in Arctic regions and biomagnification of persistent chemicals such as, PCBs and DDT means that these traditional foods from the land sea provide both benefits and some risks for northern populations. Blood levels of some organic contaminants among northern Canadian groups have been reported at much higher levels than those living in southern regions of the country (Van Oostdam et al., 2005).

Social and economic changes occurring in northern communities, along with environmental pressures (e.g. contaminants and climate change) challenge aspects of northern food security. In fact, food insecurity rates are notably higher among northern aboriginal groups than elsewhere in the country (Statistics Canada, 2005). In Nunavut, over half the population (56%) report food insecurity of some form. Rates in the Northwest Territories (28%) and Yukon (21%) were also well above the national level (Statistics Canada, 2005). In particular, people living in northern regions are more likely to experience food insecurity in terms of food scarcity.

Mental and Social Health

Living conditions are changing throughout the Arctic for indigenous as well as non-indigenous residents in relation to the shift from an economy based on hunting and gathering to more involvement and access to wage earning activities (AHDR, 2005). The societal changes and processes of modernity have been associated with a number of social and mental health issues in these regions, yet have also improved infrastructure and general health status (ACIA, 2005). Some regions report suicide rates much higher than national or regional averages, particularly among younger aboriginal males. This has been discussed in association with the changing roles of men in northern aboriginal cultures in providing for families and communities. As hunting and other traditional activities become increasingly difficult to pursue without access to economic support, the shift from hunting to wage earning employment positions or finding a balance between them, creates significant stress for members of the population (Bjerregaard and Young, 1998). This psychosocial dynamic takes the shape of an inner-struggle between modernization and traditional aspects of life in many northern aboriginal communities.

CLIMATE CHANGE AND HEALTH IN THE CANADIAN NORTH

An analysis of the climate projections for the eastern and western Canadian Arctic (divided at 102° W latitude), based on seven Global Circulation Models (GCMs) using six different emission scenarios for the future shows little difference in conditions over the coming decades for the two regions (Bonsal and Prowse, 2006). In general, model projections identify significant changes in future temperature and precipitation over the Canadian North during the current century. All scenarios show increased temperature and precipitation across the North. These increases are projected to occur during all seasons with the greatest changes taking place in winter and significantly smaller changes in the spring and fall (McBean, 2005). The most significant increases in temperature are projected in the high Arctic (particularly over the Arctic Ocean) with less of an increase closer to the northern limits of the Canadian provinces. Patterns in projected changes in precipitation are not as clear however increases similarly appear to be greater further north (Bonsal and Prowse, 2006; Kattsov and Kaller, 2005).

For both the eastern and western Arctic mean annual temperature changes are projected around 2°C (range 1-3°C) for 2020, with increases in precipitation near 5-8% (range 0-12%). In the middle of this century, temperature increases of 2-7°C (median of 4°C) are projected for both regions with an increase in precipitation of between 5 and 25% (median 15%). As one projects further into the future, greater variability is seen with the climate models. For 2080, temperature increases for the two regions are projected between 3.5-12.5°C (median 6°C) with an increase in precipitation ranging between 8 and 40% (median 20%). However, current models show significantly more variability in projections for precipitation than temperature in the Canadian North, and especially during winter months. As a result, caution is needed in using projections for some variables and for more long term time horizons.

Impacts of Climate Change on Human Health

The relationships between climate change and human health in northern populations are complex and often mediated via a number of environmental, physical, social and behavioural factors. Furgal et al. (2002) identify a list of direct and indirect relationships for northern populations between climate change and health. A similar approach is adopted in the Arctic Climate Impact Assessment (ACIA, 2005). The direct impacts are considered "those health consequences resulting from direct interactions with aspects of the environment that have changed or are changing with local climate (i.e., resulting from direct interactions with physical characteristics of the environment such as air, water, ice, land, and for example exposure to thermal extremes). Indirect impacts are "those health consequences resulting from indirect interactions mediated via human behaviours and components of the environment that have changed or are changing with local climate" (Berner and Furgal, 2005)

Direct Impacts on Human Health

Extreme Events

The direct impacts of climate on human health are primarily related to interactions with such phenomenon such as extreme weather events, temperature, and changes induced by exposure to UV-B radiation (Table 6). In 2000, following an avalanche associated with excess snow accumulation in a mountainous region of Nunavik near the community of Kaniqsualujjuaq that killed 9 people and injured 25, the Quebec Ministry of Public Security requested a review of avalanche risks and protective measures in several Nunavik and St. Lawrence north shore communities (Lied, 2000). Based on the analysis it was estimated that it was reasonable to expect a recurrence of the 1999 event every 50 years (Lied, 2000).

Table 6 Summary of potential direct climate related health impacts in Nunavik and Labrador (adapted from Furgal et al., 2002)

Identified climate related change	*Potential direct health impacts*
Increased (magnitude and frequency) temperature extremes	Increased heat and cold related morbidity and mortality
Increase in frequence and intensity of extreme wather events (e.g. storms etc.) Increase in uncharacteristic weather patterns	Increased frequency and severity of accidents while hunting and traveling resulting in injuries, death, psychosocial stress
Increased UV-B exposure	Increased risks of skin cancers, burns, infectious diseases, eye damage (cataracts), immunosuppression

Thunderstorms and high humidity have been associated with short-term increases in hospital admissions for respiratory and cardiovascular diseases elsewhere around the world (Kovats et al., 2000). In the North, some reports of respiratory distress on very hot days in the summer have been reported by Elders in Labrador and Nunavik communities in recent years (Furgal et al., 2002). According to work conducted for the ACIA, as the baseline climate changes, ongoing fluctuations are very likely to cause new extremes to be reached and the occurrence of conditions that currently create stress (e.g., summer temperatures greater than 30°C) are also likely to rise (Kattsov and Kallen, 2005).

Changes in Weather

Residents of small, predominantly aboriginal communities in all regions of the Canadian Arctic have reported that the weather has become more 'uncharacteristic' or less predictable and in some cases, that the onset of storm events occur more quickly today than in previous memory (Nickels et al., 2006; Ford et al., 2006; Huntington and Fox, 2005). In their community case study on vulnerability to environmental change in Arctic Bay, NU, Ford et al. (2006) report that 'increased storminess' was said to increase the danger of summer boating and decrease access to some hunting grounds by residents. These impacts have safety as well as associated economic impacts at the household level in terms of damaged equipment or loss of traditional food procurement. Motor vehicle (including snow machine and four wheel all terrain vehicle) injury is currently a significant cause of death and hospitalization in the NWT and Yukon, and more common among younger aboriginal males living in smaller communities (GNWT, 2004; CYFN, 2006). However whether these injuries are the result of accidents in the community or on the land, and whether or not they are associated with extreme weather events has not been confirmed.

Cold Related Injuries

The greatest warming in the Canadian North is expected during the winter months and more is projected for eastern regions (Kattsov and Kaller, 2005). As a result, winter warming may reduce cold-induced injuries such as frostbite and hypothermia among northern residents, and especially those in eastern regions. Current data does not allow an analysis or projection of the status of these health impacts in the Canadian North. However, in northern Sweden, Messner (2005) identified that a temperature rise of 1°C was associated with an increase in non-fatal acute myocardial infarctions (AMIs), by 1.5% and that a strong positive correlation existed between the Arctic Oscillation and the number of AMIs in that country.

Indirect Impacts on Human Health

Indirect health impacts to climate change are primarily related to second and third order effects of changes in temperatures and shifts in ice conditions, changes in exposure to zoonotic diseases, multiple changes and impacts to traditional food security, permafrost melting and implications for health infrastructure and the combined effects of environmental and other forms of related change on social and mental well-being (Table 7).

Table 7 Summary of potential indirect climate related health impacts in Nunavik and Labrador (adapted from Furgal et al., 2002)

Identified climate related change	*Potential indirect health impacts*
Increased (magnitude and frequency) temperature extremes	Increase in infectious disease incidence and transmission, psychosocial disruption
Decrease in ice distribution, stability and duration of coverage	Increased frequency and severity of accident while hunting and travelling resulting in injuries, death, psychosocial stress Decreased access to country food items, decreased food security, erosion of social and cultural values associated with country foods preparation, sharing and consumption
Change in snow composition (decrease in quality of snow for igloo construction with increased humidity)	Challenges to building shelters (igloo) for safety while on the land
Increase in range and activity of existing and new infective agents (e.g. biting flies)	Increased exposure to existing and new vector-borne diseases
Change in local ecology of water-borne food-borne infective agents (introduction of new parasites and perceived decrease in quality of natural drinking water sources)	Increase in incidence of diarroheal and other infectious diseases Emergence of new diseases
Increased permafrost melting, decreased sturctural stability	Decreased stability of public health, housing and trasportation infrastructure Psychosocial disruption associated with community relocation (partial or complete)
Sea level rise	Psychosocial disruption associated with infrastructure damage and community relocation (partial or complete)
Changes in air pollution (contaminants, pollens and spores)	Increased incidence of respiratory and cardiovascular diseases, increased exposure to environmental contaminants and subsequent impacts on health development

Both scientific studies and local observations report an increase in the length of the ice free season, a decrease in ice thickness and total sea ice cover throughout the North (Walsh, 2005; Huntington and Fox, 2005; Nickels et al., 2006). Sea-ice coverage has decreased by 5 to 10% during the past few decades, as has multi-year ice coverage and the thickness of sea ice in the central Arctic (Walsh, 2005). Earlier break-up and later freeze-up have combined to lengthen the ice-free season of rivers and lakes in many regions by up to three weeks since the early 1900s. Model projections indicate a continuation of these trends throughout the 21st century with the greatest losses of sea-ice projected for the Beaufort Sea (Walsh, 2005).

Flato and Brown (1996) estimate that continued warming will decrease landfast ice thickness and duration of cover by approximately 0.06m per and 7.5 days per 1°C, respectively. Ford et al. (2006) suggest that this could mean a decrease in thickness of 50 cm and duration of coverage by 2 months by 2080-2100 for the community of Arctic Bay, NU. The ice provides a stable traveling and hunting platform for northern residents and is critical to the reproduction and survival of some Arctic marine species (e.g. ringed seals, *Phoca hispida* and polar bears, *Ursus maritimus*) that are important food and cultural species. Changes in the timing of the ice season, and the security of this substrate for human use are therefore critical to safety for northern residents, aboriginal and non-aboriginal alike, that are active on the land.

Residents of communities in many Inuit communities for example, reported that the changes in ice characteristics increases danger and decreases access to hunting areas and traditional foods (Reidlinger and Berkes, 2001; Huntington and Fox, 2005; Nickels et al., 2006). An increase in the perceived number of accidents and drownings associated with ice conditions is reported in some communities today (Lafortune et al., 2004) however no review of accident data has been conducted to confirm these trends to date. Finally, aboriginal communities in all northern regions report that these ice changes have negative implications for social cohesion and mental well-being in the community because of their impacts to sharing of traditional foods and disruption of seasonal traditional land based activities (Furgal et al., 2002; Berner and Furgal, 2005).

Vector-borne and Food-borne Diseases

Climate warming during El Niño Southern Oscillation events has been associated with illness in marine mammals, birds, fish, and shellfish. Disease agents associated with these illnesses have included botulism, Newcastle disease, duck plague, influenza in seabirds, and a herpes-like virus epidemic in oysters. It is likely that temperature changes arising from longer duration climate warming will be associated with an increased occurrence of disease and epidemics in these species which can be transmitted to humans (Bradley et al., 2005). In association with warming waters, ringed seals and bowhead whales (*Balaena mysticetus*) in the Arctic Ocean north of Alaska have been found, for the first time, to be infected with parasites which also infect humans, *Giardia* spp. and *Cryptosporidium* spp. (Hughes-Hanks et al., 2005).

Many zoonotic diseases currently exist in Arctic host species such as tularemia in rabbits, muskrats and beaver, rabies in fox (Dietrich, 1981), brucellosis in ungluates, fox and bears, echinococcus in rodents or canine species (Chin 2000), trichinella and cryptosporidium and are spread via temperature mediated mechanisms (e.g. movement of animal populations, water temperature of surface waters used by Arctic populations). The most common forms of food and water-borne diseases in the NWT are giardia, salmonella and campylobacter, the latter two of which can be contracted by eating raw, poorly cooked or unpasteurized foods and the former via drinking contaminated water (GNWT, 2005). Some regions have documented significant cases of zoonotic infections in the past. For example, since 1982, 11 outbreaks involving 86 confirmed cases of trichinosis have been documented in Nunavik. Walrus meat was the source in 97% of cases (Proulx et al., 2000) but no deaths have been recorded from the disease. However, Inuit communities in the central and eastern Arctic have identified an increase in parasites in caribou over recent years and concern for the safety of consumption of this meat (Nickels et al., 2006). Charron and Sockett (2005) report that the range of the vector tick *Ixodes scapularis* for Lyme disease will extend into the Northwest Territories by 2080 based on current climate models.

Food Security

Climate warming and warming of permafrost has negative implications for ice road, all season road and airstrip security and accessibility in northern communities. Changes in these critical transportation infrastructures may influence market or store food transport from southern regions and therefore physical and economic access in small remote communities where many items are already prohibitively expensive for some households. Reports of some significant impacts to transportation infrastructure of this nature have already been documented (Transport Canada, 2003). Work by Allard et al. (2002) in Nunavik, which has no road network, reports instability of airstrips as a result of current permafrost warming. However, conversely, a longer open water season with decreasing sea ice coverage and extent will provide greater access to coastal communities throughout the year and make ship transportation more viable and thus the overall impact to the transport for market food stuffs into northern communities is difficult to predict.

Climate change has significant implications for country food security among northern residents via changes in animal distributions related to habitat shifts with warming temperatures or shifts in local community ecology, and changes in northerners' access to these species as a result of lengthened ice free seasons or increases in uncharacteristic and extreme weather events (Furgal et al., 2002). According to Brotton and Wall (1997) climatic changes could have significant impacts on the Bathurst caribou herd of the NWT, possibly reducing their numbers and in turn, reducing their harvesting potential for regional residents. In workshops conducted by Nickels et al. (2006) the majority of Inuit communities reported currently experiencing impacts on country food security associated with changes in environmental conditions. Higher winds in Nunavut and Nunavik communities were reported to make travel and hunting more dangerous by boat in the summer and therefore impact access to seals in open water and whales (Nickels et al., 2006; Ford et al., 2006). What these and other climate-related changes mean in terms of per capita shifts in consumption of wildlife species throughout Arctic communities is currently unknown but is the focus of ongoing research (Guyot et al., 2006).

In contrast to these seemingly negative impacts, increased summer temperatures and growing periods in locations such as the western Arctic enhance opportunities for small scale northern agriculture, creating an additional and potentially more cost efficient local source for some foods that are often expensive and difficult to access in northern stores. As a result, the combined effects of climate changes on total food security and health are difficult to predict.

Water Quality

The Arctic is dominated by water as ice, precipitation or in its many bodies of water however there is significant evidence supporting the concern that climate change is and will continue to impact the quantity and quality of freshwater resources in the North (Walsh, 2005). Northern residents have already expressed concern regarding the quality of water in communities during the 2001 Aboriginal Peoples Survey. The number of Inuit residents who felt their drinking water at home was unsafe to consume ranged from 9% (Labrador) to 43% (Nunavik). In the Yukon, 25% of First Nations residents reported that their water was unsafe for consumption (CYFN, 2006). However the rates of giardia in the NWT appear to be decreasing (4.7 per 10,000 in 1991 to 2.9 per 10,000 in 2002) while rates of E. Coli have remained comparatively the same over this period (GNWT, 2005). *Giardia* and *Cryptosporidium* are parasites of animal and human origin and the presence of Giardia is associated with the release into the environment of animal or human feces which, in runoff, can infect waterways that supply village drinking water. Ingestion of contaminated water can have significant gastrointestinal implications especially for infants and small children. Chlorination of water is ineffective in eliminating the cysts of these two parasites.

In northern communities water is either taken from a local lake or reservoir at higher elevation than the town site and delivered by gravity to homes, delivered via an above ground piping system (utilidor system) to and from a treatment facility, delivered by truck to individual households and stored in tanks as in most northern villages, or it is collected and brought to the house using an individual bucket haul system (Fandrick, 2005). In one of the few studies of drinking water in northern communities and their vulnerability to environmental change to date, Martin (2005) reported that approximately 30% of the Nunavik population chooses to use raw or untreated water directly from a natural source such as a local stream or brook for daily household purposes. In their examination of various water sources in and around the community, Martin et al. (2005) reported that water currently held in household tanks was of good microbiological quality and safe to drink. On the other hand, raw water samples collected from natural sources and then often stored in plastic containers inside the home was often contaminated. To date, communities throughout the North have reported changes that they associated with shifts in local climate conditions in the quantity and quality of their drinking water resources (Moquin, 2005). For example, increasing temperatures in the western Arctic are reported to be supporting increased algal and plant growth in water ways which impacts drinking water quality.

Permafrost and Health Infrastructure

Greater than 40% of Canada's land surface is on Permafrost. Half of this area contains permafrost that is warmer than –2°C and is thus likely to be impacted under current climate model predictions (Smith et al., 2005). The northwest Canadian Arctic is particularly sensitive as significant warming has already occurred in upper layers of permafrost in that region (Burn as in Couture et al., 2003). Similar observations exist for the central Arctic and although cooling was previously observed in the east, Brown et al. (2000) and Allard et al. (2002), report that upper layers in the Ungava region now also have increased by up to nearly 2°C since the mid-1990s. In general, degradation of continuous permafrost to discontinuous permafrost and the disappearance of discontinuous permafrost are projected to occur at the southern boundaries of all Arctic regions. These changes have potentially significant implications for public health infrastructure in northern communities such as waste water treatment and distribution, water distribution systems relying on pipe systems, basic housing, public buildings (such as hospitals) and emergency transportation access for remote isolated communities (Warren et al., 2005).

Infrastructure challenges already exert significant stress on public health status in some northern regions. Currently many northern residents live in conditions of overcrowding and are faced with issues regarding the quality and affordability of housing. As of 2001, 28% of residents in Labrador, 68% in Nunavik, 54% in Nunavut, 35% in the Inuvialuit Settlement Region of the NWT and 43% in the Yukon lived in overcrowded homes (Statistics Canada, 2001; CYFN, 2006). Approximately 16% of homes in the NWT and 33% in the Yukon required major repairs, as compared with the national average of 8% (GNWT, 2005; CYFN, 2006; Statistics Canada, 2001). Additionally, coastal communities appear the most sensitive to these changes as they come in addition to coastal erosion in low lying communities which further exacerbates these threats to infrastructure.

Bradley (2005) argues that many northern communities are more vulnerable to the acute impacts of climate change forces because they are isolated, have a lack of transportation and emergency response infrastructure. The degradation of permafrost has been a key factor in that it has impacted transportation infrastructure that is part of this emergency response capacity. This is particularly important in remote locations with fewer access routes and methods available to them such as those communities only accessible by air or water for such things as medical evacuations. In this sense, deformation of an airstrip because of permafrost warming, as was documented in Tasiujak, Nunavik (Allard et al., 2002) is that much more important.

Mental, Social and Cultural Well-being

Many of the impacts described above, on their own, or in concert with one another, represent forces of change to many northern residents for whom the connection with the local environment is a strong component of their mental health, culture and identity. Curtis et al. (2005) and Berner and Furgal (2005) describe climate and other forms of environmental change in northern communities as a force involved in the acculturation process of aboriginal residents. For many remote northern aboriginal communities, these forms of change are involved in overall socio-cultural and economic changes which have been associated with symptoms of psychosocial, mental and social distress such as alcohol abuse, violence and suicide. Climate-related changes and their associated impacts already being observed in some northern communities, include such things as the disruption of traditional hunting cycles and patterns in Arctic Bay (Ford et al., 2006; Nickels et al., 2006), loss of the ability of elders to predict weather and provide information to hunters and other community residents (e.g. Community of Kangiqsujuaq et al., 2005), and fear and concern over coastal erosion and damage and loss of sacred sites and infrastructure (e.g. cemeteries and homes) (Community of Tuktoyaktuk et al., 2005). These impacts represent implications for cultural, social and mental health predominantly among northern aboriginal residents. Our current understanding of the importance of environmental accessibility and stability to health status is still quite limited however some studies have shown that the ability to go on the land, travel, hunt, fish or collect berries and be safe outside of the community is a critical determinant of health for northern aboriginal people (Owens, 2005).

Potential Positive Impacts on Human Health

Despite the predominantly negative impacts associated with climate changes reported in earlier sections, some potentially positive effects on human health among northern populations are possible. Climate change leads to increased vegetation growth/cover, increased production of animal growth, the opening of new trade routes, and the potential for new or intensified forms of economic activity (ACIA, 2005). The northward progression of new animal species could further facilitate the insurance of traditional food security via the appearance of new species to hunt. As mentioned earlier, direct impacts of winter warming in some regions may include a reduction in cold-induced injuries such as frostbite and hypothermia and a reduction in cold stress. A warming climate in the Arctic could offer a variety of new economic opportunities to arctic residents, including increased maritime activity and supply opportunities, exploration and development of mineral resources, and growing interests in tourism (Health Canada, 2002). However, the net benefit or loss for northern populations is difficult to project because of the interconnected nature of these relationships.

HEALTH ADAPTATIONS

Many of the changes associated with shifts in local climates have the potential for wide-ranging impacts on human health and wellbeing, and place new stresses on the health sector. Some changes are unavoidable as is evidenced by numerous changes in the environment already being experienced in the northern regions, predominantly among remote aboriginal communities (e.g. Ford et al., 2006; Nickels et al., 2006). Although reducing the human contribution to climate change via reduction in GHG emissions could potentially slow the change, the trend towards a changing climate is itself unavoidable. It is therefore necessary to look at strategies to adapt and cope with the impacts of climate change, especially in the northern regions of Canada. Within the context of human health, the process of adaptation referred to here are actions taken (including public health actions, policies, and strategies) to minimize the negative health impacts of climate change (Health Canada, 2002). These actions can take a number of different forms (e.g. behavioural, institutional, technological, economic) and be of primary, secondary or tertiary nature (McMichael and Kovats, 2000).

The unpredictable nature of climate change in the North makes strategies for adapting complex and in need of dynamic solutions. There is no consensus on what changes will occur, and to what extent they will affect the health and well-being of northern residents. The sheer size and low population density of northern Canada means that the impacts and the severity of these impacts may differ significantly from region to region (Government of Canada, 2001). Thus, no 'one size fits all' approach to adaptation is likely to exist. For adaptation strategies to be most effective they must involve and be developed at the local level; the individuals affected by the impacts are often best informed to develop solutions (Furgal and Seguin, 2006). For example, workshops in the Inuvialuit Settlement Region in the NWT identified a number of actions individuals are already taking in response to observed changes and their impacts on human activities and livelihoods in these coastal communities (Table 8).

Some general adaptations to climate change impacts on human health discussed in this chapter include the modification of travel and hunting habits in order to overcome weather unpredictability and minimize accidents while travelling on poor ice and the altering of construction methods that take into account the melting of permafrost. Hunters in some Aboriginal communities have responded to a decrease in water quality while on the land by taking along freshwater on their hunting trips for drinking requirements (Ford et al., 2006). Cabins are being built around key hunting areas (Community of Ivujivik et al., 2005) to lessen the potential impact of weather unpredictability and reduce the number of

Table 8 Examples of environmental changes, effects and coping strategies/adaptations reported by community residents in the Inuvialuit Settlement Region to minimize negative health impacts of climate change (adapted from Nickels et al. 2002).

Observation	*Effect*	*Coping Strategy/Adaptation*
Warmer temperatures in summer	Not able to store country food properly while hunting, food spoils faster, less country foods are consumed	Travel back to community more often in summer while hunting to store country food safely (in cool temperatures) *Need:* investment more funds for hunting activities Decrease amount of hunting and storage for future with fewer places to store extra meat *Need:* re-investment in government supported community freezer program
Warmer temperatures in summer	Can no longer prepare dried/ smoked fish in the same way, "it gets cooked in the heat" Less dried/smoked fish eaten	Alter construction of smoke houses: build thicker roofs to regulate temperature Adapt drying and smoking techniques
Lower water levels in some areas and some brooks/creeks drying up	Decrease in sources of good natural (raw) drinking water available while on the land Increased risk of water borne illnesses	Bottled water now purchased and taken on trips
More mosquitoes and other (new) bitting insects	Getting bitten more Increasing concern about health effects of new biting insects not seen before	Use insect repellent, lotion or sprays Use netting and screens for windows and entrances to houses *Need:* information and education on insects and Biting flies to address current perception/fear
Changing animal travel/migration routes	Makes hunting more difficutl (requires more fuel, gear and time) Some residents (e.g. Elders) cannot afford to go hunting and consume less country foods	Initiation of a community program for active hunters to provide meat to others (e.g. Elders) who are unable to travel/hunt under changing conditions *Need:* financial and institutional support to establish program

people getting stranded in storms. Communities in Nunavik are developing a local ice monitoring program and are investigating the development, monitoring and use of locally relevant indicators for safe ice conditions such as the Sum of Freezing Degree-Days (SFDD; Tremblay et al., 2005). In Nunavut, some hunters are making more use of available technology and consulting online satellite images of the sea ice prior to travel, especially in locations such as Arctic Bay, NU during the late spring narwhal hunt (Ford et al., 2006). Hunting efficiency has also improved dramatically through the use of modern firearms and improved transportation, such as boats, snowmobiles, all terrain vehicles, and aircraft.(ACIA, 2005) At the same time, these advances have the potential to erode traditional knowledge and skills, which may then increase exposure to some environmental hazards. For instance, the loss of understanding of short-term weather changes and ice conditions could result in exposure to more dangerous conditions while traveling or hunting (ACIA, 2005). Therefore the implementation of new technologies for adaptation must be considered carefully to ensure that they support and enhance resilience and do not indirectly increase exposure and unduly increase costs.

Challenges to Adaptation

The adaptive measures taken by northern residents in the face of climate related environmental changes are dependant upon basic abilities and resources. The distribution of these capacities, like exposures to climate change and variability throughout the North, differ among and between regions and communities. These factors include the material resources available to individuals or communities, their access to technology, the information base they draw upon to understand risks and modify exposure, their regulatory or institutional power, the underlying health status of the population and their access to basic public health infrastructure and the distribution of these capacities among individuals within communities and between communities and regions.

Material Resources

The access to economic wealth among individuals and communities facilitates the access to and implementation of various technological adaptation measures in the face of climate change not otherwise possible. For example, the access to resources to hire, equip and train search and rescue personnel at the municipal level can have significant positive impacts on morbidity and mortality associated with the rescue of stranded individuals injured in natural weather related disasters or stranded in bad hunting or traveling conditions. The regional or municipal access to financial resources to fund, operate and maintain community freezers in communities can significantly aid in the adaptation of individuals to stresses on country food security associated with shifts in animal populations. The shifting ice seasons have significant impacts on iceroad networks that exist in the western Arctic and provide access to communities for the shipment of such things as market foods and other products that factor heavily in the health of northern residents throughout the winter. One potential strategy to adapt to decreased accessibility of such infrastructure is the construction of permanent all weather roads. However, as Dore and Burton (2001) estimate, the costs associated with the construction of permanent roads in northern regions are exceptionally expensive.

Access to economic resources is equally important at the individual level in adapting to climate impacts on health. As reported by Ford et al. (2006) in Arctic Bay, only some families are able to purchase critical equipment (e.g. larger boats) to adapt to changes in weather (e.g. increased storminess) and maintain a high level of hunting activity to minimize impacts to household food security. In this regard, household and individual wealth is critical in terms of adaptive capacity. A review of basic individual socio-economic indicators shows that the economic capacity of individuals in Nunavut and Nunavik is significantly less than the average in other regions of the country (Table 5). This is likely, in part, related to lower economic diversity in these regions versus others that renders these populations more vulnerable to changes in both local resource base and global economic trends and markets.

Technology

As stated above, the access to technology has been reported to aid in the adaptation to climate change impacts throughout the North. For example, the use of GPS units by younger hunters in some Nunavik and Nunavut communities is said to decrease the impacts of changing weather and ice conditions on the safety and ability to travel and hunt successfully (e.g. Ford et al., 2006; Communities of Nunavik et al. 2005; Communities of Nunavut et al., 2005). However, the adoption of such strategies also comes with a cost, in that individuals are increasing their exposure to these climate variables and therefore the net vulnerability balance between increased adaptive capacity and increased exposure is often difficult to determine. The adoption of other basic forms of technology will become critically important in the North. For example, the installation of screens in the windows of homes in ISR communities is said to help alleviate the stress of extreme indoor temperatures on hot days while protecting residents from the increased presence of biting flies and other insects (Communities of the ISR et al., 2005).

Information and Skills

When reviewing basic education statistics between regions in the North one recognizes that most northern aboriginal populations have a lower average level of formal education than that of other northerners or those living in other regions of the country (Table 5). However, it can be argued that forms of informal or traditional information and skills are just as, if not more important for adaptation in small remote communities of the Canadian North to changes in local climate conditions must also be considered.

The pressure for the evolution of a locally specific knowledge base has been exceptionally strong among the Arctic aboriginal people, driven by the need to survive off highly variable natural resources and in remote and challenging conditions. As a result, a strong understanding of weather, snow and ice conditions exists as they relate to hunting, travel, survival and natural resource availability (e.g. Krupnik and Jolly, 2002). There is an increasing awareness of the value of aboriginal knowledge and its role in adaptation to climate and other forms of environmental change in the circumpolar North and around the world (e.g. ACIA, 2005). Its value must not be forgotten when assessing the adaptive capacity of individuals and communities to the health impacts of climate change in Arctic regions. Its value is evident in the ability of aboriginal hunters to safely navigate new travel and hunting routes in response to decreasing ice stability and safety in regions such as Nunavik (e.g. Lafortune et al., 2004); in the ability of many Arctic aboriginal groups to locate and hunt species that have shifted in their migration times and routes such as geese or caribou or to locate and hunt alternate species and maintain aspects of traditional food security (e.g. Krupnik and Jolly, 2002; Huntington and Fox, 2005; Nickels et al., 2002). In general, aboriginal people show considerable resilience as a result of short-term coping mechanisms such as prey-switching in response to changing animal abundance and the development of these and long term adaptation strategies often depend on this strong local knowledge foundation (Berkes and Jolly, 2002). However the generation and application of traditional knowledge requires active engagement with the environment, close social networks in communities and respect and recognition for the value of this form of knowing and understanding the local environment as part of traditions and culture. Currently, the social, economic and cultural trends in some communities and predominantly among younger generations towards a more western lifestyle in the Arctic has the potential to erode the cycle of traditional knowledge generation and transfer and hence its contribution to adaptive capacity.

Institutional Arrangements

The devolution of powers from federal to territorial, provincial and regional scales throughout the North has had some positive impacts on the delivery of health services for Arctic populations in the past (O'Neil, 1991). The establishment of co-management regimes for resources and self-government regimes for various services which empower local scales is oriented towards giving decision-making power where it is justly situated and also most effectively placed; at the local scale where the issues are best understood. The institutional capacity of local and regional scale organizations to identify problems and make decisions using locally appropriate solutions is critical to the adaptive ability of communities. For example, greater uncertainty and threats to food security associated with increased climate variability supports the need for resilient and flexible resource procurement activities. Yet, resilience and adaptability depends not only on ecosystem diversity, but on institutional arrangements which govern social and economic systems (Adger, 2000), allowing flexibility in such things as hunting seasons with shifts in species ecology (Chapin et al., 2004; Armitage, 2005).

Public Health Infrastructure

The public health system in the Canadian North is challenged in a variety of ways in providing services and meeting the changing needs of the population. Despite spending significantly more on public health per capita than elsewhere in the country, the average life expectancy of both males and females is significantly less (Table 3). Northern regions are serviced with community health centres, regional hospitals and a medical evacuation system in which residents are flown south to better equipped centres with specialist and emergency facilities when required. The number of public health professionals and specialists per capita is significantly less in the northern Territories than other regions of the country and access to health services is identified as a challenge among many northern residents, and in particular the aboriginal people. Based on geographic location, smaller, more remote communities in regions dependant upon air travel as the link with southern regions (e.g. with no road network) are more challenged in their access to specialist and emergency facilities. The public health surveillance infrastructure has also been found in need of reinforcement and pilot projects aiming at capacity building are being put forward within the ArcticNet research program (Gosselin et al. 2006a, 2006b). This combination of factors challenges community health capacity in responding to the current and potential future impacts of climate change on human health in these regions.

Equity

Disparity among aboriginal and non-aboriginal, and remote and regionally centred populations across the North for many health and adaptive capacity indicators suggests that the more regionally centralized populations, closer to north-south transportation connections (e.g. road networks, regional airports), more heavily engaged in wage-based employment, and less dependant on local resources for household sustenance and livelihoods are less exposed and in some instances more able to adapt to changes in local climate (e.g. have access to economic resources to purchase other forms of transportation and hunting equipment, have funds to purchase market foods when land-based foods are scarce, have easier access to emergency medical services in the event of accidents). In general, these are regional centres with a predominantly higher non-aboriginal population. However, in terms of social capital, traditional skills and knowledge, and access to a diversity of environmental resources, the more remote, smaller communities are better equipped to adapt to changes and variability in local climate (e.g. have the traditional knowledge to find new hunting locations and routes, have the traditional survival skills to travel in dangerous weather, have extended social networks to spread the risk of impacts among a larger number of individuals). In general, these are communities with a predominantly higher proportion of aboriginal residents. Thus the pattern of equities in terms of adaptive capacity differs from location to location based on a variety of factors, however in the same location it appears that a number of social and economic inequities disadvantage aboriginal populations in terms of their ability to adapt to climate change impacts on health. It is important to note that to date this is solely based on qualitative observations of the data presented here and requires further investigation.

Existing Burden of Disease

The health status of northern populations are challenged in comparison with other regions of the country (e.g. Tables 3, 4; see "Current Health Status of Northern Populations"). Lower life expectancy, higher infant mortality, a higher percentage of low-weight births (Nunavut), a significantly higher number accidents and greater incidence of lung cancer as compared to the national averages characterize a northern population with challenged health status on a number of fronts. As noted by the GNWT (2005) for some indicators, the status of aboriginal residents is poorer than non-aboriginal residents and is particularly the case in smaller, more remote communities where they comprise a much larger percentage of the total population. According to Wigle et al. (2005) children and youth in the Canadian Arctic, and particularly aboriginal children, suffer from comparatively lower health status as measured by a series of indicators than children and youth in other Arctic countries and in comparison to the Canadian general population. This situation has recently been characterized as being catastrophic in Nunavik (INSPQ, 2006) where life expectancy is actually declining, notably in relationship with suicide mortality rates which are seven times higher than the rest of the province and significantly higher mortality rates for children aged 0-4 years. The underlying health status of the population may, in the future, significantly hinder some individuals' adaptive ability to environmental changes which presenting more difficult condition for health and well-being.

Socio-ecological Resilience

Previously, significant adaptive capacity among aboriginal groups was associated with a combination of strong human and social capital, social and cultural organizational flexibility and the ability to understand and respect human relationships to the land and generate, share and apply locally developed land-based knowledge. This socio-ecological resilience was a significant component of the survival of aboriginal peoples throughout the North over thousands of years which included significant variations in climatic regimes. The ability of Arctic aboriginal peoples to utilize their local resources has always been associated with, or affected by, seasonal variation and changing ecological conditions. One of the hallmarks of their success in adapting has been flexibility in technology and social organization, and the knowledge and ability to cope with climate change and circumvent some of its negative impacts. Some of these characteristics still exist in aboriginal communities today, whereas others have been eroded by social, cultural and economic shifts over previous decades.

The Risk of Maladaptation

Various definitions exist for the concept of 'maladaptation'. Smit et al. (2000) state that maladaptation is the lack of adaptation or an action that becomes increasingly worse in providing an effective response to change. Others simply remind us that maladaptation can result in negative effects that are as serious as those first being the impetus for reaction (Scheraga and Grambsch, 1998). Knowing that few adaptations are likely to have little to no associated negative implications, we can think of each along a spectrum from 'ideal adaptation' resulting in only positive outcomes to being truly maladaptative. For example, adaptive measures must be suitable with culture, history, economic capacity, current health status, etc. The health advisories released in the 1970s and 1980s to reduce consumption of traditional foods in some northern regions because of concern for exposure to environmental contaminants such as mercury and PCBs have had an undesired effect in some cases of reducing the amount of all traditional foods consumed by individuals and raising the general level of concern about food safety (Van Oostdam et al., 2005). Meanwhile, further research has shown that the benefits of consumption for physical, social and mental health outweigh the known risks associated with contaminant exposure (Blanchet et al., 2000) thus identifying the maladaptive nature of some aspects of those original behaviour changes. In terms of responses to climate change, although GPS units were rumoured to increase travel safety and adaptive capacity among young hunters in some Inuit communities, Elders and older hunters reported that it allowed younger, less experienced hunters to take more risks and get into dangerous situations without the appropriate environmental knowledge (Nickels et al., 2006).

The dramatic shift which is resulting in the loss of many aspects of a traditional life-style, that had developed over thousands of years among northern aboriginal populations can be looked at as being a form of maladaptation towards which northern populations must remain vigilant. However, after decades of using ski-doos as the principal means of winter transportation, some hunters have started to rejuvenate interest in and practice of traditional dog-sled teams for hunting and traveling (Ford et al., 2006). They are often considered more reliable and therefore safer by some. The trend towards the reliance on snow-mobiles has had serious implications in terms of human health as a 62 year old Elder from Kujjuaaq, Nunavik reports:

> *"Our bodies have changed. We used to exercise like dogs, having to run with our dogs...we used to sweat a lot. Nowadays, you hop on a ski doo and never get off and can go everywhere. That makes our tradition change too. [...] Ever since our dogs were killed, we have been just sitting on our vehicules. That's why we feel the sharp cold now". (as in Furgal et al., 2002)*

THE FUTURE OF THE CANADIAN NORTH

Basic Trends in the Canadian North and their Importance for Climate Change and Health

Significant changes taking place among Canadian Arctic populations over recent decades have implications for exposure to climate variables and their impacts on health today and in the future. Since the establishment of communities, most growth has taken place in the three main urban centres of the territories. More recently, significant amounts of growth have been associated with increases in the non-aboriginal population associated with resource development and public administration. It is expected that the North will remain to be a young population with a growing proportion of people over the age of 65, thus increasing dependency ratios (old on young) across the North. This is expected to be most significant in the Yukon and predominantly associated with an aging population (Statistics Canada, 2006a). These shifts in socio-demographic and economic status will further erode some aspects of community and individual adaptive capacity in the face of climatic changes. Similarly, shifts in culture, away from the land and towards a more sedentary lifestyle and a higher level of engagement in wage-earning employment is expected to have negative impacts on the northern community's ability to adapt to environmental change at a time when variability is increasing.

In general, many northern community economies are now a mix of traditional land-based renewable resource/subsistence activities and formal wage-earning sector activities, many of which are tied to non-renewable resource extraction. These traditional components of economic systems represent substantial inputs into the community economy and will likely to continue to be important in the future. The Conference Board of Canada (2005) estimates that Nunavut's land-based economy is worth CDN$ 40-60 million per year with an estimated $30 million being attributed to all food-oriented economic activities. Tourism which includes sport hunting camps and polar bear hunts contributes a significant proportion to the economies of small communities. Increases in tourism associated with greater access during lengthened summer periods are expected, while increased challenges to the accessibility and availability of some wildlife may change the financial contribution they represent to the total economy of northern populations. It is currently very difficult to project the net impact on houshehold incomes and thus their economic capacity to adapt.

The establishment of autonomous governments and administration throughout the North in the last 35 years is likely to represent the most substantial political change since European settlement in this area. Public administration is currently the largest secondary sector in the North. The movement towards regionally specific administrative structures are likely to represent positive impacts on health and well-being and enhanced adaptive ability.

With climate warming and diminished ice cover, it is expected that the Northwest Passage through the Canadian Arctic will become more open and feasible for ship passage in the future. This will provide opportunities for significant economic development (e.g. mineral exploration), or the potential for environmental impacts in these regions that may have far reaching implications for communities (e.g. pollution of local marine resources; Kelmelis et al., 2005). This increased access to a previously isolated region also represents a major challenge for already fragile communities having to cope with recent and new health threats coming from the westernization of their lifestyles without an adapted, well-staffed and stable public health infrastructure.

A Modern or Traditional Future?

Arctic indigenous populations have been described as existing 'between two worlds', and undergoing rapid social, cultural, political and economic transition. The social tension in some communities is often attributed, in part, to this struggle in the transition from 'old' (traditional life on the land) to 'new' (mixed economy, sedentary, fixed-place lifestyle) in a short period of time (one to two generations). The introduction of large scale economic resource development projects and their associated socio-economic benefits and impacts in small communities further accelerates and stresses this process of modernity. Today, children in northern communities grow up in an environment unlike that of their parents and grandparents. Many large northern centres now resemble southern cities in a variety of ways, however the links between traditional and modern ways of life are not always clear.

A diverse spectrum of aspirations for the future direction of their lives and their communities exists both within and between communities. It can be argued that this transition makes the health of individuals more vulnerable in the face of environmental change, and that a loss of aspects of a traditional lifestyle decreases the ability of northern populations to cope with climate change today (Ford and Smit, 2004). A shift in the pursuit of traditional activities (hunting and fishing) from a full time occupation towards a weekend event among some individuals because of their increased participation in wage-earning jobs from Monday to Friday, significantly hinders their development of traditional skills and land-based experience. In an unpredictable and changing environment this puts these individuals more at risk for such things as accidents associated with uncertain ice or weather conditions. The loss of the ability to read and predict the weather via traditional cues, similarly puts individuals at greater risk for being stranded on the land while hunting and travelling. As this local ecological and traditional knowledge is the basis for survival and successful adaptation in the natural environment, its' erosion or decreased generation and transmission may be seen as increasing vulnerability among young hunters and others. Likewise the movement of more southern migrants to the North, and their participation in land-based activities in these changing circumstances increases their exposure to environmental hazards in an already challenging physical environment.

However, there are also potential benefits in terms of health and adaptive capacity related to increased economic opportunities and modernity in the northern regions. The increased economic resources at the household level as a result of participation in development projects provide the capacity to adapt via the purchase of new equipment (e.g. skidoo or boat, GPS) and facilitates the pursuit of hunting and fishing activities in a wider variety of environmental conditions. This connection between old and new, or subsistence and market economies presents both benefits and risks. As Duhaime et al. (2002) report in Nunavik, households headed by two individuals where one was engaged in wage-earning employment, while the other practiced subsistence activities were the most productive households in terms of traditional food acquisition and consumption. This marriage in which aspects of traditional lifestyles that convey significant health benefits for individuals and households are supported or formalized via the connection to modern institutions or resource bases, may prove valuable for enhancing adaptive health capacity in northern communities in the future.

CONCLUSION

The current path for many communities is far from easy and will not always result in a happy ending in terms of climate and climate impacts on the way of life and health of its residents. Demographic pressures are added to this situation, while unemployment is already high and social stress is showing many and severe impacts. Climate change will likely accelerate this process by providing more access to the rich natural reserves of the North and putting under greater stress the current housing and municipal and public health infrastructure. Some specific problems are also likely to increase due to climate change (eg. UV exposure, food safety and security, water quality, poor indoor air quality). Current formal educational levels are comparatively low and may not facilitate an easy transition towards a more modern society, while ecosystems that deliver traditional food resources are likely to become more and more difficult to access for reasons explained in the preceding sections.

Technological fixes can help, as argued above. However, due to the lower economic status of many households, northern communities do not always have the resources to put into practice new sustainable technologies. Therefore, significant investments from all levels of government would be a significant step forward towards long-term adaptation and economic independence.

Key knowledge gaps remain, but these are likely to be addressed through existing and emerging research programs such as, ArcticNet and the initiatives that are proposed under the International Polar Year. Perhaps what is most greatly needed is the long-term commitment from all levels of government to a multi-faceted strategy to simultaneously address several key determinants of health. Climate change is an excellent opportunity to improve public health infrastructure in the North, to develop and implement long-term programs to begin addressing the health impacts of the current shift in community and personal behaviours and lifestyle, to improve the educational level of aboriginal people and enhance access to jobs (including public health professions) and other resources needed to successfully cope with climate and other forms of environmental change.

References

Adger, W.N. 2000. Social and ecological resilience: are they related? Progress in Human Geography 24(3): 347-364.

Allard, M., R. Fortier, C. Duguay and N. Barrette. 2002. A trend of fast climate warming in northern Quebec since 1993, Impacts on permafrost and man-made infrastructures. In: American Geophysical Union, 2002 Fall Meeting, Moscone Center, San Francisco, California, USA.

Arctic Climate Impact Assessment (ACIA). 2005. Arctic Climate Impact Assessment, Scientific Report. Cambridge University Press, Cambridge, UK.

Arctic Human Development Report (AHDR). 2005. Akureyri: Steffanson Arctic Institute. ISBN: 9979-834-45-5.

Armitage, D.R. 2005. Community-based narwhal management in Nunavut, Canada: Change, uncertainty and adaptation. Society and Natural Resources 18: 715-731.

Berkes, F. and D. Jolly. 2002. Adapting to climate change: socio-ecological resilience in a Canadian western Arctic community. Conservation Ecology 5(2): 16 (online) URL: http://www.consecol.org/vol15/art18

Berner, J. and C. Furgal. 2005. Human Health. In: Arctic Climate Impact Assessment. Cambridge University Press, Cambridge, UK.

Berner, J.E. 2005. Climate change and health in the circumpolar North. International Journal of Circumpolar Health 64(5): 435-437.

Bjerregaard, P., G. Mulvad and H.S. Pederson. 1997. Cardiovascular risk factors in Inuit of Greenland. International Journal of Epidemiology 26: 1182-1190.

Bjerregaard, P. and T.K. Young. 1998. The Circumpolar Inuit: Health of a population in transition. Munksgaard, Copenhagen, Denmark.

Blanchet, C., E. Dewailly, P. Ayotte, S. Bruneau, O. Receveur and B.J. Holub. 2000. Contribution of selected traditional and market foods to the diet of Nunavik Inuit women. Canadian Journal of Dietetic Practice and Research 61(2) 1-9.

Bonsal, B.R. and T.D. Prowse. 2006. Regional assessment of GCM-simulated current climate over northern Canada. Arctic. 59(2): 115-128.

Bradley, M.J., S.J. Kutz, E. Jenkins and T.M. O'Hara. 2005. The potential impact of climate change on infectious diseases of Arctic fauna. International Journal of Circumpolar Health 65(4): 468-477.

Bradley, M.J. 2005. Climate related events and community preparedness. International Journal of Circumpolar Health 64(5): 438-439.

Brotton, J. and G. Wall. 1997. Climate change and the Bathurst Caribou herd in the Northwest Territories, Canada. Climatic Change 35: 35-52.

Brown, J., K.M. Hinkel and F.E. Nelson (eds). 2000. The Circumpolar Active Layer Monitoring (CALM) Program: research designs and initial results. Polar Geography 24(3): 165-258.

Chabot, M. 2004. Kaagnituurma ! As long as I am not hungry. Socio-economic status and food security of low income households in Kuujjuaq (Nunavik Regional Board of Health and Social Services & the Corporation of the Northern Village of Kuujjuaq, Kuujjauq and Pontiac). The Canadian Review of Sociology and Anthropology, 41, 2.

Chapin, F.S., G. Peterson, F. Berkes, T.V. Callaghan, P.C. Angelstam, C. Beier, Y. Bergeron, A.-S. Crépin, K. Danell, T. Elmqvist, C. Folke, B. Forbes, N. Fresco, G. Juday, J. Niemela, A. Shvidenko and G. Whiteman. 2004. Resilience and Vulnerability of Northern Regions to Social and Environmental Change. Ambio. 33(6): 344-349.

Chapin, F.S. III, M. Berman, T.V. Callaghan, P. Convey, A.S. Crépin, K. Danell, H. Ducklow, B. Forbes, G. Kofinas, A.D. McGuire, M. Nuttall, R. Virginia, O. Young and S.A. Zimov. 2005. Polar Systems. In: Millenium Ecosystem Assessment. Island Press: Washington, D.C., USA.

Charron, D. and P. Sockett. 2005. Signs of change, signs of trouble: Finding the evidence. Health Policy Research, 11, November 27-30.

Chin, J. (ed.). 2000. Control of Communicable Diseases Manual. American Public Health Association, Washington, D.C., USA.

Community of Kangiqsujuaq, Furgal, C., Qiisiq, M., Editloie, B., Moss-Davies, P. 2005. Unikkaaqatigiit - Putting the Human Face on Climate Change: Perspectives from Kangiqsujuaq, Nunavik. Ottawa: Joint publication of Inuit Tapiriit Kanatimi, Nasivvik Centre for Inuit Health and Changing Environments at Université Laval and the Ajunnginiq Centre at the National Aboriginal Health Organization.

Communities of Tuktoyaktuk, Nickels, S., Furgal, C., Castleden, J., Armstrong, B., Binder, R., Buell M., Dillion, D., Fonger, R., Moss-Davies, P. 2005. *Unikkaaqatigiit – Putting the Human Face on Climate Change – Perspectives from Tuktoyaktuk, Inuvialuit Settlement Region*. Ottawa: Joint publication of Inuit Tapiriit Kanatimi, Nasivvik Centre for Inuit Health and Changing Environments at Université Laval and the Ajunnginiq Centre at the National Aboriginal Health Organization.

Community of Ivujivik, Furgal, C., Qinuajuak, J., Martin, D., Marchand, P., Moss-Davies, P. 2005. Unikkaaqatigiit – Putting the Human Face on Climate Change –

Perspective from Ivujivik, Nunavik. Ottawa: Joint publication of Inuit Tapiriit Kanatami, Nasivvik Centre for Inuit Health and Changing Environments at Université Laval and the Ajunnginiq Centre at the National Aboriginal Health Organization.

Communities of Nunavik (Ivujivik, Puvirnituq and Kangiqsujuaq), Furgal, C., Nickels, S., Kativik Regional Government - Environment Department. 2005. Unikkaaqatigiit: Putting the Human Face on Climate Change: Perspectives from Nunavik. Ottawa: Joint publication of Inuit Tapiriit Kanatimi, Nasivvik Centre for Inuit Health and Changing Environments at Université Laval and the Ajunnginiq Centre at the National Aboriginal Health Organization.

Communities of Nunavut (Arctic Bay, Kugaaruk and Repulse Bay), Nickels, S., Furgal, C., Buell, M., Moquin, H. 2005. Unikkaaqatigiit - Putting the Human Face on Climate Change: Perspectives from Nanavut. Ottawa: Joint publication of Inuit Tapiriit Kanatami, Nasivvik Centre of Inuit Health and Changing Environments at Université Laval and the Ajunnginiq Centre at the National Aboriginal Health Organization.

Communities of the Inuvialuit Settlement Region (ISR-Aklavik, Inuvik, Holman Island, Paulatuk and Tuktoyaktuk), Nickels, S., Buell, M., Furgal, C., Moquin, H. 2005. Unikkaaqatigiit - Putting the Human Face on Climate Change: Perspectives from the Inuvialuit Settlement Region. Ottawa: Joint publication of Inuit Tapiriit Kanatami, Nasivvik Centre of Inuit Health and Changing Environments at Université Laval and the Ajunnginiq Centre at the National Aboriginal Health Organization.

Conference Board of Canada. 2005. Economic Outlook for Nunavat. Economic Service. The Conference Board of Canada, Ottawa, ON. Canada.

Council of Yukon First Nations (CYFN). 2006. Health Status of Yukon First Nations. Council of Yukon First Nations, Whitehorse, YK.

Couture, R., S. Smith, S.D. Robinson, M.M. Burgess and S. Solomon. 2003. On the hazards to infrastructure in the Canadian North associated with thawing of permafrost. Proceedings of Geohazards 2003, 3rd Canadian Conference on Geotechnique and Natural Hazards, The Canadian Geotechncial Society, pp. 97-104.

Curtis, T., S. Kvernmo and P. Bjerregaard. 2005. Changing living conditions, lifestyle and health. International Journal of Circumpolar Health. 64(5): 442-450.

Dietrich, R.A. (ed.). 1981. Alaskan Wildlife Diseases. Institute of Arctic Biology, University of Alaska, Fairbanks, USA.

Dore, M.H.I., and I. Burton. 2001. The Costs of Adaptation to Climate Change in Canada: A Stratified Estimate by Sectors and Regions Social Infrastructure. Final report submitted to the Canadian Climate Change Impacts and Adaptation Office, Ottawa, ON.

Doskooch, B. 2006. Conservatives climate change vision, a foggy one. www.CTV.ca. June 5, 2006.

Duhaime, G., M. Chabot and A. Gaudreault. 2002. Food Consumption Patterns and Socioeconomic Factors Among the Inuit of Nunavik. Ecology of Food and Nutrition pp. 91-118.

Environment Canada. 2006. Environment Canada Department website. http://www.ec.gc.ca/climate/home-e.html (accessed August and December 2006).

Fandrick, B. 2005. Water management issues in Inuit communities. ITK Environment Bulletin, Inuit Tapiriit Kantami, Ottawa, ON, 3: 9-11.

Flato, G.M. and R.D. Brown. 1996. Variability and climate sensitivity of landfast Arctic sea ice. Journal of Geophysical Research 101: 11. 25-767.

Ford, J., B. Smit and J. Wandell. 2006. Vulnerability to climate change in the Arctic: A case study from Arctic Bay, Nunavut. Global Environmental Change 16: 145-160.

Ford, J.D. and B. Smit. 2004. A framework for assessing the vulnerability of communities in the Canadian Arctic to risks associated with climate change. Arctic 57(4): 389-400.

Freeman, M.M.R. 1996. Identity, health and social order. In: M.L. Foler, and L.O. Hansson (eds.) Human Ecology and Health: Adaptation to a Changing World. Gothenburg University, Gothenburg, Sweden. pp. 57-71.

Furgal C., D. Martin and P. Gosselin. 2002. Climate Change and Health in Nunavik and Labrador: Lessons From Inuit Knowledge. In: I. Krupnik and D. Jolly (eds.). The Earth is Faster Now: Indigenous Observations of Arctic Environmental Change, Arctic Research Consortium of the United States in Cooperation with the Arctic Studies Center, Smithsonian Institution, Fairbanks, USA. pp. 266-299.

Furgal, C.M. and J. Seguin. 2006. Climate Change, Health and Community Adaptive Capacity: Lessons from the Canadian North. Environmental Health Perspectives 114(12): 1964-1970.

Furgal, C.M. 2005. Monitoring as a community response for climate change and health. International Journal of Circumpolar Health 64(5): 440-442.

Gosselin, P., S. Owens, C. Furgal, L. Château-Degat and J.-F. Proulx. 2006a. Public Health Surveillance and Climate Change Case Study Results in Nunavik. Poster at the annual Arcticnet NCE Scientific Meeting. Victoria, B.C. December 12-15, 2006.

Gosselin, P., S. Owens, C. Furgal, D. Martin and G. Turner. 2006b. Public Health Surveillance and Climate Change Case Study Results in Nunatsiavut. Poster at the annual Arcticnet NCE Scientific Meeting. Victoria, B.C. December 12-15, 2006.

Gouvernement du Québec. 2006. Le Québec et les changements climatiques, Un défi pour l'avenir. Plan d'actions 2006-2012. Online at: http://www.asmavermeq.ca/pdf/plan_action.pdf

Government of Canada, 2001. Canada's third national report on climate change: actions to meet commitments under the United Nations Framework Convention on Climate Change. Ottawa, ON.

Government of Canada. 2006a. Government of Canada Climate Change Site. http://www.climatechange.gc.ca/ (accessed in August and December, 2006).

Government of Canada. 2006b. Managing the Federal Approach to Climate Change. 2006. Report of the Commissioner of the Environment and Sustainable

Development. Government of Canada. Available online at: http://www.oag-bvg.gc.ca/domino/reports.nsf/html/c20060901ce.html

Government of the Northwest Territories (GNWT). 2004. Injury in the Northwest Territories, A Descriptive Report. Government of the Northwest Territories, Department of Health and Social Services. Yellowknife, NT.

Government of the Northwest Territories (GNWT). 2005. The NWT Health Status Report: 2005. Northwest Territories Health and Social Services, December 2005.

Guyot, M., C. Dickson, K. Macguire, C. Paci, C. Furgal and H.M. Chan. 2006. Local observations of climate change and impacts on traditional food security in two northern Aboriginal communities. International Journal of Circumpolar Health. (In press).

Hassi, J., M. Rytkonen, J. Kotaniemi and H. Rintamaki. 2005. Impacts of cold climate on human heat balance, performance and health in circumpolar areas. International Journal of Circumpolar Health 64(5): 459-476.

Health Canada. 2002. Climate Change and Health and Well-being: A Policy Primer for Canada's North. Health Canada, Climate Change and Health Office. Ottawa, ON.

Hughes-Hanks, J.M., L.G. Rickard, C. Panuska, J.R. Saucier, T.M. O'Hara, R.M. Rolland and L. Dehn. 2005. Prevalence of *Cryptosporidium* spp. and *Giardia* spp. in five marine mammal species. Journal of Parasitology 91(5): 1225–1228.

Huntington, H. and S. Fox. 2005. Chapter 3: The Changing Arctic: Indigenous Perspectives. In: Arctic Climate Impact Assessment. Cambridge University Press, Cambridge, UK.

Institut national de santé publique du Québec (INSPQ). 2006. Portrait de santé du Québec et de ses régions 2006. Les analyses (section 1.4). INSPQ. Québec. 131 pages. Accessible en ligne le 4 décembre 2006 à : http://www.inspq.qc.ca/pdf/publications/portrait_de_sante.asp?E=p

Kattsov, V.M. and E. Kallen. 2005. Chapter 4: Future Climate Change: Modeling and Scenarios for the Arctic. In: Arctic Climate Impact Assessment. Cambridge University Press, Cambridge, UK.

Kelmelis, J., E. Becker and S. Kirtland. 2005. Workshop on the foreign policy implications of artic warming; notes from an international workshop. U.S. Geological Survey, Open-File Report 2005-1447.

Kovats, R.S., B. Menne, A.J. McMichael, C. Corvalan and R. Bertollini. 2000. Climate change and human health: Impact and adaptation. World Health Organization, Geneva.

Krupnik, I. and D. Jolly (eds). 2002. The Earth is Faster Now: Indigenous Observations of Arctic Environmental Change, Arctic Research Consortium of the United States in cooperation with the Arctic Studies Center, Smithsonian Institution, Fairbanks, USA.

Lafortune, V., C. Furgal, J. Drouin, T. Annanack, N. Einish, B. Etidloie, M. Qiisiq, P. Tookalook, and the Communities of Kangiqsujuaq, Umiujaq, Kangiqsualujjuaq and Kawawachikamach. 2004. Climate change in Northern Québec: Access to Land and Resource Issues. Kativik Regional Government, Kuujjuaq, Nunavik.

Lied, K. 2000. Évaluation des risques d'avalanche au Nunavik et sur la Côté-Nord du Québec, Canada. Rapport pour la Ministère de la Sécurité publique du Québec, le 28 avril.

Lucas, M., E. Dewailly, G. Muckle, P. Ayotte, S. Bruneau, S. Gingras, M. Rhainds and B.J. Holub. 2004. Gestational age and birth weight in relation to n-3 fatty acids among Inuit (Canada). Lipids l. 397: 617-626.

Martin, D. 2005. Quality of drinking water in Nunavik: How a changing climate affects disease. ITK Environment Bulletin, Inuit Tapiriit Kanatami, Ottawa ON. 3: 13-15.

Martin, D., B. Levesque, J.S. Maguire, A. Maheux, C.M. Furgal, J.L. Bernier and E. Dewailly. 2005. Drinking water quality in Nunavik: Health impacts in a climate change context, Final Report, Project funded by ArcticNet and ACADRE (Nasivvik), CHUL Research Centre, Universite Laval, Quebec City, Quebec, Canada. August 2005.

McBean, G. 2005. Chapter 2: Arctic Climate: Past and Present. In Arctic Climate Impact Assessment. Cambridge University Press. Cambridge, UK.

McMichael, A.J. and S. Kovats. 2000. Climate change and climate variability: Adaptations to reduce adverse health impacts. Environmental Monitoring and Assessment 61: 49-64.

Messner, T. 2005. Environmental variables and the risk of disease. International Journal of Circumpolar Health 64(5): 523-533.

Ministère de l'Environnement du Québec (Environnement Québec). 2000. Étude d'impact du projet de modification réglementaire sur l'eau potable en regard des communautés autochtones, Direction des politiques du secteur municipal, Service de l'expertise technique en eau. May.

Moquin, H. 2005. Freshwater and Climate Change. ITK Environment Bulletin, Inuit Tapiriit Kanatami, Ottawa ON. No. 3: 4-9.

Nickels, S., C. Furgal, M. Buell and H. Moquin. 2006. Unikkaaqatigiit - Putting the Human Face on Climate Change: Perspectives from Inuit in Canada. Ottawa: Joint publication of Inuit Tapiriit Kanatami, Nasivvik Centre for Inuit Health and Changing Environments at Université Laval and the Ajunnginiq Centre at the National Aboriginal Health Organization.

Nickels, S., C. Furgal, J. Castelden, P. Moss-Davies, M. Buell, B. Armstrong, D. Dillon, and R. Fongerm. 2002. Putting the Human Face on Climate Change Through Community Workshops. In: I. Krupnik and D. Jolly (eds). The Earth is Faster Now: Indigenous Observations of Arctic Environmental Change. Arctic Research Consortium of the United States, Arctic Studies Centre, Smithsonian Institution, Washington, D.C., pp. 301-333.

O'Neil, J.D. 1991. Regional health boards and the democratization of health care in the Northwest Territories. Arctic Medical Research, 1991, Suppl. 50-3.

Ouranos. 2005. Adapting to Climate Change. Ouranos Climate Change Consortium, Montreal, Qc. ISBN: 2-923292-01-4.

Owens, S. 2005. Climate Change and Health Among Women of Labrador. M.Sc. Thesis, Département de médicine sociale et préventive, programme de santé communautaire, Faculté de Médecine, Université Laval.

Proulx, J.F., D. Leclair and S. Gordon. 2000. Trichinellosis and its prevention in Nunavik, Quebec, Canada. Santé Nunavik, MSSS, Beauport, Québec.

Riedlinger, D. and F. Berkes. 2001. Contributions of traditional knowledge to understanding climate change in the Canadian Arctic. Polar Record 37(203): 35-328.

Scheraga, J.D. and A.E. Grambsch. 1998. Risks, opportunities, and adaptation to climate change. Climate Research 10: 85-95.

Smit, B., I. Burton, R.J.T. Klein and J. Wandel. 2000. An anatomy of adaptation to climate change and variability. Climate Change 45, 11: 223-251.

Smith, S.L., M.M. Burgess, D. Riseborough and F.M. Nixon. 2005. Recent trends from Canadian permafrost thermal monitoring network sites. Permafrost and Periglacial Processes 16: 19-30.

Statistics Canada 2001. Canadian Community Health Survey. Catalogue: 82-221-XIE, Statistics Canada, Ottawa, ON.http://www.statcan.ca/bsolc/english/bsolc?catno=82-221-XIE (accessed Dec 01, 2006).

Statistics Canada. 2002. The health of Canada's communities. Statistics Canada Catalogue 82-003. Supplement to Health Reports, Vol. 13.

Statistics Canada. 2005a. Canadian Environmental Sustainability Indicators. Available online on Dec. 14, 2006 at: http://www.statcan.ca/Daily/English/051214/d051214c.htm

Statistics Canada. 2005b. Food Insecurity. Statistics Canada Catalogue 82-033 XIE, Health Reports Vol. 16(3).

Statistics Canada. 2006a. Population projections for Canada, Provinces and Territories, 2005-2031. Catalogue no. 91-520-XIE, Statistics Canada, Ottawa, ON.

Statistics Canada. 2006b. Mortality, Summary List of Causes, 2003. Health Statistics Division Catalogue no. 84FO2O9XIE, Statistics Canada, Ottawa, On.

Transport Canada. 2003. Impacts of Climate Change on Transportation in Canada. Workshop Report, Canmore, AB March 2003. Report Prepared by Marbek Resource Consultants for Transport Canada, Ottawa, ON.

Tremblay, M., C. Furgal, V. Lafortune, C. Larrivée, J.P. Savard, M. Barrett, T. Annanack, N. Enish, P. Tookalook and B. Etidloie. 2006. Climate change, communities and ice: Bringing together traditional and scientific knowledge for adaptation in the North. In: R. Riewe and J. Oakes (eds), Climate Change: Linking Traditional and Scientific Knowledge. Aboriginal Issues Press, University of Manitoba. ISBN: 0-9738342-1-8.

Van Oostdam, J., S.G. Donaldson, M. Feeley, D. Arnold, P. Ayotte, C. Bondy, L. Chan, E. Dewailly, C.M. Furgal, H. Kuhnlein, E. Loring, G. Muckle, E. Myles, O. Receveur, B. Tracy, U. Gill and S. Kalhok. 2005. Human health implications of environmental contaminants in Arctic Canada: A review. Science of the Total Environment (351-352): 165-246.

Walsh, J.E. 2005. Chapter 6: Cryosphere and Hydrology. In: Arctic Climate Impact Assessment, Cambridge University Press, Cambridge, UK.

Warren, J., J. Berner and J. Curtis. 2005. Climate change and human health: infrastructure impacts to small remote communities in the North. International Journal of Circumpolar Health 64(5): 487-497.

Weller, G. 2005. Chapter 18: Summary and Synthesis of the ACIA, In: Arctic Climate Impact Assessment. Cambridge University Press, Cambridge, UK.

Wigle, D., A. Gilman, K. McAllister and T. Gibbons. 2005. Analysis of Arctic Children and Youth Health Indicators. Report prepared for the Arctic Council's Sustainable Development Working Group. http://www.sdwg.org/ (accessed August 20, 2006).

World Health Organization (WHO). 1967. The Constitution of the World Health Organization. WHO Chronicle 1: 29.

37

CHAPTER

Assessment of Human Health Vulnerability in Cuba due to Climate or Weather Variability and Change

Paulo Lázaro Ortíz Bultó[1], Luis Lecha Estela[2], Alina Rivero Valencia[1] and Antonio Pérez Rodríguez[3]
[1]Climate Center, Institute of Meteorology, Havana, Cuba
[2]Centre for Environmental Research and Services of Villa Clara, Cuba
[3]Tropical Medicine Institute "Pedro Kourí" (IPK), Havana, Cuba

The purpose of this chapter is to describe the main advances and results of research made in the development of specific methodological approaches to analyzing weather, adaptation to climate change and climate variability in a human context during the past decade. This chapter also discusses the main results of a Health Watch and Warning and Bioclimatic Prediction System developed for Cuba. Using scenarios of climate change for different diseases studied, this model may be used to estimate the impact of climate variability and change. Finally, this chapter measures vulnerability to adaptations and the expected cost of different adaptations to the impact of climate change.

INTRODUCTION

It is commonly accepted that climate and weather play a significant role in the behaviour of many diseases, some of which are among the most important causes of morbidity and mortality in Cuba. Often, these diseases strike in the form of epidemics, which may be triggered by variations in climactic conditions that make higher transmission easier (Kuhn et al. 2005).

In recent years, the complex relationships between climate, weather and human health have received much scholarly attention. However, the majority of these studies do not include the integrated effects and relationships between climate or weather conditions and disease behaviour (including vector transmission, ecological, social and epidemiological environment, etc.). There is a complex interaction between climate or weather variability, vectors, pathogens and human health, which cannot be reduced to a simple lineal relationship between meteorological elements, climate indexes and disease rates.

Concern about the potential impact climate change could have on human health began in the mid-1980s, with indications that greenhouse gases produced by human activities could influence the world's climate, resulting in the intensification of the greenhouse effect. Given the clear evidence that many health outcomes are highly sensitive to climate variations, it is inevitable that long-term climate change will have some effect on the human health. Climate variability and change will influence all natural, human, and socioeconomic systems, affecting health and many aspects of ecological and social systems. Climate conditions may create conditions that facilitate the occurrence of health crises and outbreaks of some infectious diseases (McMichael and Kovats 1999).

Climate change has been observed in all regions of not only the world, but Cuba also - seasonal changes in mean air temperature, precipitation patterns, and significant increases of extreme events are all present. The climate system's inertia will ensure that these trends continue. Even if every country immediately reduces greenhouse gas emissions, the concentration of these emissions will continue to increase for years. Climatic anomalies will also continue. It is, therefore, necessary to strengthen a coordinated framework in order to enhance international cooperation and continuity and to mitigate the most negative impacts of climate change.

One very important aspect not always considered in current investigations is the impact weather extremes has on human health. Therefore, those weather contrasts that can produce massive meteorotropic reactions must be identified. The native population's vulnerability to abrupt weather changes and the effects of extreme hidrometeorological events (such as hurricanes, floods, tornadoes and heat/cold waves) need to be studied, as does the limits of the native population's ability to adapt to climate and weather variability. At this moment, the medical community recognizes that climate and/or weather factors have a significant impact on human health and well-being. It is now necessary, however, to organize and promote more demonstrative case studies.

The potential impact of climatic change on human health is a function of a given population's exposure to changing climate conditions and its sensitivity to these changes. This sensitivity, in turn, depends on social, economic, institutional, and demographic conditions as well as other factors. These issues must be considered in order to properly understand the main effects climate change has on human health on both regional and local levels. This information, however, is not always available. Access to it, however, allows the medical community to effectively formulate adaptative procedures for the health sector. The meteorological community can contribute to diversifying and increasing society's knowledge of the impact climate change has, and the danger that it represents. This new evidence will contribute to the development of public awareness of the need for implementing the Kyoto Protocol.

The potential effects of climate or weather variability and change on population health in Cuba are discussed in this chapter. The main climatological characteristics and weather types of Cuba are described, as well as the main patterns of climate and weather-sensitive diseases that are of primary concern. Analyses of the associations between climatological anomalies, weather contrasts and disease patterns highlight the population's current vulnerability to diseases linked to climate and/or weather variability. This chapter also describes observations of a climatic trend and the most accepted effects of "El Niño" events and weather contrasts on human health.

The main effects of climate or weather variability on the health of the Cuban population are described on the basis of current non-adaptative responses, from the daily occurrence of massive health crises to seasonal disease outbreaks. This chapter also uses several examples to highlight the role bioclimactic seasonal predictions and daily biometeorological forecasts play in preventing the occurrence of disease in order to explain the structure and organization of preventive medical procedures developed to minimize the impact climate change has on human health.

Cuban research advances in the health sector made during the past decade have examined the relationship between health and climate, directly focusing on the relationship between health emergencies and daily weather changes, and considering the relationship between seasonal outbreaks of diseases and current seasonal climate variability. The information obtained through this research shows the non-adaptive, or meteorotropic, reactions of the native population to the impact of climate and weather. The new information obtained can forecast, and potentially minimize, the population's future vulnerability.

A brief analysis of the epidemiology of infectious and non-transmissible diseases is included in this chapter, as are general considerations on climate and its relationship to human health. This chapter also explains the application of complex indexes, methods and tools developed to evaluate the impact of climate variability on the health sector. The fundamentals required to assess vulnerability and the adaptation of the health sector is also discussed. The chapter also discusses issues related to the process of conducting an assessment in health sector. Finally, it presents ways to address the health-related risks of climate change.

Most climate change assessments should start by determining how populations currently cope with climate variability, particularly with weather extremes such as floods, droughts, hurricanes and the typical heat waves of temperate regions. This in turn will indicate where additional interventions are needed. Improving the capacity to cope with current climate variability will likely increase the population's resilience to climate change. The final section of this paper introduces an assessment of this problem in small island states.

Besides describing a model for considering how climate change affects human health, this chapter also describes the model's application and the different steps taken during the study of the vulnerability and adaptation assessment of Cuban samples. It also discusses a study case used to model Anomaly Variability and Climate Change on the Human Health-Assessment Risk Epidemic and Costs Estimate (MACVAH/AREEC).

The potential economic costs associated with the future impacts of climate change or strong weather contrasts were also estimated. The results derived from this study can be useful for the medical development of appropriate and more effective adaptative responses for the native population and for the implementation of new preventive health procedures to minimize the expected impacts of climate or weather variability.

THE GENERAL CHARACTERISTICS OF THE CUBAN CLIMATE

The Cuban archipelago is constituted by the island of Cuba, Youth Island and 1600 other small islands and keys. Cuba's climate results from its location in the northern portion of the tropics, near the Tropic of Cancer. The climate changes little over the year. Cuba's climate is tropical and seasonally wet, with marine influence and semi-continental features. The months from May to October are generally hot and rainy. The dry season runs from November to April. Winter is characterized by lower air temperatures and less precipitation. The rainfall amount depends on the strengthening or weakening of the North Atlantic subtropical anticyclone (Lecha et al. 1994). The most important weather changes are linked to the presence of disturbances in tropical circulation (tropical waves and hurricanes). Tropical cyclones may contribute significantly to total rainfall. In the winter, drought conditions can be severe in almost all parts of Cuba, particularly in the eastern region. Drought reduces the water available for washing and sanitation, thus increasing the risk of disease.

Climatic Trends in Cuba

From the mid-1950s until the present, the mean ambient temperature in Cuba has increased between 0.4 and 0.6°C. Minimum mean temperatures have increased approximately 1.5°C, while the maximum temperature has remained almost constant. These warmer temperatures are associated with an increase in winter precipitation and a decrease in summer precipitation. The increase in winter precipitation can be linked to an increase in the frequency of extreme extratropical events, particularly after the 1970s. Table 1 summarizes the main climate trends observed in Cuba during the 1990s.

Table 1 Main climate trends observed in Cuba during the 1990s

Increase in mean environmental air temperature, primarily due to increases in minimum temperature (1.4°C)
Decrease in diurnal variation temperature (Oscillation) (2°C)
Increase in precipitation in the dry season and decrease in the wet season
Later start of the wet and dry seasons, and a lag in the summer precipitation
Increase in extreme weather events: e.g. droughts, floods, and other dangerous meteorological events
Stronger hurricane seasons
More frequent extreme events, e.g. ENSO [**warm events** (1991-1993, 1994-1995, 1997-1998, 2002-2003) and **cold events** (1994, 1996, 1998-1999, 1999-2000)]

The Effects of "El Niño" Events in Cuba

The frequency of climate anomalies has increased in the past few decades. The National Climate Center at the Meteorological Institute has a prediction model for the Multivariate ENSO Index (PMEI) (Ortíz and Rivero 2003b). Designed by Ortíz, this model obtains very good results, forecasting the occurrence of "El Niño" or "La Niña" events three months in advance. Positive values are associated with warmer events and negative values are associated with cold events. Winter trend anomalies in the 1980s and 1990s (Ortíz et al. 2006) are shown in Figs. 1 and 2.

The climate or weather variability can be expressed in various temporal scales (WHO 2003). The "El Niño" Southern Oscillation (ENSO) has been identified as a significant element, since it has contributed to the rise in climate and weather variability in Cuba. ENSO events cause significant

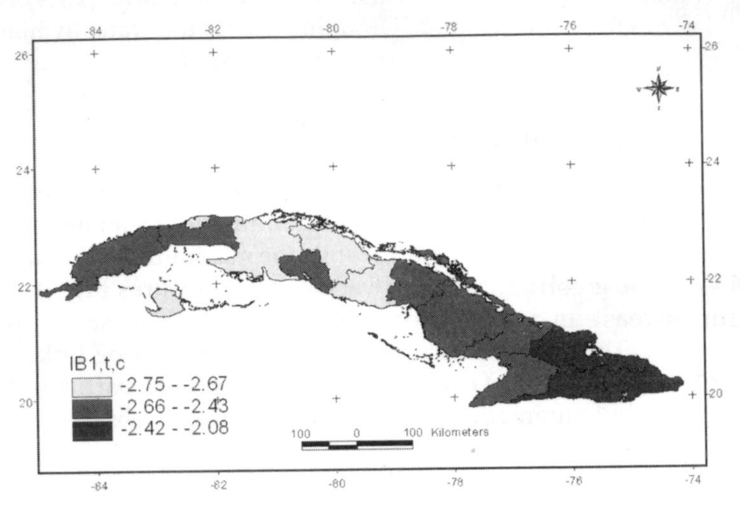

Fig. 1 Winter trend anomalies in the 1980s using the $IB_{1,t,c}$ index
Colour image of this figure appears in the colour plate section at the end of the book.

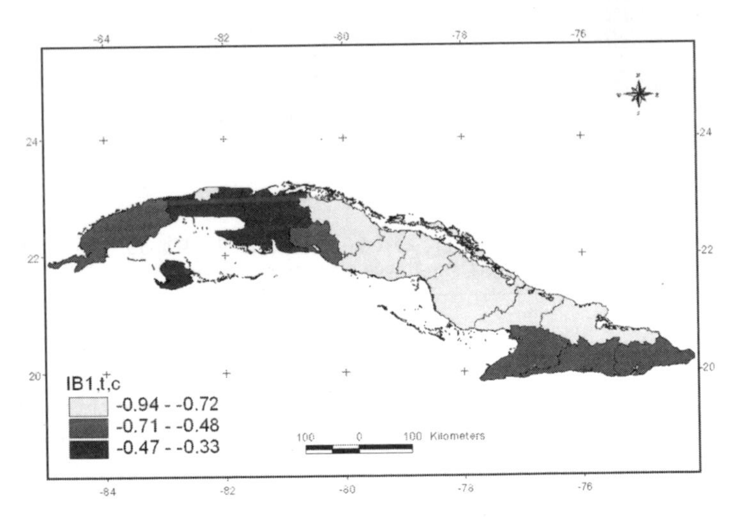

Fig. 2 Winter trend anomalies in the 1990s using the $IB_{1,t,c}$ index
Colour image of this figure appears in the colour plate section at the end of the book.

anomalies in the behaviour of atmospheric circulation patterns, resulting in positive rainfall anomalies and higher than normal minimum temperatures during the winter months (the dry season), and severe weather events of increased frequency (Cárdenas 1998). These events also cause a broad spectrum of contrasting daily weather types, which result in strong meteorotropic effects on the native population (Lecha 1993, 1998).

Many regions can be affected by an increase in vector density and transmission potential when rainfall increases (WHO 2003). Ecosystem impacts are significant, resulting in a high level of *Aedes aegypti*. Temperature also affects the behaviour of both vectors and humans, increasing the probability of transmission. Increases in temperature decrease the incubation period of disease-bearing mosquitoes, for example. The number of cases of diarrhoea are also considerably increased by poor sanitary conditions.

Climate and Weather-sensitive Diseases in Cuba: Epidemiological Behaviour

The Current Situation for Infectious Diseases

Public health is a high priority in Cuba. Reliable disease surveillance began in 1960. In 1997, the most prevalent diseases were Acute Respiratory Injections (ARIs), Acute Diarrhoeal Diseases (ADDs), bronchial asthma (BA), Viral Hepatitis (VH), and chicken pox (V); rates were 43,905.4; 8,996.9; 8200; 238.5; and 222.9 per 100,000, respectively. The prevalence of BA was 8.6% in urban areas and 7.5% in the rural zones. Other diseases of importance included gonorrhoea and syphilis, with rates of 304.3 and 142.2 per 100,000. There were low rates of meningococcal (3.05 per 100,000), bacterial (9.12 per 100,000), and viral meningitis (26.4 per 100,000). There were no reported cases of poliomyelitis, diphtheria, whooping cough, measles, rubella, mumps, or neonatal tetanus as a result of vaccination programs carried out since the early sixties. In 1998, the Health National System reported that there were 1,783 medical care institutions that offered medical assistance to 100 per cent of the population (Gutierrez 1998).

Dengue fever was first identified in Cuba in 1943, although it may have caused an epidemic in 1902. In 1977, dengue serotype 1 (DEN 1) was introduced and quickly spread throughout the country. During the resulting epidemic, which lasted until 1978, 553,132 cases were reported. The first great epidemic of dengue hemorrhagic fever (DHF) in the Western Hemisphere occurred in Cuba in 1981, with 344,203 cases of dengue fever, 10,312 of DHF, and 158 deaths (Guzman et al. 1990; Kourí et al. 1989, 1997). Dengue serotype 2 (DEN 2) was the causative agent. In response, a vector control program was initiated with support from all levels, including direct actions by the government. In addition, a surveillance program was implemented, including the establishment and improvement of diagnostic laboratories. These programs had very good results, with no autochthonous cases reported until 1997. With the exception of the capital (Havana, population of 2.25 million) and the cities of Santiago de Cuba and Guantánamo, Cuba was declared free of *Aedes aegypti*. Since the initiation of vector-control programs, *Aedes aegypti* has been detected, and quickly eliminated, particularly along the highway that unites these cities. The main difficulties in Havana have been the size of the city and the heterogeneity of the population. In Santiago and Guantánamo, the primary problem has been the lack of a constant supply of drinking water, which compels the population to store water in containers that serve as breeding sites for the vector. As a result, these cities have experienced epidemics in recent years. In 1997, Santiago de Cuba was affected by an epidemic in which 17,114 clinical cases were reported, of which 205 were DHF. There were 12 deaths (Kourí et al. 1997). The next epidemic was in 2000 in Havana, when there were 138 cases of dengue. Another epidemic occurred in 2001-2002 in Havana, with nearly 12,000 cases.

ARI, ADD, bacterial meningitis (BM), viral meningitis (VM), dengue fever (DF), and BA were included in the vulnerability assessment because these diseases are known to be climate sensitive and because they have relatively high burdens of disease.

The Current Situation for Non-transmissible Diseases

Table 2 describes the main causes of mortality in Cuba in 2005.

Table 2 Main causes of mortality in Cuba during 2005.

	Deaths	*Mortality rate (100,000 inhabitants)*
Heart diseases	22,223	197.4
Cancer	18,959	168.4
Vascular brain diseases	8,787	78.1
Accidents	4,447	39.5

Other important causes of mortality were chronic respiratory diseases, reporting 3,004 deaths (rate 28.6), Diabetes Mellitus, reporting 1981 deaths (rate 17.6) and hepatic cirrhosis, reporting 1,153 deaths (rate 10.2). It is important to note that the epidemiological behaviour on mortality and morbidity is similar to developed countries.

General Aspect to Focus the Studies of Climate Change and Health

Human health is an integrating theme of climate variability and change. Population health is affected by climate, particularly by climatic effects acting through natural disasters, climate-sensitive diseases. Climate also affects different sectors, (i.e., agriculture, water resources and others) with an indirect impact on Human Health. This results in spatial and temporal changes in a huge variety of health risks, from heat waves to floods and landslides, to malaria and malnutrition, and more indirectly through disruption to human societies, employment and livelihoods. Health is therefore both a key climate-sensitive sector in its own right, and also provides an important justification for addressing climatic impacts on other sectors (Fig. 3).

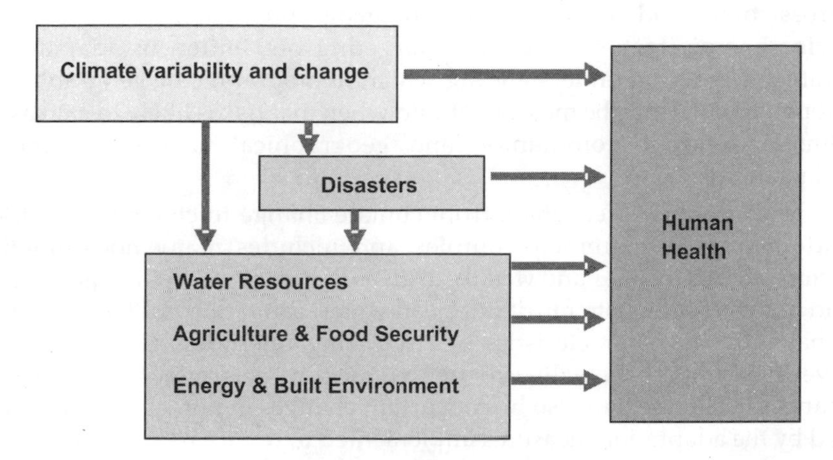

Fig. 3 Health as an integrating issue in climate variability and climate change (Corvalán 2006).

Realistically assessing the potential health impact of climate variability and change requires understanding both the vulnerability of a population and its capacity to respond to new conditions (Ebis 2006). The vulnerability of human health to climate change is a function of:

- **Sensitivity,** which includes the extent to which health or the natural or social systems on which health outcomes depend are sensitive to changes in weather and climate (the exposure-response relationship) and the characteristics of the population, such as the level of development and its demographic structure;
- **Exposure** to weather or climate-related hazards, including the character, magnitude, and rate of climate variation;

- **Adaptation measures** and actions in place to reduce the burden of a specific adverse health outcome (the adaptation baseline), the effectiveness of which determines, in part, the exposure-response relationship.

Populations, subgroups, and systems that cannot or will not adapt are more vulnerable, as are those that are more susceptible to weather and climate change. Understanding a population's capacity to adapt to new climate conditions is crucial to realistically assessing the potential health effects of climate change. In general, the vulnerability of a population to a health risk depends on factors such as population density, level of economic development, food availability, income level and distribution, local environmental conditions, health status, and the quality and availability of health care.

These factors are not uniformly distributed across a region or country or across time and differ based on geography, demography, and socioeconomic factors. Effectively targeting prevention or adaptation strategies requires understanding which demographic or geographical subpopulations may be most at risk and when that risk is likely to increase. Thus, individual, community, and geographical factors determine vulnerability.

The cause-and-effect chain from climate change to changing disease patterns can be extremely complex and includes many non-climatic factors such as income and wealth, distribution, provision of medical care, and access to adequate nutrition, clean water, sanitation epidemiological behaviour, and characteristics of circulating microbes. Therefore, the severity of impacts actually experienced shall be determined not only by changes in climate, but also by concurrent changes in non-climatic factors and by the adaptation measures implemented to reduce negative impacts.

Steps in Vulnerability and Adaptation Assessment

The book "Methods of Assessing Human Health" (Ebis 2006) proposes seven steps towards developing an assessment of impact. The steps are as follows:

1. Determine the scope of the assessment;
2. Describe the current distribution and burden of climate-sensitive health determinants and outcomes;
3. Identify and describe current strategies, policies, and measures designed to reduce the burden of climate-sensitive health determinants and outcomes;
4. Review the health implications of the potential impacts of climate variability and change in other sectors;

5. Estimate the future potential health impacts using scenarios of future changes in climate, socioeconomic, and other factors;
6. Synthesize the results;
7. Identify additional adaptation policies and measures to reduce potential negative health impacts.

Key issues for ensuring that an assessment is informative, timely, and useful include stakeholder involvement, an adequate management structure, and a communication strategy.

Three broad categories of health impacts are associated with climatic conditions: impacts that are directly related to weather/climate; impacts that result from environmental changes that occur in response to climatic change; and impacts resulting from consequences of climate-induced economic dislocation, environmental decline, and conflict (McMichael et al. 2001).

Changes in the frequency and intensity of heat events and extreme rainfall events (i.e. floods and droughts) will directly affect population health. Indirect impacts will occur through changes in the range and intensity of infectious diseases and food- and water-borne diseases, as well as changes in the prevalence of diseases associated with air pollutants and aeroallergens.

Climate and Weather Assessment on Health in Multiples Scales

Local, regional (national), and global scales are interconnected in supporting and facilitating action on climate change. Data from multiple scales and researchers are needed in order to understand the complex relationships between climate, weather and human health.

The Data. Innovative approaches to health and climate assessment are needed and should consider the role of socio-cultural diversity present among countries. This requires both qualitative and quantitative data, and the collection of long-term data sets on standard health outcomes at comparable temporal and spatial scales. The data favours improve on forcasting models and other application in the health sector.

Researchers. Researchers are needed to evaluate community-based assessments and systematic responses to the issues the impact of climate change will raise in all countries and regions.

Climate variability may influence human health through three interconnected ways: distribution and quality of water, life cycle of disease vectors and host/vector relationships, and ecosystem dynamics of predator/prey relationships.

Some cases with high priority diseases identified in the small island states

1. Diseases identified: malaria, dengue, diarrhoea disease/typhoid, heat stress, skin diseases, acute respiratory infections, viral hepatitis, varicella (chicken pox), meningococcal disease and asthma, toxins in fish and malnutrition.
2. The possibility of dust-associated diseases with the annual atmospheric transport of African dust across the Atlantic is unique to the Caribbean islands.
3. In addition to weather and climate factors, consideration of social aspects, such as culture and traditions, are important factors in disease prevalence.

Methods and Tools for the Studies of Climate Variability and Change in the Health Sector

Research methods used so far include predictive modelling, analogue methods and early effects. Predictive models include biological models (e.g. malaria), empirical statistical models (e.g., temperature-mortality relationships), the use of the complex index simulation variability climate change and other processes (e.g., relationship climate index and diseases) and integrated assessment (IA) models. A balance between empirical analysis and scenario-based methods is needed to integrate the different methods through, for example, IA methods. The outcome of an assessment may not necessarily be quantitative for it to be useful to stakeholders.

A variety of methods and tools are available to assess climate change vulnerability in the health sector; few, however, are available on CD-ROM or downloadable from websites. Both quantitative and qualitative approaches have been taken within national assessments of the potential health impacts of climate change. The three key issues to be addressed are: (1) estimating the current distribution and burden of climate sensitive diseases, (2) estimating the future health impacts attributable to climate change, and (3) identifying current and future adaptation options to reduce the burden of disease. Guidance and direction for each issue is discussed briefly.

Disease-specific Models

Predictive models of the health impacts of climate change use different approaches to classify the risk of climate-sensitive diseases. For malaria, results from predictive models are commonly presented as maps of potential shifts in distribution attributed to climate change. The models are typically based on climatic constraints on the development of the vector and parasite; they produce maps that identify potential geographic areas of risk, but do not provide information on the number of people who may be at risk within these areas. Few predictive models incorporate adequate assumptions about other determinants of the range and incidence of disease, such as land-use change or prevalence of drug resistance for malaria, or adaptive capacity.

In Cuba's case, scenarios of regional climate change and other changes are used as inputs into a model on climate and health. The model's spatial combination is combined with a Generalized Autoregressive Conditional Heteroscedasticity (GARCH) model, with exogenous variables for a model on climate and health. The Cuban model is called MACVAH/ AREEC.

MACVAH/AREEC Models

Model to Anomaly Variability and Climate Change on the Human Health-Assessment Risk Epidemic and Costs Estimate Models (MACVAH/ AREEC): This model describes anomaly variability and climate change and its impact on human health. When scenarios of climate change and health models form the input, the model outputs proposals for diseases, generating maps of risks for epidemics in Cuba using GIS. Finally, the model estimates the impact of costs to variability and change. The spatial correlation explains each disease's capacity to disseminate, while the range of the correlation describes the epidemic's trend.

The climate and health model was based on Spatial Models combined with Generalized Autoregressive Conditional Heteroscedasticity (GARCH) models with a dummy variable. In this case, the socio-economic and climate index complex is used as a dummy variable in the models.

Finally, the parameter estimates for each particular model calculates the impacts according to the following expression:

$$I_1 = \frac{C_0}{1 - \sum_{i=1}^{k} a_i} \quad (1)$$

$$I_2 = \frac{C_1}{1 - \sum_{i=1}^{k} a_i} \quad (2)$$

$$I_m = \frac{C_0 + C_1}{1 - \sum_{i=1}^{k} a_i} \quad (3)$$

where $I_{1:}$ I_2 and I_m are the long-run effect of climate change on the given diseases and C_0 is the coefficient that describes the magnitude, the signal climate change on diseases is C_1, This value coefficient describes the effect economic condition with change on diseases and the I_m expresses the combination of impact from climate change and economic conditions, described by the values C_0 and C_1.

The models consider the incidence of acute respiratory infections, acute diarrhoea disease, viral hepatitis, varicella (chicken pox), meningococcal disease, streptococcal pneumonia, viral meningitis, malaria and dengue. Ecological data include the larval density and biting density per hour of mosquito vectors, as well as the number of houses where larval activity was observed. Socioeconomic data include the percentage of houses without potable water, the percentage of houses with dirt floors, the adult (age 16 and above), illiteracy rate, monthly birth rates and a monthly index based on the number of houses where a focus of *Aedes aegypti* mosquitoes was observed.

These complex indexes (Ortiz 1998, 2004, 2006) were applied to characterize climatic variation by regions. They were used to build maps of climatic risk across the country and to determine periods of high risk for various diseases. This model describes future changes to disease patterns, according to the scenarios and information available in the studies. As one measure of the uncertainties, the equation of conditional variance of stochastical factor in each specific model was considered.

The MACVAH/AREEC model was used in the country studies of Panama, the Dominican Republic, Bolivia and Paraguay with satisfactory results.

Other aspects to consider in studies of the effects of climate variability and change on health are the limitations and sources of uncertainty, which can be due to problems with data, models, unknown relationships between climate and health, as well as other sources of uncertainty, such as ambiguously defined concepts or terms and inappropriate spatial or temporal units, among others.

Advantages of the Models

The models include different signals of disease variability and climate complex indexes that describe the interaction between both groups of processes. Projection case number of diseases, output maps of futures transmission potential (risk maps) and the models can be applied to air-borne diseases, water-food borne diseases and vector-borne diseases; although these models require spatial variability.

Limitation of the Models

The major limitation of this model was the need for information to estimate the parameters and the required interdisciplinary interpretation of the results.

Vulnerability and Adaptation

Assessing the potential health impacts of climate variability and change requires understanding both the vulnerability of a population and its capacity to respond to new conditions.

Adaptation includes the strategies, policies, and measures (hereafter referred to as adaptation options) undertaken now and in the future to reduce the burden of climate-sensitive health determinants and outcomes. Adaptation can be anticipatory (actions taken in advance of climate change) or responsive, and can encompass both spontaneous responses to climate variability and change by affected individuals and planned responses by governments or other institutions (Smith et al. 2001). An example of a public health adaptation is an early warning system for extreme events.

The primary goal of building adaptive capacity is to reduce future vulnerability to climate variability and change. Increasing the adaptive capacity of a population shares similar goals with sustainable development: both increase the ability of countries, communities, and individuals to effectively and efficiently cope with the challenges of climate change (Ebis 2006).

Estimating the Current Distribution and Burden of Climate and Weather-sensitive Diseases

During the early 1990s significant biometeorological research with non-transmissible diseases were carried out in Cuba. This research was entered into a large database with daily biometeorological information from selected Cuban meteorological stations related to the daily occurrence of five non-transmissible diseases (cardiovascular, bronchial asthma, high blood tension, headaches and head strokes).

The Cuban health system has a complete organization to collect and evaluate the daily occurrence of significant diseases. With the aid of this infrastructure, all of the cases of these non-transmissible diseases reported at Emergency Rooms were included. Daily information was also collected from selected family doctor's consults, taking into account the previous classification and monitoring of patients.

Estimating the Future Potential Health Impacts Attributable to Climate Change

Estimates of possible future health impacts of climate change must be based on an understanding of the current burden and recent trends in the incidence and prevalence of climate-sensitive diseases, and of the associations between weather/climate and the health outcomes of concern. In most countries, the ministry of health, hospitals, and similar sources can provide data on disease incidence and prevalence on scales needed for analysis. These sources can also provide information on whether or not current health services are satisfying demand. The current associations between climate and disease need to be described in ways that can be linked with climate change projections. Adverse health outcomes associated with interannual climate variability such as El Niño could also be considered (Kovats et al., 2003b).

Once the current burden of disease is described, models of climate change or qualitative expert judgments on plausible changes in temperature and precipitation over a particular time period can be used to estimate future impacts. Health models can be complex spatial models or can be based on a simple relationship between exposure and response. Models of climate change should include projections of how other relevant factors could change in the future, such as population growth, income, fuel consumption, and other relevant factors. Projections from models developed for other sectors can be incorporated, such as projections for flood risk, changes in food supply, and land-use changes.

The exercise of attributing a portion of a disease burden to climate change is in its early infancy. Analysis should consider both the limits of epidemiologic evidence and the ability of the model to incorporate the non-climatic factors that also determine a health outcome. For example, the portion of deaths due to natural climatic disasters that can be attributed to climate change shall reflect the degree to which the events can be related to climate change. For vector-borne diseases, other factors such as population growth and land use may be more important drivers of disease incidence than climate change.

Three sets of approaches are described: (1) comparative risk assessment, (2) disease-specific models, and (3) qualitative assessment.

Qualitative Assessment

Potential future health risks of climate change can be estimated from knowledge of the current burden of climate-sensitive diseases, the extent of control of those diseases, and how temperature and precipitation can affect the range and intensity of disease. For example, is highland malaria a current problem? What is the extent of that problem? How well is the disease controlled during epidemics? How could the burden of disease be affected if temperature increased so that the vector moved up the highlands? Similarly, future risks can be estimated from relationships used in the WHO Global Burden of Disease project.

Identifying Current and Future Adaptation Options to Reduce the Burden of Disease

Adaptation includes the strategies, policies, and measures undertaken now and in the future to reduce potential adverse health effects. Individuals, communities, and regional and national agencies and organizations will need to adapt to health impacts related to climate change (Adger et al. 2005). At each level, options will range from incremental changes in current activities and interventions, to translation of interventions from other countries/regions to address changes in the geographic range of diseases, to development of new interventions to address new disease threats. The degree of response shall depend on factors such as who is expected to take action; the current burden of climate-sensitive diseases; the effectiveness of current interventions to protect the population from weather- and climate-related hazards; projections of where, when, and how the burden of disease could change as the climate changes (including changes in climate variability); the feasibility of implementing additional cost-effective interventions; other stressors that could increase or decrease resilience to impacts; and the social, economic, and political context within which interventions are implemented (Yohe and Ebi 2005; Ebi and Burton, submitted).

Because climate change shall continue for the foreseeable future and because adaptation to these changes shall be an ongoing process, active management of the risks and benefits of climate change need to be incorporated into the design, implementation, and evaluation of disease control strategies and policies across the institutions and agencies responsible for maintaining and improving population health. In addition, understanding the possible impacts of climate change in other sectors could help decision-makers identify situations where impacts in another sectors, such as water or agriculture, could adversely affect population health.

For each health outcome, the activities and measures that institutions, communities, and individuals currently undertake to reduce the burden of disease can be identified from (1) review of the literature; (2) information available from international and regional agencies (WHO, the Pan American Health Organization, UNEP, and others) and national health and social welfare authorities (ministries of health); and (3) consultations with other agencies and experts that deal with the impacts of the health outcome of concern. Ideally, the effectiveness of adaptation measures should be evaluated.

Many of the possible measures for adapting to climate change lie primarily outside the direct control of the health sector. They are rooted in areas such as sanitation and water supply, education, agriculture, trade, tourism, transport, development, and housing. Inter-sectorial and cross-sectorial adaptation strategies are needed to reduce the potential health impacts of climate change.

Relationship between Weather Variability and Health Outcomes

During the early 1990s significant biometeorological research with non-transmissible diseases were carried out in Cuba. This research was entered into a large database with daily biometeorological information from selected Cuban meteorological stations related to the daily occurrence of five non-transmissible diseases (cardiovascular, bronchial asthma, high blood tension, headaches and head strokes).

The Cuban health system has a complete organization to collect and evaluate the daily occurrence of significant diseases. With the aid of this infrastructure, all of the cases of these non-transmissible diseases reported at Emergency Rooms were included. Daily information was also collected from selected family doctor's consults, taking into account the previous classification and monitoring of patients.

Between 1991 and 1995, the Cuban Institute of Meteorology's research project "Effects of weather and climate on human health under the conditions of the humid tropics" was developed. An essential part of the investigation was the demonstration of the catalytic influence of certain local weather types and types of synoptic situations (TSS) in the daily occurrence of health crises.

To evaluate the impact of weather on health, the daily Emergency Room reports of six chronic non-transmissible diseases were reported: cardiovascular diseases, arterial hypertension, acute vascular brain strokes, bronchial asthma in children and adults, migraines and some other types of respiratory diseases.

One of the main results obtained during the years 1996-1998 was the development of a Health Watch and Warning System (SAAS in Spanish), which works starting from a program of calculation of the interdaily (24 hours) contrasts of the partial density of the oxygen in the air (Lecha and Delgado 1996). This program was elaborated on the basis of a specific application of a Cuban Geographical Information System (TeleMap 1994).

However, in those years it was not possible to develop an objective forecast of the partial density of the oxygen content in the air. This is because it was not possible to access the original data source of the objective forecast models available. The baroclinic model available to the Cuban Meteorological Service was not even operational for terms longer than 48 hours. Until recently, this significant practical limitation affected the implementation of the operative biometeorological forecast.

At the present time (Moya and Estrada, personal communication) an operational model was developed to forecast the daily occurrence of precipitation for all of Cuba's territory. The model application runs on a program that is compatible with the database of the Global Forecast System (GFS) located on the internet (http://nomad5.ncep.noaa.gov/ncep_data/index.html). The predicted values of the variables that are necessary to make the calculations of the partial density of the oxygen content are taken from this site (Lecha 1996).

Through cooperative efforts, the feasibility for generating an operative output of this program for its biometeorological applications was evaluated. For data, version 2 of the SAAS program uses air temperature, atmospheric pressure and surface humidity from 12 to 168 hours in advance (one week). The model can be initialized every 12 hours.

The preliminary results are satisfactory. When the program is running on the internet, it is automatically compatible with GFS databases. It calculates the predicted values of the partial density of oxygen in the air and its differences for 24-hour terms, generating as output graphic sequences of 14 maps (model initialization every 12 hours) of the interdaily contrast of the partial density of oxygen in the air (Fig. 4).

The meteoro-pathological response of the local population depends on the magnitude and sign of the partial oxygen density difference in 24 hours. Remarkable increases are more related to bronchial asthma crises and heart diseases, while decreased results are associated with migraines, high blood pressure and brain-vascular diseases. The actual model output represents the reduction of the partial density of oxygen of the air (hypoxia) in a red and the increase (hyperoxia) in blue.

The model calculations are made with the same resolution of GFS database that would be implemented in all workspaces in WMO regions.

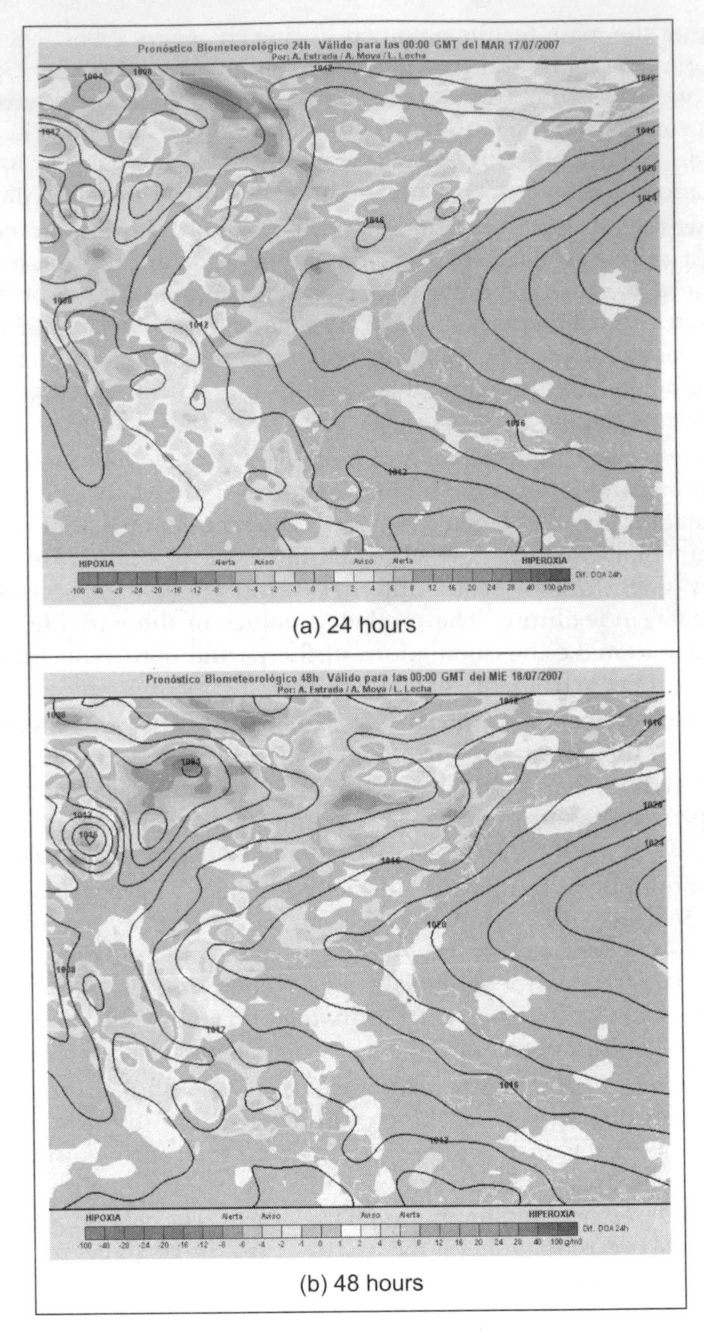

Fig. 4 Demonstrative output maps from the model SAAS version 2.0 initialized the day March 16, 2007 at 0000 GMT (19:00 hours, local time in Cuba).

Colour image of this figure appears in the colour plate section at the end of the book.

The Cuban Meteorological Service workspace is referred to from 10 to 50 degrees North latitude and from 50 to 105 degrees West longitude.

Recently, the Centre for Environmental Research and Services of Villa Clara and the Cuban Meteorological Service set up a Website in order to offer the operational service of biometeorological forecasts to a group of Cuban medical institutions on-line. The service is available at http:// pronbiomet.villaclara.cu

The medical and meteorological counterparts work in co-ordination in order to validate the daily results of the biometeorological forecasts and to guarantee the infrastructure needed to apply this information to the Emergency System. Moreover, the working team is involved in the development of organizational procedures and preventive actions for the population in order to diminish the potential health impacts associated with future significant biometeorological events.

Observed Climate and Health Relationships in Cuba

Given the clear evidence that many health outcomes are highly sensitive to climate variations, it is inevitable that long-term climate change will have some effect on the health of the global population. Climate variability and change will influence all natural, human, and socioeconomic systems, thus affecting not only health, but also many aspects of ecological and social systems. Climate is one factor that may create conditions that facilitate the development of some disease-causing microorganisms (McMichael and Kovats 1999)

Viral Hepatitis A

Viral hepatitis type A is a water-food borne disease. This virus has a high resistance to extreme environment conditions, contributing to viral persistence and the possibility of community spread (Piatkin and Krivochein 1981). In Cuba, viral hepatitis type A is seasonal, increasing from August to October during the baseline period of 1961-1990. However, with recent climate variability and change, winter seasons are warmer and rainier, resulting in the advance of peak transmission in the months from March to June of each year. These new seasonal conditions are shown by the range of values of indexes. $IB_{1,t,C}$ and $IB_{3,t,C}$; $IB_{1,t,C}$ is highly positive and values of $IB_{3,t,C}$ are moderately positive (Ortiz et al. 2006).

These climatic patterns favour contamination of drinking water due to the overflow or waste of black water, producing contaminated drinking-water wells, and a quick increase of some vectors, like flies and cockroaches, when poor sanitary conditions are combined warm and humid conditions; this is shown in Fig. 6. Similar behaviour and mechanisms are observed with diarrhoea diseases, although different specific agents are involved. Climate anomalies can increase the incidence of waterborne diseases, which are most likely to occur within communities that do to not have adequate drinking-water supplies and sanitation systems (WHO 1996).

Acute Respiratory Infections

During the 1961-1990 baseline, two peaks characterized ARIs in March and October (Fig. 5). Currently, as a consequence of increasing climate anomalies (e.g. drought and warmer winters), a new peak is now observed in June when the rainy season is delayed (Fig. 6). Low temperatures during the winter season and close contact between persons also may be possible causes of this increase. These changes are shown in the response of the combination of the climatic indexes $IB_{1,t,C}$ and $IB_{2,t,C}$, with a high range of $IB_{1,t,C}$ and a low range of $IB_{2,t,C}$, characterizing warmer and drier summer seasons.

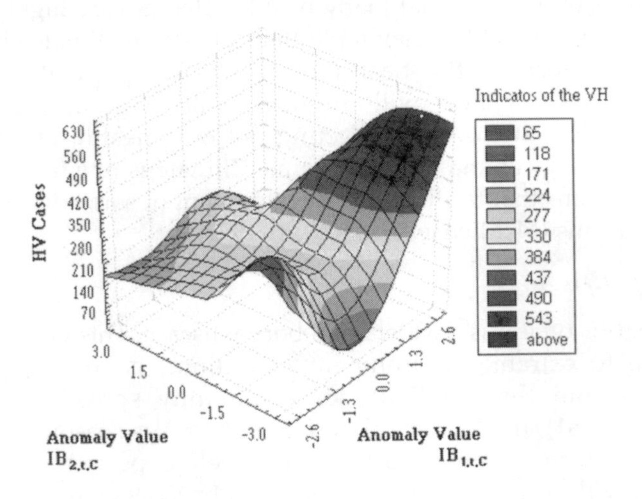

Fig. 5 Association between climate variability and viral hepatitis according to the indexes.

Colour image of this figure appears in the colour plate section at the end of the book.

There are multiple causative agents of ARIs, with the most frequent being those of viral origin. Droughts, cold winds, and abrupt temperature variation during the winter season, combined with an increase in dusty conditions, can infect the mucous membranes of the respiratory passages, which can facilitate contracting an ARIs (San Martín 1963). In addition, close personal contact during winter months can contribute to the spread of ARIs.

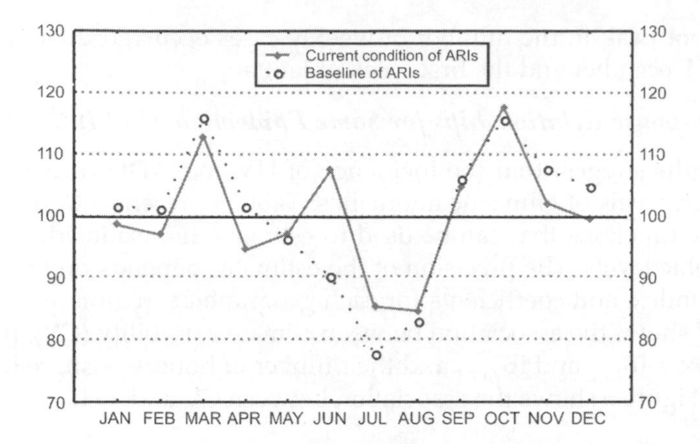

Fig. 6 Observed impact of CV on the seasonal pattern of ARIs

Chicken pox also is transmitted person-to-person. During the baseline period, the seasonal peak was in March (the end of winter). Currently, the peak is observed in April, a month characterized by high CV. High CV may result in infiltration of the upper respiratory tract, increasing viral transmission, particularly among infants and children. The climate patterns are characterized by a combination of moderate values of $IB_{1,t,C}$ with high values of $IB_{3,t,C}$ (dry and high contrasting conditions).

Meningitis Caused by Streptococcus pneumoniae

The main cause of bacterial meningitis in Cuba since 1999 has been *Streptococcus pneumoniae*. *Streptococcus pneumoniae* is a bacterial agent common in the upper respiratory tract, with CV apparently contributing to the infection, particularly in children under 5 years old and the elderly. The disease occurs most often between January and April, although there is a regional difference in the pattern of *Streptococcus pneumoniae*. The central region has higher solar radiation and more climate variability, which is shown by high values of $IB_{2,t,C}$ and low values of $IB_{3,t,C}$. This, combination of physical-geographical characteristics and socioeconomic conditions ($IB_{4,t,C}$) may explain the high incidence of diseases all year around.

Viruses and bacteria quickly mutate, thus allowing for environmental adaptation (McMichael and Kovats 1999). Climate variability and change may be additional stresses that increase mutation rates of different microorganisms, thus increasing emerging and re-emerging diseases.

Some authors (PAHO 1997) have suggested that pneumococcal infections might increase in the winter. In Cuba, the number of weekly cases increased approximately four-fold from summer to winter, and a prominent peak in the number of weekly cases occurred during the last week of December and the first week of January.

Dose-Response Relationships for Some Epidemiological Indicators

Our results suggest that the incidence of HV and ADDs are associated with high levels of climactic anomalies. Table 3 presents stratified dose-response functions that can be used to estimate disease incidence for all geographic levels. The precision of the estimates depends on the disease, climate index, and coefficients for each geographical region or local area. Figure 7 shows the association between climate variability (CV), based on the indexes $IB_{1,t,C}$ and $IB_{2,t,C}$, and the number of houses positive for *Aedes aegypti*. Figure 6 shows the association between CV and VH.

Table 3 Function dose-response. Impact of climate variability in some diseases

Diseases	*Impact level*	*Coefficient estimate for the function dose-response*
ADDs	High	= 1109 (CIB3) × susceptible population in the study region
	Means	= 458.9(CIB3) × susceptible population in the study region
	Low	= 311.8 (CIB3) × susceptible population in the study region
VH	High	= 31.42 (CIB3) × susceptible population in the study region
	Means	= 27.18 (CIB3) × susceptible population in the study region
	Low	= 18.77 (CIB3) × susceptible population in the study region

CIB3: values of the change of the one $IB_{t,3,C}$ according to ranges.

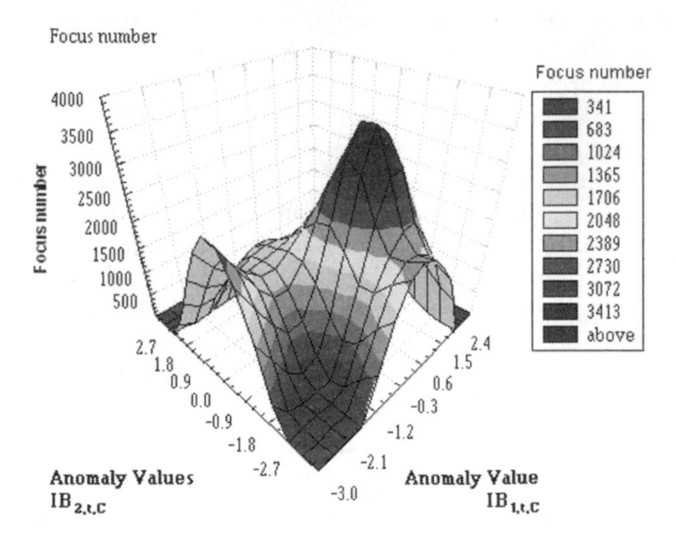

Fig. 7 Association between climate variability and the number of positive houses (hotspot) of the *Aedes aegypti* by climate variability according to indexes. *Colour image of this figure appears in the colour plate section at the end of the book.*

IMPACTS OF SCENARIOS OF CLIMATE VARIABILITY

To create scenarios of climate variability, Cuba was divided into three geographic regions: the western region (which includes the provinces of Pinar del Río, La Habana, Matanzas, and Juventud Island); the central region (which includes the provinces of Cienfuegos, Villa Clara, Sancti Spíritus, Ciego de Avila and Camaguey); and the eastern region (which includes the provinces of Las Tunas, Granma, Santiago de Cuba and Guantánamo); eighteen subregions, and eight zones according to the structure of the relief and the characteristics described by climate index ($IB_{1,t,c}$) where i = 1, 2, ...

Baseline data for 1961-1990 were analyzed by decade and compared with conditions from 1991 to 2000. This allowed identification of climate variability in different regions of the country. It also indicated that major variability was related to the anticyclone, and was found in the mean latitudes during the winter and, to a lesser degree, in mountainous regions and Youth Island. Less variability was found inland, in the eastern region, and along the southern coast (Ortíz and Rivero 2003b, Ortíz et al. 2004, 2006)

Analyses were carried out from the rainier period to the drier period using all possible combinations. The results were used to describe and quantify the magnitude of climate variability in space/time using complex climatic indexes. Climate variability was stratified on the basis of historic information on the effects of QBO and NAO, certain phases of which increase hot and dry weather during the warm season (Cardenas 1998, Enfield 1998). ENSO results in more warm and rainy conditions during the cold season (Ortíz and Rivero 2004). The different combinations of climate variability resulted in the following scenarios: one, positive values of NAO with MEI in the warm phase and West-East QBO; and two, negative values of NAO with the other parameters constant, e.g. dry season described by $IB_{1,t,c}$ values in Figs. 8 and 9. These figures show two extreme scenarios of variability with different levels of anomalies.

Using this type of analysis offers one tool for the development of surveillance systems to identify, control and/or adapt activities to reduce projected health impacts (Table 4).

Scenarios of Climate Change

Climate scenarios were based on the HadCM2 (Hadley Center Model) general circulation model using different concentrations of CO_2 (one and two times CO_2). The outputs were used to obtain climate variability rates (Mitchell et al. 1995) that were used as input to the Bultó indexes (Ortiz et al. 1998, Ortiz and Rivero 2004). Using the scenarios of climate variability, climate change scenarios generated to 2010, 2020 and 2030.

Table 4 Projected impacts in human health, period (2011-2021)

Diseases	*Projected impacts*	*Transmission way*
Bronchial asthma	Decrease of the number of cases in dry period.	**Air-borne diseases**
Acute respiratory infections	New epidemic outbreaks with picks in the rain period. Increase of the risk in adults, due to the demographic trends.	
Meningococcal diseases	Increase in the months of the dry period season and in the rain period from August to October	
Viral meningitis	Increase of the cases and variation of the trend, with increase in the months of the rain period. (with significant increase in the months Jun-Jul and Sep-Oct)	
Meningococcal meningitis	Change in the geographic distribution in the country. Variation of the circulating agents	
Bacterial meningitis	Small increment of the cases in the country.	
Varicella (chicken pox)	Advance in the season of appearance of the seasonal rise and continuation of a high incidence in the year.	
Viral hepatitis Type A	Increase in the months of the dry period.	**Water-food borne diseases**
Acute Diarrhoeal Diseases	Increase in the months of the dry period and displacement of the seasonal pick of May for Jul-Aug.	
Dengue	More frequency of epidemic outbreaks and changes in the space pattern and time (new affected regions and increase of the epidemic period).	**Vector borne diseases**
Malaria	Increase of the risk and probable reemergence of the diseases in the country. Spread to new areas and increase in the exposure period.	

Under both scenarios, the forecast climate conditions projected an increase of ARI and of ADD by oral/food transmission in 2015 (Figs. 10 and 11). A new outbreak of ARI was projected for June. An increase in incidence of ADD was projected in the first months of the year, with seasonal displacement from May to Jul-Aug. Climate conditions in the winter season were projected to be warmer and rainier, and the rainy season was projected to be drier and hotter, which may then influence the incidence of ADD.

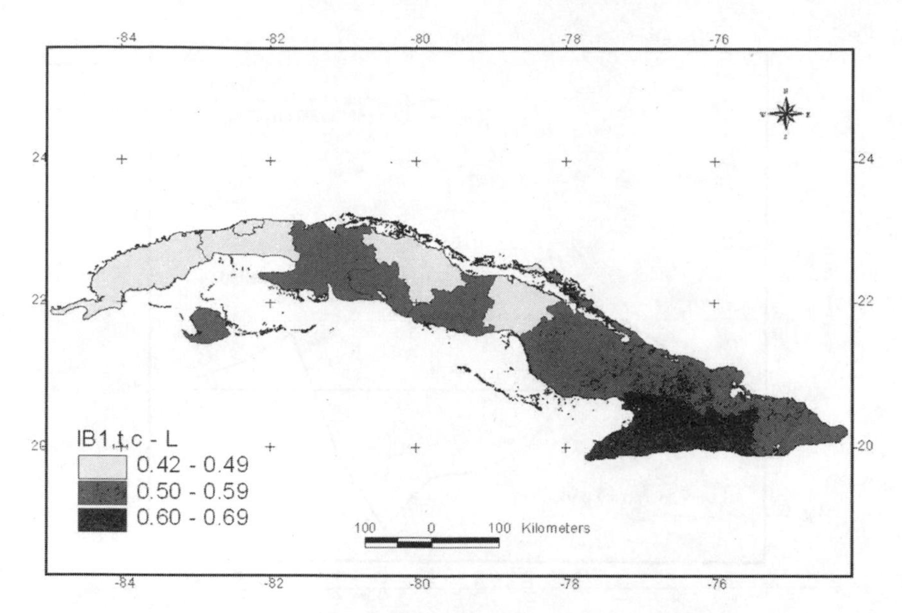

Fig. 8 Scenario of climate variability. Low sensibility range < 0.70 (change per decade)
Colour image of this figure appears in the colour plate section at the end of the book.

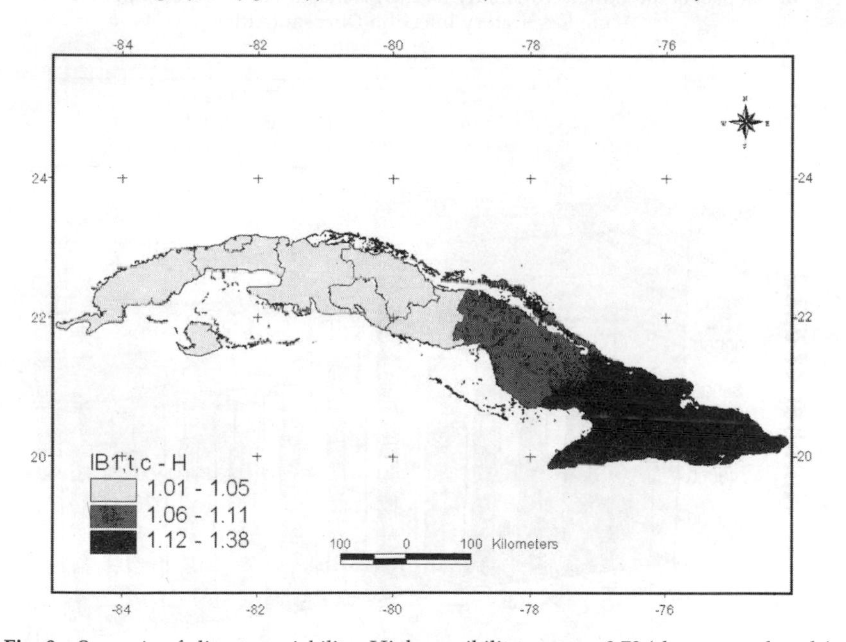

Fig. 9 Scenario of climate variability. High sensibility range > 0.70 (change per decade)
Colour image of this figure appears in the colour plate section at the end of the book.

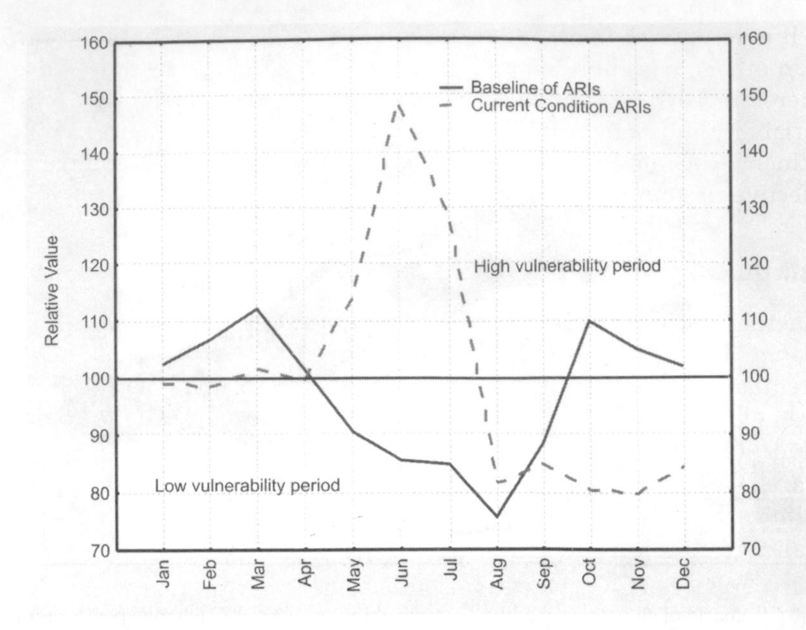

Fig. 10 Impact of the climate variability on the behaviour of the seasonal pattern of the Acute Respiratory Infection Diseases (ARIs)

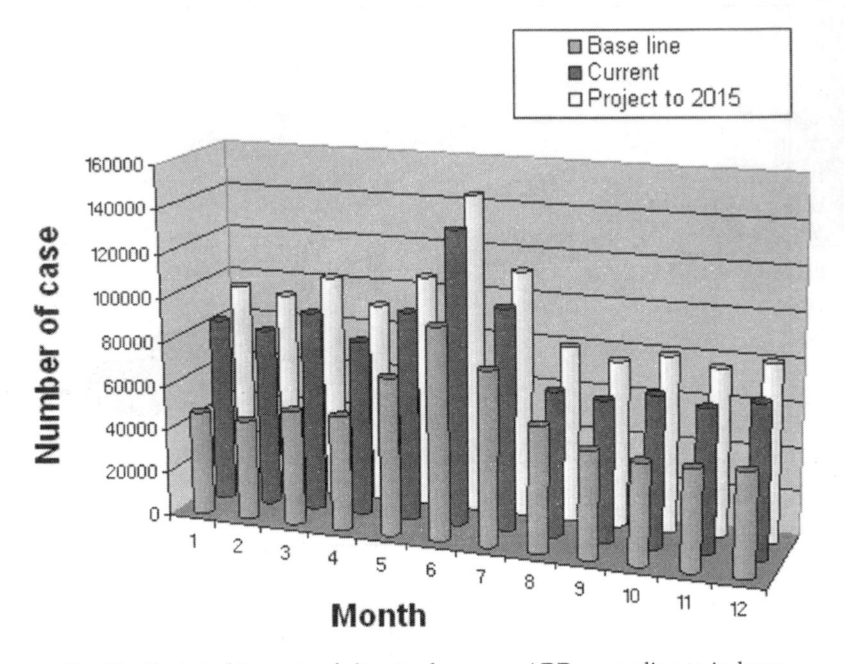

Fig. 11 Projected impacts of climate change on ADDs according to indexes

It is important to note that some Bultó indexes suggest more impact than others, based on the epidemiological characteristics of the disease. Therefore, each health outcome is likely to respond differently to climate variability and change. It is important to understand that many factors can influence the rate and intensity of these diseases, such as the complexity of effective community response.

Adaptation Measures in Cuba

Whether or not the projected health impacts of climate variability and change are actually experienced will depend on the measures used to attenuate or prevent these impacts. Adaptation includes the policies, strategies and measures designed and implemented to reduce potential adverse health effects. Increasing the adaptive capacity of a population shares similar goals with sustainable development: both increase the ability of individuals and communities to cope with changes and challenges (Burton and Marten 1999).

Experience in Cuba has shown that primary health care is the right level for the implementation of preventive measures to reduce population vulnerability, particularly when considering the multiple factors that are related to climate-sensitive diseases. In addition to strengthening these programs, it is important to strengthen the links between the health sector and other sectors.

In general, the vulnerability of a population to climate-related health risks depends on important aspects of the local environment. The level of material resources, the effectiveness of the government and civil institutions, the quality of the public health infrastructure and access to relevant local information on extreme weather threats (Haines and Patz 2004, Woodward et al. 1998) are essential to the development of effective adaptation responses to reduce current and future vulnerability in the community. It is necessary to identify and prioritize strategies, policies, and measures to address climate variability and change (Table 5).

The Importance of Forecasting as an Anticipatory (or Proactive) Adaptation Measure in the Human Health Sector

Projections of disease outbreaks afford decision-makers the opportunity to proactively initiate activities to reduce the impact of outbreaks. Recent advances in seasonal forecasting are generating new opportunities to minimize the impact of climate variability on health (WHO 2004). For this reason, using climactic indexes, along with forecasting models, can alert

Table 5 Some examples of adaptation measure to CV and climate change in Cuba

Adaptation options	*Current activities*	*Future activities*
To strengthen primary health care of the public health system.	Health promotion and preventive measures designed to reduce population vulnerability	Continue developing health promotion and preventive programs, increasing community participation on health. Increasing the participation of local governments and other sectors in developing the best conditions of life.
Measures to improve the health surveillance system.	Provide warning system and forecast of the main sensitive disease to all levels of the National Public Health System. Emission to Bulletin with areas of risk and vulnerability.	Incorporate new diseases and other risk factors (ecological and social) in the forecast models that contribute to diminish levels of uncertainty To forecast for different levels.
Immunization program for the groups of high risk.	Increase the use of vaccines against some community diseases.	Enhance vaccination programs for ARIs and haemophilic influenzae to achieve their successful control, etc.
Improvement of the sanitary conditions.	Increase of sanitary demands in all fields (communal, drinking water, garbage, sewage, foods and others) Maintain contingency plans	Continue developing program of education about environmental care with the participation of all social sectors and community
Stronger of the research working, education and capacity decision make and peoples in general	Research projects on climate and human health and their effects on diseases that climate-sensitive	Develop new projects including Sahara dust and vegetation index with participation from other countries that contributing understanding of different effect the weatherand climate in the diseases on region

authorities of possible changes in epidemiological behaviour, either immediately or in the near future (Ortiz and Rivero 2003a). Further, this approach can be used to project how changing weather patterns might alter the range and intensity of climate-sensitive diseases.

Other aspects to consider are that the predictions present different functions to decision-makers: they may act as an experiment and analysis tool a support tool for understanding, or an early warning system. An example of a public health adaptation is an early warning system for anomalous and extreme events.

Early warning systems based on climate forecasts and environmental observations (Fig. 12) illustrate the degree of certainty associated with data from prediction and surveillance activities. Climate forecasts and ongoing environmental observations can be combined with knowledge of disease aetiology to create disease early warning systems. Effective early warning systems can be used to inform surveillance systems to help reduce the impact of an epidemic and cope with variability and climate change.

Figures 13, 14 and 15 show the projection of climate indexes in time and spatial scales as well as the risk level to variability according to the $IB_{t,3,C}$. The temporal risks for each region of the country can be projected by linking disease incidence with demographic data and climate indexes. Decision-makers can use these results to plan anticipatory adaptation (proactive adaptation) measures, such as early warning systems (WHO 2003, 2004).

For example, under some climatic conditions, an increase in ADDs, ARIs and the number of *Aedes aegypti* would be expected; the latter could result in a high risk for dengue transmission in the May to July period (Fig. 16).

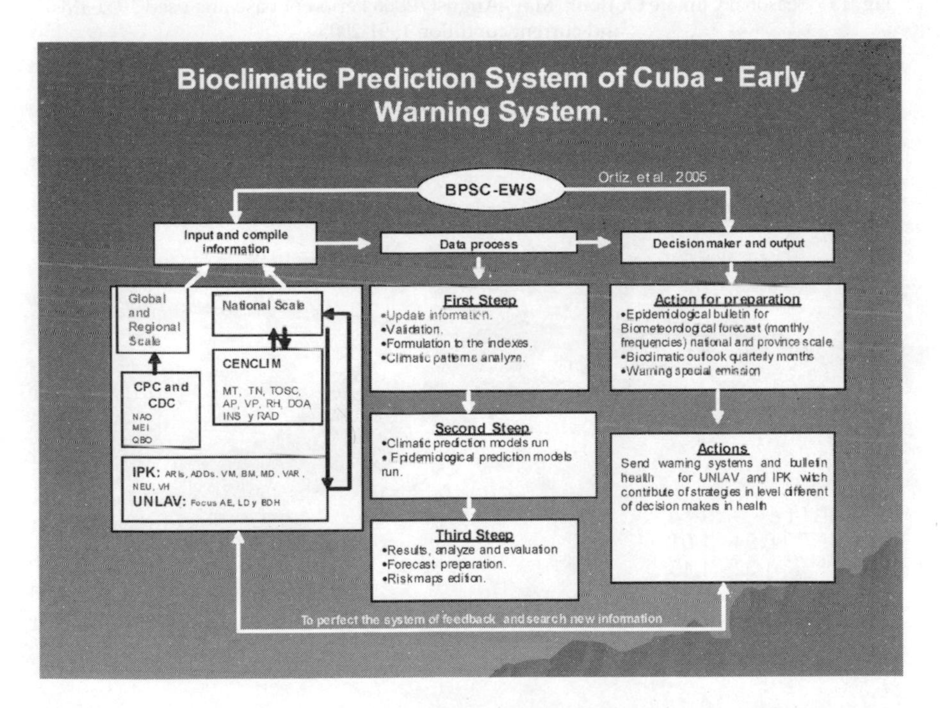

Fig. 12 Scheme of Bioclimatic Prediction System and Early Warning System (EWS) for Cuba

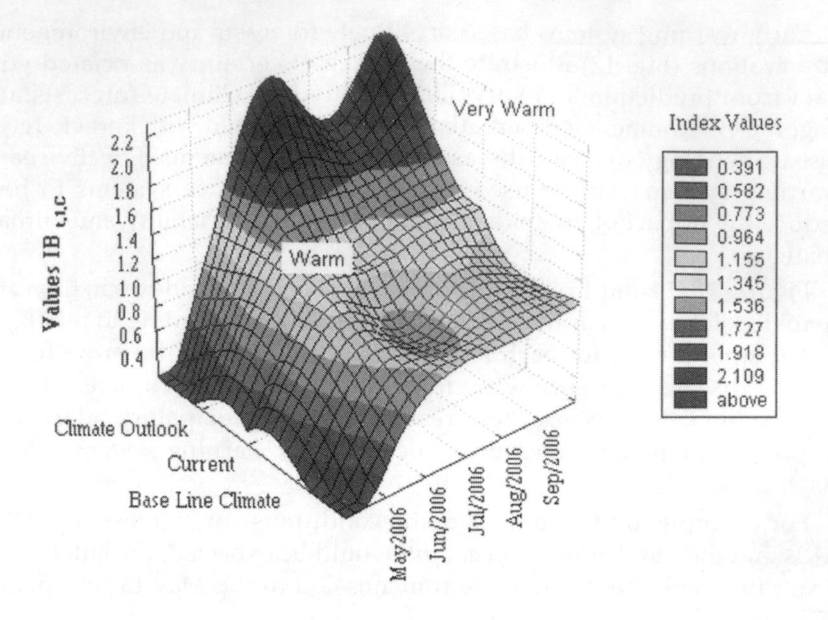

Fig. 13 Seasonal Climate Outlook. May–August/2006 Period of base line used 1961-1990 and current condition 1991-2005

Colour image of this figure appears in the colour plate section at the end of the book.

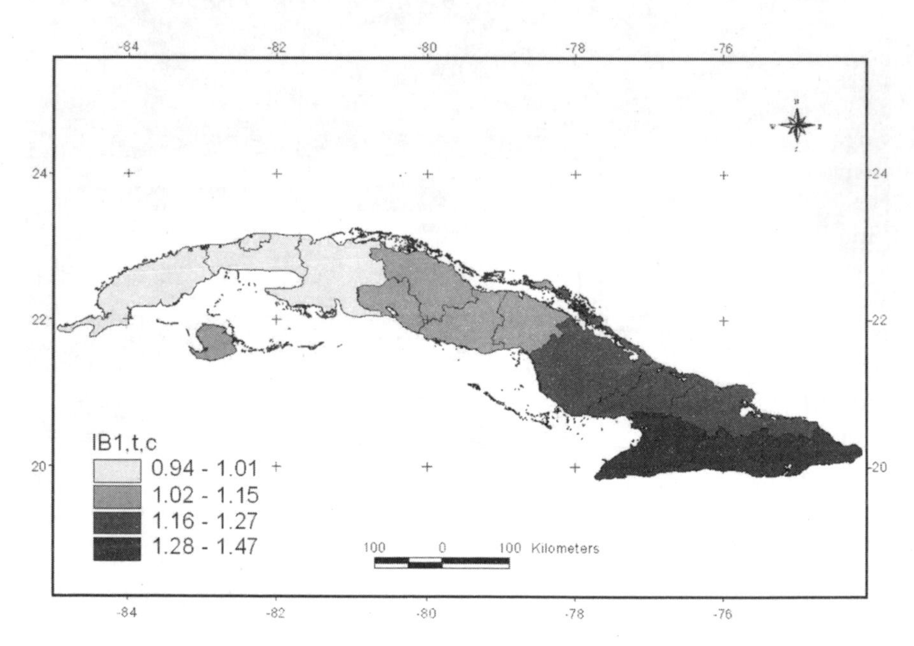

Fig. 14 Seasonal Climate Outlook (May–August/2006) according to $IB_{t,1,C}$.

Colour image of this figure appears in the colour plate section at the end of the book.

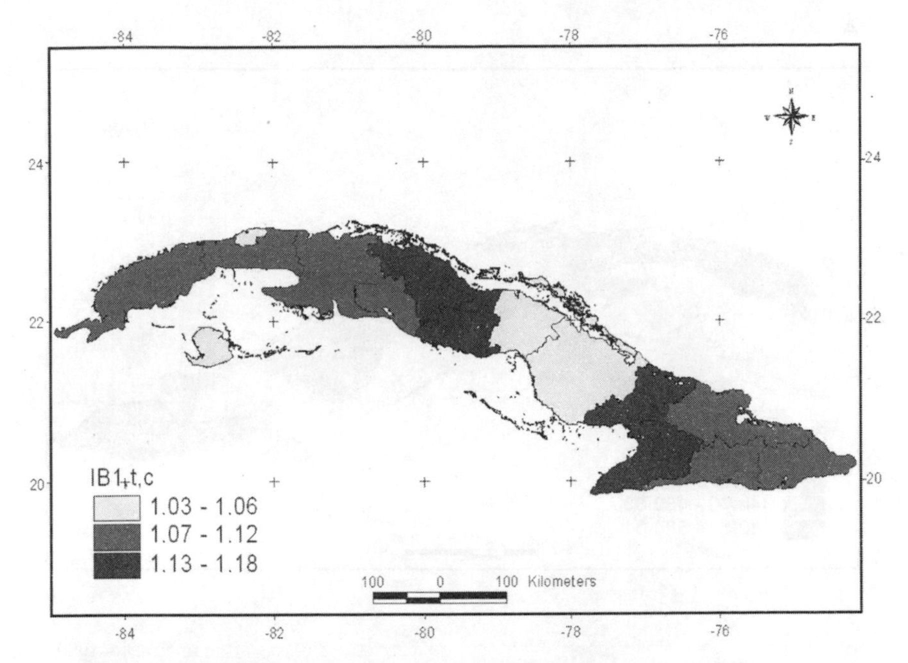

Fig. 15 Climate outlook according to $IB_{t,1,C.}$ August/2006
Colour image of this figure appears in the colour plate section at the end of the book.

Economic Impacts of Climate Variability and Change

Analyzing the economic impact of the effects of climate variability and change on human health is a complex and difficult undertaking. We used statistical data on the costs of hospitalization, treatments, and urgent care services, to assess the economic impacts (WHO 2003). To estimate the costs of morbidity attributable to climate variability, we first needed to determine how many cases are attributable to climate variability. For each disease selected for analysis, we determined any changes in disease trends due to climate variability.

It was then possible to determine the levels of disease risk, including projected increased numbers of cases, by using the dose-response functions, stratified by climate indexes. Finally, the costs associated with excess cases over baseline were estimated (Tables 6 and 7).

CONCLUSIONS

- These sections show that human health is an integrated theme of climate variability and change. Population health is affected by climate and particularly by climatic effects acting through natural

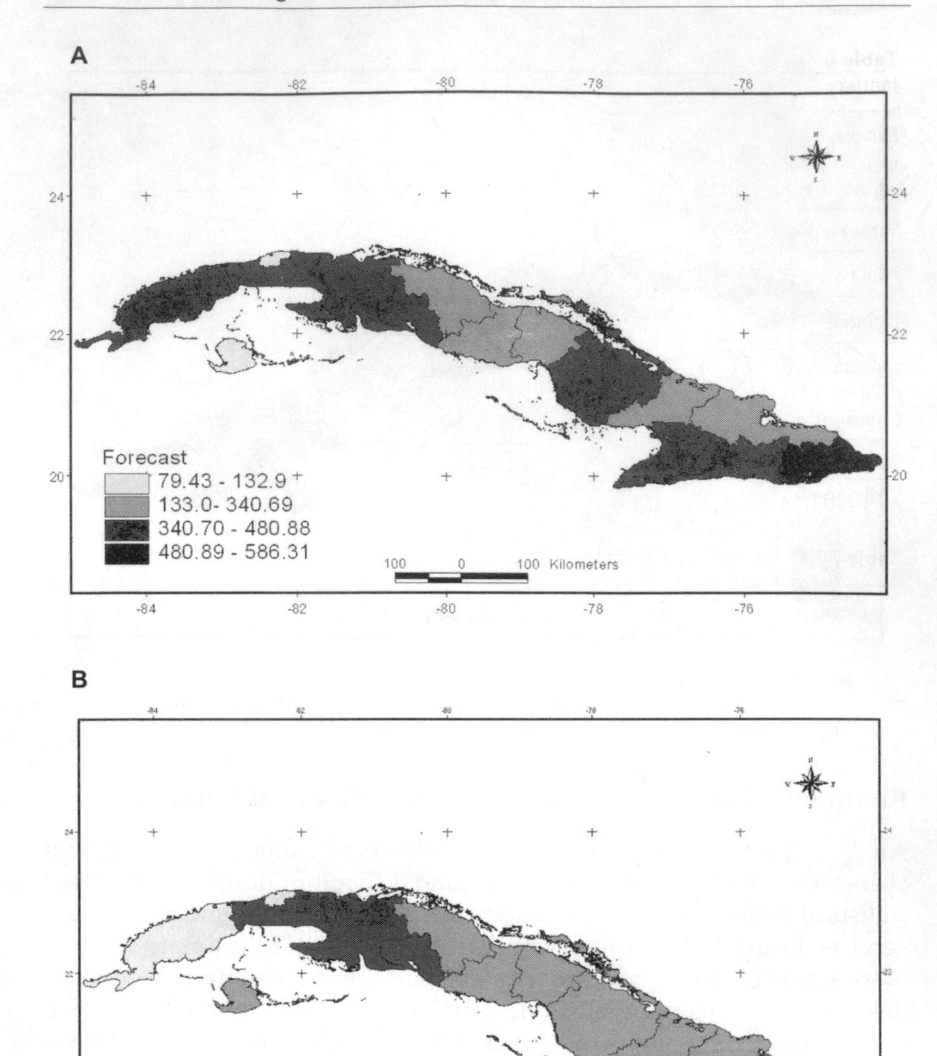

Fig. 16 Rate of per 100,000 habitants, expectation attentions by Acute Diarrhoeal Disease (A) and Acute Respiratory Infections (B) August/2006.
Colour image of this figure appears in the colour plate section at the end of the book.

Table 6 Estimate health cost (in millions US$) associated with climate variability from January 2001 to March 2002 (Ortíz et al. 2006).

Diseases	*Cost of attention*	*Hospitalization costs*	*Loss of work activities*	*Treatment cost*	*Cost of emergency services*	*Total cost*
Viral Hepatitis	8 874.06	8 657.10	917 50.00	5 505.00	1 236.79	116 022.95
ADD	373 073.6	175 067.95	547 059.20	76 064.60	36 463.4	1 207 728.75
Dengue	-	-	-	3 745 605.66	-	3 745 605.66
Streptococcus pneumoniae Meningitis *	-	231 318.00	-	-	-	231 318.00
Total cost						**5 300 675.36**

* All cases require hospitalization.

Table 7 Economic cost (in million US$) according to scenario 2010 (Ortíz et al. 2006).

Diseases	*(IC)*	*Cost of (IC)*	*Hospitalization Costs*	*Patient's Income*	*Total cost*
ARI	332 615	44 054 856.75	99 784	34 045 302.96	78 100 159.71
ADD	137 378	26 835 418.52	41 213	9 046 253.50	35 881 672.02
VH	11 027	1 433 510.00	3 308	1 966 837.56	3 400 347.56
V	19 353	2 563 111..32	-	-	2 563 111.32
MD	3 001	-	3 001	2 400 800.00	2 400 800.00
Dengue*	1 220		1 226 222		1 226 222.00
Streptococcus pneumoniae Meningitis *	100	-	100	814 500	814 500.00
General cost					**124 386 812. 64**

IC: Increase of cases. *All cases are hospitalized.

disasters, climate-sensitive diseases and through climate-sensitive sectors such as agriculture, water, or human environments.

- These results demonstrate that studies of climate and health are necessary to increase our knowledge of the effects of climate on human health; such information is important for decision-makers, and for reducing the socio-economic impacts of climate variability and change.

- This study is innovative in the development of complex climate indexes to reflect climate anomalies at different scales, and to explain the mechanisms and relationships between climatic conditions and diseases. Our results suggest that some diseases not previously thought to be climate-sensitive (VH, chicken pox, bacterial and viral meningitis, and others) do vary with the identified climactic indexes. The disease risks vary by geographical region, as described by the indexes. Therefore, climate prediction can be used to inform the design and development of prevention activities to reduce the burden of climate-sensitive diseases, thus increasing adaptive capacity to climate variability. Anticipatory prevention is better than reacting once a disease outbreak has occurred.
- In the Latin American and Caribbean region is increasing its understanding of the potential health impacts of climate variability and change by identifying those vulnerable to climate variability and long-term climate change (cyclones, floods, and droughts) on small islands.
- Health is therefore both a key climate-sensitive sector in its own right, and an important justification for addressing the impact of climatic on other sectors.
- The main roles for climate information in operational health decisions are:
    1) Identification of climatically suitable or high-risk areas for particular diseases.
    2) Early Warning Systems for climate-sensitive diseases can vary over time.
- These results demonstrate that studies of climate and health are necessary to increase our knowledge of the effects of climate on human health; such information is important for decision-makers in order to reduce the socio-economic impact of climate variability and change in the region.
- This study is innovative in the development of complex climate indexes to reflect climate anomalies at different scales, and to explain the mechanisms and relationships between climatic conditions and diseases.
- Based on our experience with studies in vulnerability and adaptation assessment, it is clear that climate prediction can be used to prepare for climate variability and extreme events caused by climate change. This preparation includes an estimation of costs.

- Our experience also demonstrates that interdisciplinary collaboration and the sharing of information, experience, and research methods between sectors are critical for effective policy formulation and the development of support tools for decision-makers.
- The results of this study evidence a clear non-lineal relationship, in a differentiated way, between the changes of the climatic variations and the changing patterns of behaviour of diseases.

References

Basilevsky, B. 1994. Statistical Factor Analysis and Related Methods. Theory and Applications. John Wiley & Sons, Inc.

Burton, I. and Van A. Marten. 1999. Come Hell or High Water. Integrations Climate Change Vulnerability and Adaptation into Bank World. World Bank Environment Department papers. World Bank, Washington, D.C., USA. Paper No. 72.

Cárdenas, P.A. 1998. Papel de los Índices Teleconectivos y del ENOS en la predictabilidad de la lluvia en Cuba. La Habana, Cuba. Technical Report 099-01.

Chan, N., K. Ebi, F. Smith, T. Wilson and A. Smith. 1999. An integrated assessment framework for climate change and infectious diseases. Environmental Health Perspectives 107, 5: 329-338.

Dillon, W.R. and M. Goldestein. 1984. Multivariate Analysis. Methods and Applications. John Wiley & Sons, Inc.

Enfield, D.B. 1998. The dependence of Caribbean rainfall on the interaction of the Tropical Atlantic and Pacific Oceans.

Gutiérrez, T. 1998. Impact of the climatic change and measures of adaptation in Cuba. INSMET. National Project "The Global Changes and the Evolution of the Environment Cuban". Technical Report 112.

Guzmán, M.G., G. Kourí, J. Bravo, M. Soler, S. Vázquez and L. Morier. 1990. Dengue hemorrhagic fever in Cuba, 1981: A retrospective seroepidemiologic study. American Journal Tropical Medicine Hygiene 42: 179-184.

Haines, A. and A.J. Patz. 2004. Health effects of climate change. American Medical Association 291, 1: 99-103.

Hair, J.F., R.E. Anderson, R.L. Tatham and W.C. Black. 1999. Análisis multivariante. Prentice-Hall. Madrid, España.

Kourí, G., M.G. Guzmán, J. Bravo and C. Triana. 1989. Dengue hemorrhagic fever/ dengue shock syndrome: lessons from the Cuban epidemic. Bulletin WHO 67: 375-380.

______. 1997. Ministerio de Salud Pública de Cuba. Dengue en Cuba. Boletín Epidemiológico. Organización Panamericana de la Salud. 18: 7.

Kovats, R.S., K. Ebi and B. Menne. 2003. Methods of assessing human health vulnerability and public health adaptation to climate change. WHO, Health Canada, UNEP, WMO, Copenhagen. (Health and Global Environmental Change) Vol. 1.

Lecha, L. 1993. Estudio bioclimático de la provincia de Cienfuegos. Edit. Academia, La Habana, Cuba.

Lecha, L. 1998. Biometeorological classification of daily weather types for the humid tropics. Int. Jour. Biomet. 42(2): 77-83.

Lecha, L. and T. Delgado. 1996. On a regional Health Watch & Warning System, 94-107. In: The Proceedings of the 14th Int. Congress of Biometeorology, Part 2, Vol. 3. Ljubljana, Slovenia.

Lecha, L., L. Paz and B. Lapinel. 1994. El clima de Cuba. Edit. Academia, La Habana, Cuba.

Lorenz, E.N. 1956. Empirical orthogonal functions and statistical weather prediction. MIT department of Meteorology, Statistical Forecast Project (Available from Dept of Meteorology, MIT Massachusetts Ave Cambridge, MA 02139). Report 1, 49.

McMichael A.J. and S. Kovats. 1999. El tiempo el clima y la salud. [In Spanish]. Boletín de la Organización Meteorológica Mundial.Vol 48, No. 1, pp. 16-21.

Michael, H.J. and M.J. Trtanj. 1999. La predicción climática para la Salud Humana. Boletín de la OMM. Volumen 48, 1: 32-34.

Mitchell, J.F.B., R.A. Davies, W.J. Ingram and C.A. Senior. 1995. On surface temperature, greenhouse gases and aerosols: models and observations. J Climate 10: 2364-2386.

Ortiz, B.P., et al. 2000. Principios metodológicos para la evaluación de impacto de la variabilidad y el cambio climático en la salud humana. Un enfoque estadístico [In Spanish]. Rev. Meteorología Colombiana. No. 3, pp. 75-84.

Ortiz, B.P., E.M. Nieves and A.V. Guevara. 1998. Models for setting up a biometeorological Warning System over a population areas in Havana. In: J. Breute, H. Feldmann and O. Ulmann (eds.). Urban Ecology. Springer-Verlag, Berlin, Germany, pp. 98-91.

Ortiz, B.P., A. Rivero, A. Perez, N. Leon, M. Diaz and C.A. Perez. 2006. Assessment of Human Health Vulnerability to Climate Variability and Change in Cuba. Environmental Health Perspectives. Volume 114, Number 12, December.

Ortiz, B.P., A. Rivero, A. Pérez, N. León and C.A. Pérez. 2004. The climatic variability and their effects in the variations of the patterns space-time of the diseases and their economic impact. Technical Report 82. Research Climate Center. INSMET. National Program the Global Changes and the Evolution of the Environment Cuban.

Ortiz, B.P. and A. Rivero. 2003a. Un modelo AR-ARCH(p) para el comportamiento de la lluvia por regiones, usando algunos índices de circulación y el índice multivariado del ENOS (MEI) [In Spanish]. Rev. Meteorología Colombiana. No. 7, pp. 11-19.

Ortiz, B.P. and A. Rivero. 2003b. Modelo para el pronostico del índice multivariado del ENOS (PMEI) [In Spanish]. Revista Cubana de meteorología. Diciembre. Volume 10, No. 2, pp. 39-44.

Ortiz, B.P. and A. Rivero. 2004. Índices climáticos para la determinación y simulación de las señales de la variabilidad climática en diferentes escalas espacio temporales [In Spanish]. Revista Cubana de Meteorología. Vol. 11, No. 1, pp. 41-52.

PAHO. 1997. División de Prevención y Control de Enfermedades, Programa de Enfermedades, Programa de Enfermedades Transmisibles, HCP/HCT, PAHO, Resurgimiento del dengue en las Américas. [In Spanish]. Boletín Epidemiológico. Organización Panamericana de la Salud. 18:1-6. Piatkin RD and Krivochein YS. 1981. Microbiología con Urología e Inmunología. Segunda Edición Editorial MIR. Moscú, 1981. 582 pp.

San Martín, H.F. 1963. Salud y Enfermedad [In Spanish]. (Ed. Confederación Médica Panamericana). Tomo 1. La Habana-Cuba.

TeleMap. 1994. Manual de Usuario. TeleMap ver 2.1. Software para las geociencias. ICH, La Habana, Cuba.

WHO. 1996. Climate change and human health. In: A.J. McMichael, et al. (eds.). Geneva, Switzerland.

WHO. 2003. Climate Change and Human Health: Risks and Responses. In: A.J. McMichel, D.H. Cambpbell-Lendrum, C. Corvalán, K.L. Ebi, A. Githeko, J.D. Scheraga, et al. (eds.). WHO/WMO/UNEP. Geneva. Chapters 4-5.

WHO. 2004. Using Climate to Predict Diseases Outbreaks: A review. Geneva: World Health Organization (WHO/SDE/OEH/04.01).

WMO. 2001. Informe final Abreviado de la Decimotercera Reunión de la comisión de Climatología [In Spanish]. Anexos 1 y 2. WMO. Génova.

Woodward, A., S. Hales and P.Weinstein. 1998. Climate change and human health in the Asia Pacific region: Who shall be the most vulnerable? Climate Research 11: 31-38.

Section IX

Beyond 2012: The Next Phase

38

CHAPTER

Climate Commitments: Assessing the Options

Daniel Bodansky*
Emily & Ernest Woodruff Chair of International Law
University of Georgia
School of Law, Athens, GA 30602, USA
E-mail: Bodansky@uga.edu

I. INTRODUCTION

The question of commitments lies at the heart of the climate change debate. Ever since climate change first emerged as a political issue in the late 1980s, attention has focused on quantified 'targets and timetables' as the principal type of commitment—the model used with great success in the 1987 Montreal Ozone Protocol. Although legally binding targets and timetables for greenhouse gas mitigation could not be agreed in the 1992 UN Framework Convention on Climate Change (due primarily to opposition by the United States), they became the centerpiece of the Kyoto Protocol—and the lightning rod for its opponents.

In considering the way forward—either under Kyoto or beyond it—a central question will be the type (or types) of mitigation commitments to employ. Should quantified emission limitation targets continue to be the principal type of climate commitment and, if so, should these targets be of

*This paper is reproduction of paper published in "Beyond Kyoto: Advancing the International Effort againt Climate Change", Pew Center on Global Climate Change 2003; with permission of Pew center

the kind found in Kyoto—that is, fixed targets, pegged to historical emission levels? Or should international climate policy strike out in a different direction by adopting different types of targets, for instance, or by focusing on technology standards or commitments on research and development? The often-tortuous history of the climate change negotiations demonstrates that politics more than policy will determine the answer to these questions.

This Chapter examines the structure of future climate mitigation commitments—that is, the different forms future commitments might take.[1] Part II addresses the function and importance of mitigation commitments. Part III identifies the range of options with respect to three key variables: what types of commitments, when, and by whom? Part IV then proposes criteria for assessing these options. Part V evaluates some of the leading proposals for future commitments.

II. WHY COMMITMENTS?

The importance of commitments may seem self-evident. However, the rejection of the Kyoto Protocol by the United States, and the reluctance of developing countries to assume binding emission limitation targets (at least until industrialized countries have taken action), make it useful to consider at the outset: What is the function of commitments? Are they essential, or could the climate change problem be addressed either through the application of pre-existing legal obligations, or through voluntary measures as the Bush Administration has proposed?

The nature of the climate change problem, as well as the history of international environmental cooperation more generally, suggest the need for commitments. The existence and implications of purported legal obligations, such as the duty to prevent transboundary pollution and the polluter pays principle, are the subject of endless debate among scholars and states. Although these principles reflect strong moral imperatives—and may even have the status of international law—in the absence of courts that could apply and enforce them, they are unlikely to be of significant use in changing states' behavior. Instead, states are likely to address climate change only if they believe it is in their interest to do so. That is why climate change negotiations have focused on "commitments,"

[1]For a discussion of two related issues—first, how to distribute the burden of mitigation commitments (based on wealth, historical emissions, per capita entitlements, or some other criteria), and, second, what the trajectory or end point of commitments should be—see Ashton and Wang (2003) and Pershing and Tudela (2003), respectively. This paper focuses on mitigation commitments and does not address the equally important issue of adaptation commitments.

requirements that a state itself assumes, rather than on "obligations," a broader term that includes norms externally imposed.

The role of commitments derives from the "collective action" nature of the climate change problem. Like other collective action problems, climate change mitigation poses a fundamental dilemma. Because most of the benefits of climate change mitigation do not accrue to the country taking action, but are instead shared by the international community as a whole, individual countries have little incentive to do anything on their own.[2] Even when the global benefits justify the costs, the country engaging in mitigation usually receives only a fraction of the total benefits. So, from its individual perspective, the costs of mitigation are likely to exceed the benefits. Of course, if the costs of reducing emissions are sufficiently low, countries might be willing to go ahead anyway, for example, to show leadership or for public relations purposes. But significant investments to reduce greenhouse gas (GHG) emissions will be in a country's individual self-interest only if they are reciprocated by other states—only if a country's actions are part of a bargain involving significant action by others to address climate change.

International commitments serve as the glue that helps hold a cooperative regime together. Before taking potentially costly actions to address climate change, states need to be confident that others will do their part as well. International commitments are the means by which countries bind themselves to one another to take mutual action.

What does it mean to say that a country "commits" itself to undertake mitigation actions? In one sense, virtually all international commitments are voluntary. Given the absence of an international legislature that can impose obligations on states, international obligations in general depend on a state's consent. But, by making a commitment (for example, to reduce GHG emissions), a state agrees to limit its future freedom of action; it promises to behave in a certain way or to achieve a certain result. While its acceptance of a commitment is voluntary, its fulfillment of the commitment is not.

International commitments fall along a spectrum. Some are political, such as the aim in the UN Framework Convention on Climate Change (UNFCCC) to return developed country emissions to 1990 levels by the year 2000; others are legal, such as the reporting requirements in the UNFCCC and the targets and timetables in the Kyoto Protocol and the

[2]In this respect, mitigation differs from adaptation. Most of the benefits of adaptation accrue directly to the country undertaking the adaptation measures. (They are, in this respect, what economists refer to as "private" rather than "public" goods.) Thus, so long as the benefits outweigh the costs, countries have an incentive to undertake adaptation measures regardless of what other states do.

Montreal Ozone Protocol. In the absence of effective institutions to interpret and enforce international law, the distinction between political and legal commitments can often seem illusory. Most international commitments—even "legally-binding" ones—depend on the good faith of states and on the diffuse costs of developing a reputation for breaking one's promises, which makes it more difficult to enter into mutually-advantageous deals in the future. But, in general, casting a commitment in "legal" form signals a greater level of seriousness by states, raises the costs of violation, and sets in motion domestic legal implementation mechanisms. That is why, even in the absence of any realistic prospect of being sanctioned for non-compliance, countries are usually reluctant to accept legally binding commitments and why the decision to do so in the Kyoto Protocol was so controversial and difficult.

Of course, no level of commitment can fully assure that a country will uphold its end of the bargain. Some countries may view their treaty commitments as aspirational rather than absolutely binding. But, compared to a strictly voluntary system, commitments provide states with greater confidence that other states will not simply say one thing and then do another. This not only promotes action by states, but provides a signal to the market that helps drive changes in private behavior. Moreover, if mechanisms can be agreed to impose specific sanctions for violations, this further raises the costs of non-compliance and thus provides additional assurance to states that others will comply with their commitments. Indeed, given the potentially high short-term costs of mitigating climate change, many analysts believe that both legally binding commitments (in contrast to voluntary actions), and a strong compliance system (with strict penalties to deter free riders) are essential.[3]

III. KEY VARIABLES

The problem of developing climate change commitments can be expressed by the following question: who *will have* what *commitments* when*?*

All three of these variables—who, what, and when—raise important, interdependent issues.

What Types of Commitments?

Specifying the content of a commitment has both formal and substantive dimensions:

[3]Barrett (2002); Victor (1999).

Binding vs. Non-binding

To begin with, there is the issue of the legal form of a commitment—in particular, whether it will be legally binding or political. This is not simply an either-or choice; a range of options present themselves:

Non-binding "commitments" Although perhaps strictly speaking a misnomer, a "commitment" can be expressed in non-legally binding language, as a recommendation ("should" rather than "shall") or an aim. The emissions target for developed countries in the UNFCCC (to return emissions to 1990 levels by the year 2000) was contained in the commitments section of the treaty, but was stated as an "aim" rather than a legal requirement.

One-way ("no-lose") commitments This is a variant of the previous option. An aim, although non-binding, could have legal consequences in the sense that, if bettered, it can provide a country with certain legal benefits. For example, if a country reduced its emissions by more than its non-binding target, then it could sell the surplus emissions to other countries.[4] Project baselines established under Kyoto's Clean Development Mechanism (CDM) are, in essence, one-way "commitments," since a country (or firm) faces no penalty if its project exceeds a baseline, but receives certified emission reduction credits if the project reduces emissions below the baseline.

Legally binding commitments A commitment can also be expressed in binding language ("shall"), like the targets and timetables in the Kyoto Protocol. It is important to note that this is a separate question from whether the commitment is subject to enforcement through a compliance system (considered below). Most international commitments do not have any specific compliance mechanisms.[5] Nonetheless, they are legally binding and must be complied with by those states that accept the commitment (in much the same way that one is bound by one's solemn promises, whether or not enforcement machinery exists).

Enforceable commitments A binding commitment can be subject to a mandatory compliance system, with authority to respond to violations, such as the dispute settlement system adopted under the World Trade Organization. This would provide the greatest assurance of compliance but would also present the greatest worry for states that are on the fence about whether to undertake mitigation commitments. The Marrakech Accords, which set forth detailed rules to operationalize Kyoto, establish a compliance procedure, including consequences for non-compliance. But the binding character of these consequences remains an open question.

---

[4]Philibert and Pershing (2001).

[5]The World Trade Organization dispute settlement system is one important exception.

Choice of Policy Instrument

The substantive content of commitments can involve an equally wide variety of policy instruments:

Emission targets An emission target is an obligation of result: it requires regulated entities (for example, countries or firms) to achieve a particular level or rate of emissions, but allows them flexibility as to how they will achieve that result.[6] Emissions targets can be specified in various ways: fixed or indexed, absolute or conditional, and economy-wide or sectoral.

Absolute targets—Until recently, most of the attention in the climate change regime has focused on fixed, countrywide emissions targets, pegged to an historical base-year emissions level (generally, 1990 emissions). The Kyoto Protocol, for example, requires industrialized countries to achieve predetermined, fixed levels of emissions for the 2008-2012 commitment period.[7] In this respect, the climate change regime has followed the approach used in several other international environmental regimes, including those addressing acid rain and stratospheric ozone depletion.

Indexed targets—Because emissions depend on a wide range of variables that are difficult to anticipate in advance (economic growth, weather, technological change, etc.), an emission target can be pegged to one or more of these variables, rather than defined in fixed terms, like the Kyoto targets.[8] Thus far, most of the literature has focused on tying emissions targets to a country's GDP so that the permitted level of emissions would be larger or smaller, depending on whether the economy grows or shrinks. The Bush Administration's carbon intensity target[9] and the proposed Argentine target[10] are both examples of indexed GDP-based targets.

Conditional targets—In contrast to the Kyoto targets, which apply come what may, a target could be formulated in conditional terms: if the specified conditions are not satisfied, then the target either would not apply at all or would be modified in some fashion. One option is to make commitments conditional on a state's achievement of a minimum level of

[6]Another way to say this is that approaches using emissions targets flow from outputs (i.e., emissions) to inputs (i.e., the activities that cause emissions), rather than vice versa. See Heller and Shukla (2003).

[7]Although the provisions on sinks in the Marrakech Accords have modified these targets, and Kyoto's flexibility mechanisms allow countries considerable leeway in how they meet their targets.

[8]The non-binding target in article 4.2(a) of the UNFCCC implicitly acknowledged a wide variety of circumstances that may cause emissions to vary.

[9]U.S. Global Climate Change Policy: A New Approach, Feb. 14, 2002, available at http://www.usgcrp.gov/usgcrp/Library/gcinitiative2002/gccstorybook.htm.

[10]Bouille and Girardin (2002).

wealth. (For example, per capita GDP could be used as a "graduation criterion" for the assumption of commitments by developing countries.) In addition, conditional targets—like indexed targets—could help alleviate fears that a fixed emission target might become an economic straitjacket. A conditional target that has received particular attention in this regard is the so-called "safety valve" approach.[11] In essence, a safety valve defines a conditional target in negative terms: the target applies unless the cost of compliance exceeds a specified level, in which case the target is relaxed through the issuance of additional emission allowances.[12]

Sectoral targets—A target can also be specified on a narrower basis than total national emissions. For example, targets could be specified for particular sectors or industries that are particularly important, politically easier to address, or comparatively insulated from international competition. Sectoral targets could be binding or "no lose," fixed or indexed. In essence, proposals to expand the CDM to apply to entire sectors rather than particular projects[13] would involve setting no-lose, sectoral emission targets: if a developing country failed to meet its sectoral target, it would face no consequences; but reducing emissions below its target would generate emission reduction credits that the country could sell.

Financial targets Rather than focus on emissions, a target can be specified in financial terms, as an amount to be devoted to climate change mitigation, either domestically or internationally. Both the UNFCCC and the Marrakech Accords set forth collective financial commitments that apply to Annex II countries as a whole, rather than individual targets for each state.

Policies and measures In contrast to a target-based approach, a commitment regarding policies and measures (PAMs) is an obligation of conduct rather than an obligation of result: it requires countries to act in certain ways, but does not require them to achieve any particular level of emissions or financial contribution. During the negotiation of the Kyoto Protocol, the European Union pushed for the inclusion of commitments related to policies and measures, but due to strong resistance from the United States, the Protocol includes only an illustrative list of possible PAMs, without requiring states to adopt them.[14] Examples of PAMs include:

[11]Kopp et al. (1997); McKibben and Wilkoxen (1997).

[12]The safety valve has been characterized as a "hybrid" approach because it mixes a quantity-based instrument (if the safety valve price is not exceeded, then the quantitative target must be met) with a price-based instrument (if the safety-valve price is reached, then additional emissions are allowed at that price). IEA (2002).

[13]Samaniego and Figueres (2002).

[14]Kyoto Protocol, art. 2.1.

Technology and performance standards—An international commitment can address the use of emission-reduction technologies. For example, it could specify mandatory standards relating to appliance efficiency, residential insulation, or the use of renewable or other non-emitting energy sources.[15] The international commitment can either require the use of particular technologies (which would tend to lock in those technologies) or set forth a performance standard (for example, relating to energy efficiency) that allows private entities flexibility as to the choice of particular technologies. Among the relatively few examples of international technology standards are the construction, design, and equipment standards for oil tankers set forth in the Marine Convention (MARPOL) including, for example, segregated ballast tanks.[16]

Taxes—An international commitment can provide for a common or harmonized tax on GHG emissions. So long as a country had the required tax in place, it would satisfy its international commitment, regardless of the actual level of emissions reduction achieved.[17]

Subsidy removal—An international commitment can require countries to remove specified subsidies, for example, on energy production or consumption. The Kyoto Protocol includes in its illustrative list of PAMs for developed countries "the progressive reduction and phasing out of subsidies."[18] Subsidies are a problem not only in industrialized countries: the International Energy Agency estimates that removing energy subsidies in just eight developing and transition countries would reduce their CO_2 emissions by 17 percent and global emissions by 4.6 percent.[19]

Emissions trading—An emissions commitment can be coupled with a PAM requiring countries to implement a domestic emissions trading program with specified features (including possible linkages with other national programs and with an international emissions trading system, or a safety-valve device).[20] The European Union directive on emissions trading represents an effort of this kind: it sets forth the parameters of a required emissions trading system for EU member states.

Technology R & D and incentives—To address the low rates of investment in research and development concerning emission-reducing technologies,[21] a commitment might require states to devote additional resources for R & D, as well as for deployment of existing and new

[15]Barrett (2002).

[16]Mitchell (1994).

[17]Cooper (1998); Nordhaus (2001).

[18]Kyoto Protocol art. 2.1(a)(v).

[19]China, India, Indonesia, Iran, Kazakhstan, Russia, South Africa, and Venezuela.

[20]McKibben and Wilkoxen (1997).

[21]Margolis and Kammen (1999).

technologies.[22] For example, countries could commit to various forms of participation in an international hydrogen initiative. The agreement on the international space station is one illustration of an international agreement focusing on cooperative research, development, and deployment.

Since a targets-based approach and a PAM-based approach are often seen as competitors, it is worth emphasizing that they could complement one another: a target could be used to specify the overall result to be achieved, while PAMs could specify the means for reaching that result. Indeed, in some cases the relationship could be even stronger. As some commentators have noted,[23] an international target- and trading approach would be most cost-effective if combined with national PAMs ensuring that domestic trading systems are complementary.

When will Commitments Apply?

Another critical question is the timing of commitments. The international negotiations thus far envision a dynamic process beginning with the relatively modest but important reporting requirements in the UNFCCC, to be followed by specific mitigation commitments in subsequent protocols. A future agreement could set forth a more detailed road map for the evolution of commitments over time.

There are two important elements to timing: first, when will a commitment take effect, and second, how long will it last?

When does the Commitment Begin?

In contrast to most treaties, which set forth commitments that take effect immediately upon the agreement's entry into force, the Kyoto Protocol establishes a commitment period beginning more than ten years after its adoption. The intent was to avoid economic disruption by giving countries and firms time to adjust to the Kyoto targets. Even so, many economists argue that, if the United States had stayed in the Kyoto system, the Kyoto targets would have cost more than necessary by requiring premature capital retirement.[24] According to this view, an even longer-term target, timed to coincide with ordinary patterns of capital turnover, would have been more economically efficient. If a commitment is too far off in the future, however, it may lack credibility; it may raise concerns that, given the lack of stability in international politics, the commitment is likely to be changed before it ever takes effect. An intermediate approach

[22] Barrett (2002).

[23] Hahn and Stavins (1999).

[24] Aldy et al. (2001). For a discussion of rates of capital turnover, see Lempert et al. (2002).

is suggested by the Montreal Ozone Protocol, which provides for the gradual phasing-in of commitments, so that the commitments start relatively soon, but do not reach their full stringency until later, in order to give individuals and industry time to adjust.

What is the Duration of a Commitment?

In most international environmental regimes, commitments have an indefinite duration; they continue in effect until the parties modify or terminate them. The Kyoto Protocol, in contrast, defines an emission target for only a five-year period, ending in 2012. This is sometimes justified as providing necessary flexibility. The rationale is that, given the significant uncertainties relating to climate change, the international regime should consist of a series of rolling commitment periods, which allow commitments to be continually redefined to take account of improved scientific and economic understanding. But indefinite commitments also could build in flexibility (for example, a carbon intensity target that increases in stringency over time) or could provide for periodical review with a view to possible adjustments. Most international environmental agreements have flexible amendment procedures, so that commitments can be periodically updated in response to new problems and new information. Similarly, the international trade rules and tariff rates set forth in the GATT/WTO regime are not time-limited. But this has not meant that they are carved in stone; instead, the trade regime has undergone major changes through periodic negotiating rounds. The real effect of making commitments with a limited duration is to reverse the ordinary presumption of continuity. In other regimes, commitments continue until they are changed; in the Kyoto Protocol, they lapse unless they are renewed. This allows states to preserve much more freedom, but at the cost of making the regime less predictable, and necessitating repeated negotiations, each of which could prove politically difficult.

Who will be Subject to Commitments?

Individuals/Private Entities

Although the climate change regime has, thus far, sought to establish obligations only for states—for example, relating to emissions targets, financial contributions, and reporting—an international commitment could conceivably apply directly to individuals, private entities, or sub-national entities such as cities. International criminal law, for example, establishes basic duties on individuals (for example, not to commit torture or genocide), the violation of which results in international criminal liability.[25] Although individual criminal responsibility seems clearly inappropriate for climate-related activities, other forms of individual liability are possible. For example, an international emissions tax could apply directly to producers or consumers of fossil fuels. Similarly, some have suggested that, given the withdrawal of key countries such as the United States from the Kyoto Protocol, the international climate regime should establish emission reduction obligations for multinational corporations.

It should be emphasized, however, that attempting to impose obligations directly on individuals or private entities would pose very difficult issues of implementation and enforcement—particularly with respect to individuals and firms located in countries that do not participate in the international regime and that therefore could not be enlisted for enforcement purposes.[26] There are, at present, no examples of international environmental regimes that apply directly to individuals.

States

Given the difficulties of imposing obligations directly on individuals, most international regimes define commitments for states and rely on them to translate these into obligations for individuals and firms under their jurisdiction.

Because of the global nature of the climate change problem, the natural tendency is to include all countries in an international climate change regime. All countries have a duty to participate because of their contribution to climate change, and they all have a right to participate because they will all be affected by it. The UNFCCC takes this approach: it is open to any state and defines at least minimal obligations for all participants. At the same time, it recognizes that the same level of commitment is not appropriate for all states. It therefore sets forth differentiated obligations, based on the principle of common but differentiated responsibilities and respective capabilities.[27]

In establishing new commitments, a key question will be whether they apply equally to all states, or whether differentiation is appropriate.

[25]The Genocide and Torture Conventions—in which the United States participates—both define crimes for which individuals can be held responsible. The newly created International Criminal Court will have jurisdiction to prosecute individuals directly for commission of crimes against humanity.

[26]International criminal law is generally based on the idea of universal jurisdiction: any state can proscribe and punish violations, regardless of where they occur. A similar approach could be used for climate change, although it would be sure to draw objections from non-participating states, such as the United States, which have objected to the new International Criminal Court on similar grounds.

[27]UNFCCC, art. 3.1.

Kyoto's mitigation commitments all take the same form, for instance, but apply only to developed countries and vary in stringency among them. Commitments could also be differentiated by form (some countries have absolute or binding targets, while others have indexed or no-lose targets); by timeframe (as in the Montreal Protocol, which gave developing countries an additional 10 years to phase out ozone-depleting substances);[28] or by conditionality (applying when a country has met a criterion such as a specific level of per capita GDP or emissions).

The criteria that might be used to determine who should participate in a climate regime, or to differentiate commitments among the participants, include the following:

Big current emitters Relatively few countries contribute significantly to climate change—15 countries, for example, account for 75 percent of global CO_2 emissions.[29] Mitigation commitments by these big emitters could largely address the climate change problem. Moreover, limiting membership in the regime to countries with mitigation commitments could simplify the negotiating dynamic significantly.

Big historical emitters Alternatively, commitments might vary depending on a country's historical contribution to the climate change problem. Here, the rationale for differentiation would be the idea that countries with high historical emissions are responsible for the current problem and have a duty to fix it—including through reductions in their current emissions. This is the essence of the so-called "Brazilian proposal" for allocating the burdens of addressing climate change.[30]

Rich countries Commitments could vary depending on a country's wealth and therefore its capacity to respond to the climate change problem.

Like-minded states A future climate regime could be limited to like-minded states, which are willing to undertake a certain level of commitments and have shared views about international implementation mechanisms such as emissions trading. Again, the idea would be to create a more favorable negotiating dynamic by conducting negotiations initially among countries with shared goals, bringing other countries in later.

[28]The timetable specified in the Montreal Protocol for industrialized countries to phase out their use of ozone-depleting substances applies conditionally to developing countries, if their per capita consumption of ozonedepleting substances exceeds a specified level.
[29]IPCC (2001), sec. 10.1.2.1.
[30]UNFCCC (2002).

IV. ASSESSMENT CRITERIA

Potential commitments need to be evaluated from both a policy and a political perspective. In some cases, synergies may exist between different assessment criteria: a climate policy that is equitable or cost-effective may in the long run be more environmentally effective. But, often, different assessment criteria will be in tension. Ensuring predictability in the costs of mitigation measures, for example, comes at the expense of predictability concerning environmental effects. More broadly, there are strong tensions between the basic goals of policy optimization and political feasibility. Formulating a sound climate change policy is not so difficult; nor is formulating a politically acceptable one. The challenge is to devise a policy that is both sound and acceptable.

Policy Criteria

What commitments are optimal from a policy perspective? There are five key criteria: environmental effectiveness, cost-effectiveness, equity, dynamic flexibility, and complementarity.

Environmental Effectiveness

Ultimately, the purpose of mitigation commitments is to reduce dangerous climate change. The bottom-line test of commitments is their effectiveness, over the long run, in preventing (or at least limiting) climate change.

An important contributor to environmental effectiveness is, of course, stringency—all other things being equal, a stronger commitment should produce a greater environmental result than a weaker one. But all other things are rarely equal and, as a result, environmental effectiveness is not solely a function of stringency. Other important factors include:

Leakage To the extent that the climate change regime is not global, private entities can avoid the impacts of commitments by shifting their operations to a non-party state. As a result, more stringent targets could actually be counterproductive, both by discouraging countries from joining and by causing emitting activities to shift to states without commitments.[31]

Stimulating technological change Some types of commitments may be more effective, over the long run, in inducing technological change. For example, many policy analysts argue that market-based approaches, such as "cap-and-trade" or taxes, are more effective in promoting ongoing

[31]For more on leakage, see Aldy et al. (2003).

technological change than technology standards, which lock in a particular technology and fail to provide incentives for further change.[32]

Changes in public attitudes, awareness, and learning Over the long run, addressing climate change will likely require changes in public attitudes and behaviors. To the extent a commitment can help do so—for example, by raising public awareness—this would be an extra benefit.

Enforceability Given the nature of the climate change problem, countries will be tempted to violate their commitments, since the near-term economic benefits of violation (reduced compliance costs) will typically outweigh the near-term environmental costs (greater climate change). For this reason, climate change commitments may be effective in changing behavior only if they can be adequately monitored and enforced.[33]

Cost-effectiveness

Since countries have only a finite level of resources to devote to climate change and other competing needs, commitments need to get the most "bang for their buck"; they need to reduce each unit of emissions at the lowest possible cost. Most economists agree that market-based approaches—such as emissions trading and taxes—are best from this perspective.[34] The more flexibility market participants have to seek out and utilize low-cost reduction options, the greater the economic effectiveness. That is why the Kyoto Protocol provides not only "where" flexibility (countries may receive credit for emission reductions in another country where the reductions can be made more cheaply), but also "what" flexibility (countries can choose the domestic policies and measures that make most sense for them) and "when" flexibility (countries can time their reductions over a five-year commitment period, and can bank surplus reductions for use in future commitment periods). As discussed above, many economists argue that even longer commitment periods would be desirable, to give companies more flexibility in timing their emission reductions to take advantage of regular capital replacement cycles and additional R & D.[35]

Equity

Commitments should treat participants fairly. As discussed in the equity paper in this report, this is important not only in determining which

---

[32]Wiener (2001).

[33]Barrett (2002).

[34]Aldy et al. (2003).

[35]Lempert et al. (2002).

commitments are politically acceptable; it is also an important end in itself. Whereas environmental and economic effectiveness can both be judged in absolute (objective) terms, equity is by its nature relational. The question is whether a commitment (or set of commitments) is sufficiently equitable to be perceived as such by all participants.[36]

Dynamic Flexibility

Given the likelihood that commitments will periodically need to be revised in light of new scientific and economic information, a commitment would ideally be formulated in a manner that allows revisions as needed. For example, both targets and taxes have a form that can be scaled up or down, becoming more stringent or lax as the circumstances warrant.

Complementarity

The withdrawal of the United States from the Kyoto Protocol opens up the possibility of a fragmented climate regime, with different country groupings adopting different types of commitments. In that case, an important factor in assessing possible commitments would be the feedbacks, complementarities, and potential linkages between commitments in different regimes. For example, if one group of countries adopted commitments involving policies and measures and another group adopted binding emissions targets, it could be difficult for the two regimes to interact. Similarly, if the two groups both adopted "cap-and-trade" regimes—one based on absolute, fixed targets and the other on indexed targets—trades between the regimes, although possible, might be difficult, and need to wait until emission reductions had been achieved and verified.[37]

Political Criteria

From a political perspective, there are two key criteria: whether a particular type of commitment can be negotiated, and whether it can be implemented.

What Commitments can be Negotiated?

In considering future commitments, the question is not simply which commitments are optimal, but which are negotiable. Most of the options

[36] Ashton and Wang (2003).

[37] Trading between systems using absolute and relative targets might also be possible through use of a gateway as in the United Kingdom trading system or a commitment period reserve. For a discussion of the possibility of trading between systems using absolute and relative targets, see Haites (2002); IEA (2002).

for mitigation commitments discussed above have been proposed at one time or another. But none has been able to command a stable consensus.

In some cases, an option may not be negotiable due to domestic political factors in particular countries. For example, carbon taxes are likely to be unacceptable to the United States in the foreseeable future, regardless of which party is in power.

But several more general considerations also affect the negotiability of mitigation commitments, including the following:

Continuity with Kyoto A commitment's continuity with Kyoto could cut both ways in terms of political acceptability. On the one hand, most countries now have a substantial investment in the Kyoto process, so a commitment's continuity with that process would be a point in its favor. At the same, Kyoto has become a negative icon for many in the United States, and is likely to remain a non-starter even once a new administration takes office. In terms of this particular criterion, indexed or conditional targets could conceivably square the circle: they are compatible with the architecture established by Kyoto, including the emissions trading mechanism;[38] but they are more flexible than the fixed, absolute targets in Kyoto, and thus could credibly be characterized as a different approach from Kyoto.

Economic predictability For countries as widely different as the United States and China, a primary concern with Kyoto-style commitments has been the possibility of high compliance costs. Although some economists estimate that the costs of compliance would be low—and that an emissions target for China could even be economically advantageous, given its potential to reduce emissions cheaply and to sell surplus credits to countries with higher mitigation costs—compliance costs depend on many unpredictable variables such as rates of economic and population growth and of technological change, which make economic estimates highly uncertain.[39] From a political standpoint, economic predictability may be as or more important than economic efficiency. Countries want to know in advance what they are undertaking and whether it makes political and economic sense.

Compatibility with sustainable development priorities Most developing countries perceive climate change mitigation and economic development to be in competition with one another: money invested in mitigation is money diverted from economic development. In the long

[38]See *supra* note 37.

[39]Estimates of U.S. compliance costs, for example, differed by more than an order of magnitude, from about $5 billion to over $400 billion per year. See Weyant and Hill (1999); EIA (1998).

run, developing countries will undertake climate change mitigation only if they see synergies with sustainable development goals, for example, through the promotion of energy efficiency, renewable energy, and sustainable land use.[40] So, to the extent that they can be crafted in a manner that advances a country's development goals, climate change commitments will be more attractive.[41]

What Commitments can be Implemented?

To be effective over the long run, commitments need to take into account the capabilities and limitations of the institutions on which implementation and compliance will depend. The importance of institutional capacity is by now well understood in the context of technology transfer: the "best" available technology is not necessarily best for a country lacking the capacity required to use the technology effectively. Instead, technologies that better fit a country's capacities may be more appropriate. At the international level, where institutions are notoriously weak, the issues of implementation and enforcement deserve particular attention. A commitment may make perfect policy sense in the abstract, but, unless it takes account of the practical realities of implementation, a gap is likely to emerge between promise and performance.

Factors relevant to implementation include the following:

Ease of monitoring Different types of commitments vary widely in terms of the ease with which they can be monitored and verified. Some analysts attribute the success of the international oil pollution regime to its reliance on construction and design commitments that are easy to verify (by direct inspection of ships when they are in port),[42] rather than on discharge standards. In the climate change context, national emissions of carbon dioxide can be estimated with a high degree of confidence, but emissions of other gases and removals by sinks are considerably more uncertain. Indexed targets introduce additional complexities, since they require monitoring not only of emission levels but also the variable to which emission allowances are pegged.[43]

Predictability of compliance Most implementation of international commitments takes place at the national level, through national law, so commitments adopted internationally need to be capable of domestic legal application. One criticism of obligations of result, such as targets and

[40]Heller and Shukla (2003).
[41]Winkler et al. (2002).
[42]Mitchell (1994).
[43]IEA (2002), at 139 (GDP measurement is relatively inaccurate in many developing countries).

timetables, is that, because compliance depends on changes in behavior by firms and individuals (as is the case with climate change), it is difficult for a country to predict accurately whether it will achieve the required result. By contrast, obligations of conduct, such as equipment standards, tend to be easier to implement at the national level: if a country engages in the required conduct (for example, by requiring firms to install the specified equipment), then it is in compliance.

V. OPTIONS FOR FUTURE COMMITMENTS

The following represent some of the most frequently discussed options for future climate change mitigation commitments. Three caveats are in order. First, these options are, of course, not the only possibilities. Instead, they represent a range of approaches chosen to illustrate many of the general issues regarding mitigation commitments. Second, the assessments of the various options identify the most prominent advantages and disadvantages of each approach, rather than applying the assessment criteria discussed above in a systematic manner. Finally, these options could be combined in various ways; they are not mutually exclusive. For example, an agreement might commit states not only to an emissions target, but also to efficiency standards and funding for research and development. Or it might set forth different types of commitments for different categories of countries—a binding emissions target, say, for industrialized countries, and a non-binding one for developing countries. Or it might set forth an evolutionary pathway, with different types of commitments kicking in at different times.[44]

Absolute, Sequentially-negotiated National Emissions Targets

The Kyoto Protocol sets forth fixed national emission targets for the 2008-2012 period. The idea is that the first five-year commitment period will be followed by other commitment periods, to be negotiated on a rolling basis. Kyoto-style targets, if applied to all significant emitters, would have several benefits:

- *Environmental effectiveness* Fixed targets, if complied with, provide the greatest environmental certainty.
- *Cost-effectiveness* Fixed targets can be cost-effective if combined with emissions trading (as in the Kyoto Protocol) and with "when

[44]For example, as the text discusses, developing countries might start with non-binding emissions targets and more towards more binding targets over time, as they satisfy specified graduation criteria.

flexibility" (either through a longer commitment period or through provisions for banking and borrowing).

- *Equity* Fixed targets (like targets generally) can be differentiated among countries to meet equity concerns.
- *Dynamic flexibility/scalability* Fixed targets (like targets generally) can be adjusted up or down to take account of new information.
- *Continuity with Kyoto* For countries that support Kyoto, fixed targets would provide the greatest continuity.

At the same time, absolute targets also have several significant drawbacks:

- *Difficulties of negotiating* The costs of achieving a fixed national emissions target are uncertain, and depend on many factors (such as rates of economic growth and technology change) that are difficult to predict. Although absolute targets can allow considerable flexibility in implementation (as illustrated by the Kyoto mechanisms), they represent a legal straitjacket in the sense that, once agreed, they do not provide for changing circumstances. This rigidity could make iterative negotiation of fixed short-term targets difficult.
- *Perceived incompatibility with development priorities* Absolute targets are particularly problematic for developing countries and countries with rapidly growing economies, since they are seen as representing a potential constraint on economic growth. Of course, targets could build in "headroom" to allow developing country emissions to grow. Unless economic and emissions growth can be predicted reliably, however, setting fixed targets for developing countries involves a difficult balance between targets that are too loose (and possibly create surplus allowances, above business-as-usual emissions, often referred to as "hot air"), and targets that are too strict and inhibit development.

Indexed National Targets

Indexed targets have some of the same advantages and disadvantages as fixed targets. On the positive side, they are cost-effective if coupled with trading, which appears difficult but not impossible; they can be differentiated between countries[45] and made more or less stringent as the circumstances warrant; and they could provide continuity with Kyoto. In addition, they provide greater flexibility than fixed targets by allowing

[45]Differentiation would be possible on the basis of not only the stringency of the target, as with fixed targets, but also the variable to which targets are indexed.

emissions to vary depending on whether the economy (or whatever variable emissions are pegged to) grows or shrinks. This can prevent the creation of "hot air" due to an economic downturn, but comes at the expense of environmental certainty. Indeed, if economic growth is sufficiently high, permitted emissions may even go up rather than down. And although the increased flexibility of indexed targets mitigates the problem of economic uncertainty, it does not eliminate it altogether.

Sectoral Targets

Sectoral targets (either fixed or indexed) have the benefit over economy-wide targets of allowing states to proceed incrementally. Rather than attempt to develop a target that makes sense for the entire economy, states can address emissions in a step-by-step manner, starting with a more limited set of activities in sectors such as energy or transportation. That is why many national strategies for addressing GHG emissions take a sectoral approach. Moreover, in some cases, more is known about emissions in one sector than another, so sectoral targets may help ease monitoring concerns. Finally, sectoral targets would make it more difficult for countries to give preferential treatment to particular sectors and, in that respect, could help ease competitiveness concerns.

But addressing emissions on a sectoral basis comes at a price. If states are restricted as to which types of emission reductions 'count' internationally, they may be unable to take advantage of the most cost-effective options. Even if targets are developed for all sectors with significant GHG emissions, separate sectoral targets prevent countries and firms from making tradeoffs across sectors, doing more in a sector where emissions can be reduced more cheaply and less in another sector where reductions are more expensive.[46] Allowing such tradeoffs not only makes economic sense; it may also make targets more negotiable by giving countries flexibility to focus on those sectors where they can reduce emissions with the least economic and political pain. Sectoral targets also could distort competitiveness and give rise to complex equity issues if different circumstances prevail in the same sector in different countries.

Hybrid Targets (Safety Valve)

Hybrid targets, advocated primarily by economists,[47] were put on the table informally by Brazil in 2000, during the negotiations that culminated

[46]Although trading across sectors could mitigate this concern, if trading were fully allowed, then the sectoral targets would, together, amount to an overall national target.
[47]Kopp, Morgenstern, and Pizer (1997); McKibbin and Wilkoxen (1997).

in the Bonn/Marrakech Accords. Hybrid targets have a number of desirable features:

- *Economic predictability and negotiability* By ensuring that the costs of mitigation commitments cannot rise above a predetermined level, hybrid targets remove one of the principal obstacles to the negotiation and acceptance of emission reduction targets.[48]
- *Equity* Although the safety valve level would need to be the same globally (otherwise the country with the lowest safety valve price could continue selling permits until the global trading price equilibrated at its safety valve level), commitments could still be differentiated through the emission reductions targets. (With a hybrid target, a country's costs are a function of both the safety valve price and the stringency of its emission target.) Thus, the safety valve, like fixed targets, is compatible with the application of equity criteria.
- *Scalability* A hybrid target could be scalable through its safety valve price as well as its emission reduction targets. To facilitate planning by business (which is currently difficult due to uncertainty about the stringency of targets after Kyoto's first commitment period), the safety valve price could have an automatic escalator, which would apply unless the parties decided otherwise.

Of course, the economic predictability of hybrid targets comes at the expense of environmental predictability—the principal strength of fixed emission reduction targets. This has an obvious downside: if mitigation costs prove high and the safety valve kicks in, then the level of actual emission reductions would be less than under a fixed target. But there are risks either way. Just as we have no assurance what level of reductions a given price will buy, we have no assurance how much a particular emissions reduction will cost. The difference is, the economic risks of excessive costs are near-term, while the environmental risks of insufficient reductions in emissions are longer-term and may be correctable through stronger measures later. Moreover, economic predictability could even provide an environmental benefit: with a guaranteed ceiling on costs, countries might be willing to accept more ambitious targets, leading to greater emissions reductions if costs prove low.

In addition to environmental uncertainty, a hybrid target would be likely to face issues of political acceptability in countries opposed to the introduction of new taxes, since the safety valve would operate, in effect,

[48]In rejecting the Kyoto Protocol, for example, the Bush Administration identified potential harm to the U.S. economy as one of Kyoto's two fatal flaws.

like a tax. Agreement could also prove difficult on a safety valve price as well as on what to do with any money raised from the sale of additional permits. (Would the money go to an international fund and, if so, who would control the fund, or would it be spent domestically?) In addition, if the safety valve price were set relatively low, it could limit incentives for technological research and innovation, by giving companies an easy way out if costs prove high.

Non-binding ("No-Lose") Targets for Developing Countries with Graduation Criteria

No-lose targets have been proposed primarily as a means of providing incentives for developing countries to accept emission targets.[49] Over the long run, developing countries may need to accept binding targets as their economies develop. No-lose targets could serve as a useful transitional device, possibly in conjunction with criteria that define when a developing country would graduate from a non-binding to a binding target. During the transitional period, no-lose targets could be combined with legally binding commitments in various ways. For example, under a "dual commitment" approach, a relatively weak but legally binding commitment could be combined with a stricter one-way commitment that, if surpassed, would allow a country to engage in emissions trading.[50] Given the high variability of economic growth rates in developing countries, an indexed rather than fixed target could be used to prevent the target from becoming too easy or too hard.

Efficiency/Technology Standards

The difficulties involved in negotiating, monitoring, and enforcing emission targets have made technology standards more attractive, even to some economists who, as a rule, criticize such standards as inefficient.[51] Technology standards—for example, relating to energy efficiency—could be negotiated by governments or through public-private partnerships. One advantage is that they could have a significant environmental impact, even in the absence of universal acceptance, through tipping effects. As Scott Barrett explains: "If enough countries adopt a [technology] standard, it may become irresistible for others to follow, whether because of network effects, cost considerations...or lock-in."[52] If so, technology standards

[49]Philibert and Pershing (2001).

[50]Kim and Baumert (2002); Philibert and Pershing (2001).

[51]Barrett (2002), at 398.

[52]Id., at 395.

would be essentially self-enforcing, and would not involve the compliance issues raised by emission targets. Moreover, trade rules may allow countries that accept a technology standard to exclude from their markets products that fail to meet the standard, putting additional pressure on non-participants to join the technology regime.[53] Finally, technology standards are comparatively easy to monitor, since in most cases they simply require inspection to make sure that the proper equipment is being used.

At the same time, technology standards have a number of significant drawbacks that have limited their appeal in the climate change negotiations thus far. They depend on governments being able and willing to pick technologies based on sound technical considerations (rather than on the basis of which technologies are produced domestically or are backed by a politically powerful lobby). They lock in technologies and do not provide an incentive for further innovation. They limit flexibility by prescribing not just a result, but how countries must achieve it. For these reasons, among others, over the last decade, environmental policy has tended to move away from command-and-control regulation towards market-based approaches.

R & D Commitments

If emission reduction technologies such as hydrogen fuel cells or carbon capture and storage became practicable and economic, this could go a long way towards overcoming the existing barriers to climate change mitigation. But recent studies indicate that, despite the high profile of the climate change issue, investments are going down overall in mitigation-related research and development.[54]

International commitments by states to provide funding for research and development are not unprecedented. For example, the international space station is the product of an agreement providing for multilateral cooperation and funding.[55] Voluntary approaches have also sometimes proven successful. Twenty-one countries including the United States currently contribute to the Consultative Group on International Agricultural Research, which funds research centers around the world.[56] So, while some countries such as the United States may be wary of any new financial obligations, financing of R & D might prove attractive,

[53] Charnovitz (2003).
[54] Margolis and Kammen (1999).
[55] Barrett (2002), at 394.
[56] See http://www.cgiar.org/index.html.

either as an alternative to more stringent types of mitigation commitments or, at a minimum, as an add-on.

VI. CONCLUSIONS

In developing new mitigation commitments, the toolbox available to policymakers contains a wide range of options. In this respect, the climate change debate has grown considerably more sophisticated over the past decade.

In moving forward, it is unlikely that one size will fit all: different mitigation commitments will prove more or less attractive to different countries. The question will be whether to undertake the extremely difficult political task of negotiating a unitary system or to accept—at least for the short- to medium-term—a more variegated set of commitments, under either a single regime based in the UNFCCC or multiple regimes at the bilateral, regional, and global levels.

In general, the various types of possible commitments are complementary to one another rather than mutually exclusive, both within and between countries. National and international climate policy could consist of a mix of different types of emission targets for different countries and sectors, as well as technology standards and R & D commitments.

But to the extent that commitments vary between countries, international climate change policy will face several important challenges: first, to ensure that the various commitments add up to a sufficient level of effort overall; second, to ensure that the mix of commitments across countries is, broadly speaking, equitable; and third, to promote linkages between different national programs and, if there are multiple international regimes, between those regimes. None of these tasks is insuperable, and careful policy analysis can help elucidate the possible solutions. But, in the end, the successful resolution of these issues will depend on mustering greater political will among states to address climate change.

References

Aldy, J.E., P.R. Orszag and J.E. Stiglitz. 2001. Climate Change: An Agenda for Global Collective Action, Pew Center on Global Climate Change, Arlington, VA, USA.

Aldy, J.E., R. Baron and L. Tubiana. 2003. "Addressing Cost: The Political Economy of Climate Change." In: Beyond Kyoto: Advancing the International Effort Against Climate Change, Pew Center on Global Climate Change, Arlington, VA, USA.

Ashton, J. and X. Wang. 2003. Equity and Climate: In Principle and Practice. In: Beyond Kyoto: Advancing the International Effort Against Climate Change, Pew Center on Global Climate Change, Arlington, VA, USA.

Barrett, S. 2002. Environment and Statecraft. Oxford University Press.

Baumert, K.A., O. Blanchard, S. Llosa and J.F. Perkaus. (eds.). 2002. Building on the Kyoto Protocol: Options for Protecting the Climate, World Resources Institute, Washington, D.C., USA.

Bouille, D. and O. Girardin. 2002. Learning from the Argentine Voluntary Commitment. In: Baumert et al. 2002, ch. 6.

Charnovitz, S. 2003. Trade and Climate: Potential Conflicts and Synergies. In: Beyond Kyoto: Advancing the International Effort Against Climate Change, Pew Center on Global Climate Change, Arlington, VA, USA.

Cooper, R. 1998. Toward a Real Treaty on Global Warming, Foreign Affairs 77(2): 66-79.

Energy Information Agency (EIA). 1998. What Does the Kyoto Protocol Mean to U.S. Energy Markets and the U.S. Economy, http://www.eia.doe.gov/oiaf/kyoto/pdf/sroiaf9803.pdf

Hahn, R.W. and R.N. Stavins. 1999. What Has Kyoto Wrought? The Real Architecture of International Tradable Permit Markets, Resources for the Future Discussion Paper 99-30.

Haites, E. 2002. Linking Domestic and Industry Greenhouse Gas Emission Trading Systems. Paper prepared for EPRI, IEA and IETA.

Heller, T.C. and P.R. Shukla. 2003. Development and Climate: Engaging Developing Countries. In: Beyond Kyoto: Advancing the International Effort Against Climate Change, Pew Center on Global Climate Change, Arlington, VA, USA.

IEA. 2002. Beyond Kyoto: Energy Dynamics and Climate Stabilization, Paris. France

IPCC. 2001. Climate Change 2001: Mitigation, Working Group III Summary for Policymakers, Cambridge University Press. Cambridge, NY.

Kim, Y.-G. and K.A. Baumert. 2002. Reducing Uncertainty through Dual-Intensity Targets. In: Baumert et al. ch. 5.

Kopp, R., R. Morgenstern and W. Pizer. 1997. Something for Everyone: A Climate that Both Environmentalists and Industry Could Live With, Resources for the Future, Washington D.C., USA.

Lempert, R.J., S.W. Popper and S. Resetar. 2002. Capital Cycles and the Timing of Climate Change Policy, Pew Center on Global Climate Change, Arlington, VA, USA.

Margolis, R.M. and D.M. Kammen. 1999. Underinvestment: The Energy Technology and R & D Policy Challenge, Science 285: 690-692.

McKibbin, W.J. and P.J. Wilkoxen. 1997. A Better Way to Slow Global Climate Change, Brookings Policy Brief No. 17, Washington, D.C., USA.

Mitchell, R. 1994. Intentional Oil Pollution at Sea: Environmental Policy and Treaty Compliance, MIT Press, USA.

Nordhaus, W. 2001. After Kyoto: Alternative Mechanisms to Control Global Warming, Paper presented to joint session of the American Economic Association and the Association of Environmental and Resource Economists, Atlanta, Georgia, Jan. 4.

Pershing, J. and F. Tudela. 2003. A Long-term Target: Framing the Climate Effort. In: Beyond Kyoto: Advancing the International Effort Against Climate Change, Pew Center on Global Climate Change, Arlington, VA. USA.

Philibert, C. and J. Pershing. 2001. Considering the Options: Climate Targets for All Countries, Climate Policy 20: 1-17.

Samaniego, J. and C. Figueres. 2002. Evolving to a Sector-based Clean Development Mechanism. In: Baumert et al. 2002, ch. 4.

UNFCCC. 2002. Scientific and Methodological Assessment of Contributions to Climate Change: Report of the Expert Meeting, Doc. FCCC/SBSTA/2002/INF.14, 16 October.

Victor, D.G. 1999. Enforcing International Law: Implications for an Effective Global Warming Regime, Duke Environmental Law and Policy Forum, Vol. 10.

Weyant, J. and J. Hill (eds.). 1999. Costs of the Kyoto Protocol: A Multi-Model Evaluation, Energy Journal, Special Issue.

Wiener, J. 2001. Policy Design for International Greenhouse Gas Control. In: Michael A. Toman (ed.). Climate Change Economics and Policy: An RFF Anthology.

Winkler, H., R. Spalding-Fecher, S. Mwakasonda and O. Davidson. 2002. Sustainable Development Policies and Measures: Starting from Development to Tackle Climate Change, In: Baumert et al., ch. 3.

39

CHAPTER

Flexible Options for Future Action

Cédric Philibert[*,**]
Principal Administrator, Energy Efficiency and Environment Division
International Energy Agency, 9 rue de la Féddération
75739 Paris Cedex 15, +331 40 57 67 47
E-mail: Cedric.philibert@iea.org

1. INTRODUCTION

The ultimate objective of the United Nations Framework Convention on Climate Change (UNFCCC) is to stabilize atmospheric concentrations of greenhouse gases "at a level that would prevent dangerous anthropogenic interference with the climate system." This exact level and the timeframe to achieve stabilization have been left undecided, but should "be sufficient to allow ecosystems to adapt naturally to climate change, to ensure that food production is not threatened and to enable economic development to proceed in a sustainable manner." To reach such an objective, the Kyoto Protocol is at best a beginning. New steps will be needed.

The relevant conversations have begun under the auspices of the UNFCCC after the Conference of the Parties held in Montreal in December 2005, following two distinct processes - the ad hoc working group on the second commitment period of the Kyoto Protocol, and the broader Dialogue on future action. The Bali Conference has somewhat linked these two processes. Where could they lead? And what future international

[*]The views expressed here do not necessarily represent those of the IEA, the OECD or their member states.
[**]OECD/IEA (2007) retains the copyright of this chapter.

framework for action against climate change could best help this achievement?

Some analysts believe that the next steps could simply extend the Kyoto agreement beyond 2012, see it joined by Parties who have not signed it yet, and give quantified objectives to emerging economies. Others propose entirely different types of agreements. It is probably wiser to further elaborate on the basic structure of the Kyoto Protocol - quantified objectives and emissions trading - and add new features. These features - partial indexation of emission targets on economic growth, price caps, and, for developing countries, non-binding targets - may help countries adopt relatively more ambitious targets than otherwise. As will be shown, they would do so in better addressing the uncertainties on abatement costs. They would provide greater incentives to participate and comply, or at least reduce the disincentives to participate. They would thus help to simultaneously broaden and deepen climate change mitigation action - two moves often suggested contradictory.

This contribution is in four parts. The first part considers the advantages and limitations of the Kyoto Protocol. The second part considers some of the radical alternatives that have been suggested as possible substitutes. The third part considers some options to increase the flexibility in future Kyoto-like arrangements with a view of attracting broader participation. The fourth part assesses the various dimensions of flexibility and shows that the options for future commitments may increase the overall economic efficiency of the action to mitigate climate change in a context of uncertainty.

2. ASSESSING KYOTO

The Kyoto Protocol is based on quantified emission objectives made flexible through emissions trading. It indirectly addresses emissions from developing countries through the Clean Development Mechanism. The most important question is if and how its structure could be expanded to cap all or almost all emissions at a global level. This may prove difficult unless the difficulties resulting from uncertain abatement costs are fully acknowledged.

2.1 The Advantages of Emissions Trading

The main strength of the architecture of the Kyoto Protocol is in quantified objectives and emissions trading. Most greenhouse gases have no direct local environmental effects; they rapidly mix in the atmosphere, and where they are emitted does not matter. Emissions trading, therefore, does not modify the environmental effect of the targets; but it lowers the costs of emissions reductions, which, depending on the level of stabilization chosen, may be considerable (IEA 2002). This, in turn, is good for the environment, especially as climate change is a long-term issue. Though usually defined as the capacity to reach a given objective at the lowest possible cost, cost-effectiveness can also offer the greater environmental benefits for a given cost—the cost that our societies are willing to pay to mitigate climate change.

Another advantage is that emissions trading, if implemented at the domestic level as well as at the international level, offers governments the flexibility to fine-tune the balance between free allocation and auctioning. This could improve the acceptability of the new regulations to incumbent emitters on the one hand, and maximize social welfare through revenue recycling, on the other. Finally, emissions trading allows international negotiations to focus on an acceptable distribution of efforts, which need not be cost-effective from the onset. This is a key for equity.

However, to be fully cost-effective an agreement would need to include all (major) emitting countries, allowing the abatement to take place wherever they cost less and preventing leakage. As is well known, developing countries have not been given any quantified objectives on their emissions in the Kyoto Protocol, which President Bush rejected. The Kyoto Protocol only caps about a third of global emissions.

2.2 The Clean Development Mechanism

To some extent, the Clean Development Mechanism (CDM) instituted by the Kyoto Protocol may substitute for quantified objectives by developing countries and give access to cheap reduction opportunities. However, the CDM is impeded by substantive transaction costs, resulting from the need to assess each project, prove it is additional to what would have happened otherwise, and to define an appropriate baseline. Relaxing the additionality criteria may augment neither the efficacy of the CDM nor its possible benefits for developing countries (Asuka and Takeuchi 2004). Arguably, however, allowing emission reductions originating from 'programmes' undertaken in developing countries, as decided at Montréal, may facilitate agreeing on much larger projects and reduce unitary transaction costs.

Still, what the CDM can accomplish remains to be seen. Nuclear power has been excluded, carbon dioxyde capture and storage not (yet?) included. Energy efficiency improvements have difficulties finding their way through the procedures for demonstrating additionality. Renewable energy projects often are two expensive. Afforestation projects are limited, and halting deforestation is not considered. The CDM thus seems to ignore some of the most important areas for emission reductions.

2.3 Extending and Expanding the Kyoto Protocol

For its detractors, the Kyoto Protocol provides too little environmental benefits at too high costs. This may be difficult to prove right or wrong for uncertainties abound on both the benefit and cost sides. Even the real effects on global emissions of the Kyoto Protocol itself are not known with precision - not to mention the dynamics it may create for the future. The control of greenhouse gases (GHG) in some countries may increase emissions in other countries as production escapes the new regulation (leakage) while the global dissemination of less carbon-intensive technologies developed in response to emission controls (spill-over) may reduce these same emissions. Grubb et al. (2002), for example, believe that Kyoto's spill-over effects will more than offset Kyoto's leakage. In any case the direct effects of Kyoto on climate change can only be small, because climate change is a problem of a 'stock' nature; what drives climate change is the slow build-up of atmospheric GHG concentrations over decades and centuries.

A more serious criticism comes from the acknowledgement that the Protocol currently caps only a third of global emissions - although there is some irony in this criticism when it comes from those who have refused to have their own emissions capped by it. Some analysts thus argue that one should simply keep Kyoto— as it is today (e.g. Oppenheimer and Petsonk 2004). It would progressively become a broader, more global agreement, as developing countries develop and reach some thresholds in per capita income (multistage approach). Or developing countries could be incorporated sooner but with large amounts of surplus emission rights, which may or may not result from the adoption of a global rule for emission allocation, such as convergence towards equal per capita allocation (Meyer 2000, Aslam 2002). Industrialized countries which have resisted participating in the first period of the Kyoto Protocol would possibly be given more lenient targets in subsequent ones.

The problem with 'progressive' approaches is that they are hardly compatible with low concentration levels if, ultimately, necessary. This is due to the late entry into the system of most developing countries (Berk and den Elzen 2001), but also to the less stringent targets given to some others. As Socolow (2006) recalls, *"much of the world's construction of long-lived capital stock is in developing countries. Unless energy efficiency and carbon efficiency are incorporated into new buildings and power plants now, wherever* *they are built, these facilities will become a liability when a price is later put on CO_2 emissions".*

The problem with the convergence option is that it may first provide a large amount of excess allowances to developing countries. In any case, uncertainties on business as usual emission levels may prevent developing countries to adopt firm and fixed emission targets - unless they were given sufficient amount of allowances to cover the highest emission growth scenarios. This would not lead to a very effective global framework, as industrialized countries would need to buy large amounts of surplus allowances before triggering any real emission reductions in developing countries (IEA 2002). One lesson from the Kyoto Protocol is that providing hot air to some countries to help others accept tough targets may not work; for example, the 'blank check' to Russia was one of the reasons invoked by the US administration to reject the Kyoto Protocol.

Later on, allocation based on per capita convergence may bind the emissions of developing countries at much lower per capita levels than those previously enjoyed by citizens of industrialized countries. Arguably, some technology spill-over will reduce the peak of energy intensity reached by new-comers in their industrial development, as happened in the past (Martin 1988). Nevertheless, this constraint on emissions might be perceived by developing countries as an unfair constraint on their economic development itself (Chen and Pan 2003).

In sum, keeping Kyoto unchanged while only playing with the 'numbers,' i.e., the size of the respective allowances, produces the following dilemma: ensure broad participation with weak targets or undercut the goal of broad participation in setting ambitious targets that not all countries will accept.

2.4 The Problem with Fixed Targets

There might be a more fundamental reason that makes a mere extension of the Kyoto-style quantified objectives to all major emitters so difficult. It has to do with their fixed and binding nature, which gives birth to the criticism of them being necessary arbitrary. This criticism has power - and indeed some legitimacy.

An economically-efficient agreement would not only be cost-effective, it would also ensure that environmental benefits outweigh abatement costs. It would provide maximum net benefits if it can get as close as possible to an optimum level of abatement is undertaken - when the marginal abatement cost equals the marginal environmental benefit. This is naturally very difficult in the case of climate change, where so many uncertainties abound about abatement costs, on the one hand, and policy benefits, i.e. the value of avoided climate damages, on the other. However, the economic theory shows that in the context of uncertain costs not all possible arrangements are equally effective.

If abatement costs were known with certainty, then a quantified objective would define a price, or a price-based instrument (e.g. a carbon tax) would define a global quantity. As abatement costs are uncertain, quantity and price instruments are not equivalent. A price instrument would offer certainty on the marginal cost incurred, but not on the actual level of abatement. A quantity instrument would offer certainty on the level of abatement, but not on the costs incurred.

Which instrument is preferable to mitigate climate change? The stock nature of the problem makes the marginal policy benefits roughly constant over any credible policy interval. That is, avoided marginal climate damages might be high or low, but the first tonne of carbon dioxide that is not emitted in any given year is likely to bring about the same benefit than the last one.

Naturally, it is always possible that there are thresholds in GHG concentrations leading to 'non-linear' responses from the climate system, or 'climate surprises'. Recent scientific studies have tended to identify critical temperature changes for some climate change impacts, such as: less than 1°C for coral bleaching, 1°C for the disintegration of Greenland ice sheet, 1-2°C for broad ecosystem impacts with limited adaptive capacity, 2°C for the disintegration of West Antarctic Ice Sheet, 3°C for the shutdown of thermohaline circulation. But only probability density functions can yet express the link between GHG concentration levels and these temperature changes (Schneider and Lane, 2006). It seems even less possible to identify any 'tipping point' in response of the climate system to small variations in emission trends over a decade or two, which have relatively little short term impact on the evolution of concentration levels.

By contrast, the cost of abating the first tonne is minimal, while the cost of abating 'the last one' (of course, depending on the depth of the cuts) might be very high and possibly higher than the marginal benefit it provides. Therefore, price instruments, which spontaneously adjust the emission cuts to the reality of the costs, should be preferred over quantity instruments. In other words, the certainty provided by quantitative targets on emissions in any given year has little value but may cost too much (Newell and Pizer 2003, Pizer 2002, Philibert 2006).

Governments need not be fully aware of this economic literature to show – for many of them at least – reluctance in adopting sufficiently ambitious (if any) emission-reduction objectives. Abatement costs are obviously uncertain, as they depend from uncertain economic growth, technology developments and evolutions of relative energy prices. Cost uncertainties give birth easily to cost controversies forcefully fuelled by various interest groups, which will eventually prevent some governments to step in.

In sum, the Kyoto Protocol, despite its inherently limited short term effects on climate change, offers some very interesting features for the future international architecture of climate change mitigation. However, the conjunction of the stock nature of the climate problem and of the uncertainties surrounding abatement costs make any arrangement based on fixed quantitative goals less than fully economically efficient as well as unlikely to be universally accepted as a means to cap emissions.

3. RADICAL ALTERNATIVES

Many proposals have been made for succeeding the first commitment of the Kyoto Protocol (Bodansky 2004, Philibert 2005a). While some build upon the Kyoto structure, others are more radical alternatives. The most often quoted radical alternatives seem to be commitments on policies and measures, carbon taxes, and 'technology protocols.' This section considers them in turn.

3.1 Policies and Measures

An existing obligation in the UNFCCC commits all Parties to undertake policies and measures that help mitigate climate change. Identifying specific policy requirements may be a logical extension from existing commitments. One possible approach would be to invite developing and/or developed countries to identify a set of win-win policy reforms, according to their national circumstances. Developing countries, for example, would look for 'sustainable development policies and measures' corresponding to their own sustainable development objectives (Winkler et al. 2002), then identify whether they lead to emission reductions below business-as-usual levels, and then seek to have them financed by industrialized countries through the Convention process.

In the course of the negotiations leading to the Kyoto agreement, however, developing countries have proven very reluctant to make commitments on policies and measures seen as contradictory to their sovereignty. It may be difficult to ensure that a wide set of policies and measures provide cost-effective emission reductions. The international financing of the latter could more easily leverage both public and private financing through emissions trading than through other mechanisms in the Convention. This does not mean that the efforts to identify and implement policies that respond to local sustainable development objectives while reducing (at least in relative terms) GHG emissions are useless; quite to the contrary, they are very useful, in both industrialized and developing countries. Emissions trading should add to these efforts – not replace them.

3.2 Carbon Taxes

Carbon taxes offer perhaps the most convincing alternative to the Kyoto framework from a theoretical perspective, especially under the form of harmonized domestic taxes advocated by Nordhaus (2002). Their political economy, however, remains difficult. At the domestic level, taxes are usually unpopular and raise profitability concerns for industry if some competitors in other countries do not face the same additional costs. Taxes offer little flexibility to governments to accommodate these concerns while maintaining their environmental effectiveness.

At the international level, uniform tax rates are required for reasons of cost-effectiveness, but the resulting distribution of costs may be unacceptable, especially by developing countries, likely to ask for side-payments. In sum, carbon taxes can be – and already are – useful as part of domestic policy packages, but making them the centrepiece of any future international strategy may prove extremely difficult.

3.3 Technology Protocols

Technology protocols have been suggested as a possible alternative to the Kyoto Protocol, in particular by Barrett (2003), who believes that Kyoto lacks credible incentives for participation and enforcement mechanisms. His proposal would involve collaborative research and development in developing new technologies, follow-up protocols establishing technology standards, a multilateral fund to help spread the new technologies to developing countries, a short-run system of pledge-and-review, and a further protocol for adaptation assistance.

Clearly, although various behavioural changes might help achieve stabilization of concentration, deep technology changes will be required. Policies and measures specifically designed to 'push' research and development might bring an invaluable contribution to such technical change. Dissemination of new technologies, however, is unlikely to be rapid enough in the absence of long-term price signals that only economic instruments, such as either taxes or tradable permit schemes, would provide (Philibert 2003). Could technological standards substitute for price signals in providing for rapid dissemination of innovation?

Barrett recognizes that such an approach would not be cost-effective and thus only a second best. But, he argues, the setting of standards "often creates a tipping effect. If enough countries adopt a standard, it may become irresistible for others to follow, whether because of network effects, cost considerations (as determined by scale economies), or lock-in." This may not always be the case, however. Let us suppose some industrialized countries adopt a standard that would, for example, force energy-intensive industries, the power sector, and refineries to give up fossil fuels or capture and store the carbon dioxide. It is not easy to figure out why this would obligate or incite the rest of the world to follow suit even if this entails huge costs.

Would new multilateral funds make the difference? Maybe - but can new funds leveraging only public money do more than mechanisms, such as emissions trading, leveraging potentially both public and private money? Also, some of these technologies may be disseminated by their own virtues as they become fully cost-effective thanks to economies of scale and learning curves. The technology spill-over effects may be similar to the Kyoto case. Finally, the Intergovernmental Panel on Climate Change (IPCC) made clear that energy efficiency improvements at the end-user level, likely to provide the bulk of short-term affordable emission reductions, require 'hundreds of technologies' (Moomaw and Moreira 2001). Should one then negotiate hundreds of protocols?

In sum, international technology collaboration is useful but already exists, notably through 40 International Energy Agency Implementing Agreements. It could be strengthened, and standards might be one area for improvement (Philibert 2004). Technology collaboration certainly should accompany or be part of future climate agreements. It remains doubtful that it should be the centrepiece. And it may not be easy to 'integrate' in a single action framework countries following a 'technology-only' strategy and countries accepting quantified emission limitations, for fairness in setting efforts and fears of distorting economic competition may require difficult comparisons of either efforts or likely results (Philibert 2005c).

4. TRANSFORMING KYOTO

Transforming Kyoto into a superior agreement would mean finding ways to make the agreement global and more effective in addressing cost uncertainty. These points are linked; it would probably be easier to get developing countries involved in a global emissions trading regime on the basis of assigned amounts that would be exactly set on their business as usual, unabated emission trends, if these could be known with certainty. Thus, they would have everything to gain and nothing to lose from accepting targets. Similarly, the difficulties for some industrialized countries to accept their Kyoto targets are in part due to the difficulty of estimating the resulting costs with certainty and without controversy. This section considers the following five options for future quantitative emission commitments: banking and borrowing, dynamic targets, targets with price caps and non-binding targets.

4.1 Banking and Borrowing, Long Periods

As climate change is a long term issue what matters is somehow the overall amount of GHG emitted over a long period of time. Increasing the cost-effectiveness in achieving a given cumulative objective would require a greater 'when flexibility' (so nicknamed by reference to the 'where flexibility' that either emission trading or taxes provide). In the context of emissions trading schemes, where flexibility could come from the ability to bank surplus emission allowances and borrow them from the future, or (very) long commitment periods. Banking is fine but would only be effective after initial periods where market players can accumulate allowances to face price spikes. Creating the conditions for this would delay action. Borrowing has proven effective in domestic policies but seems problematic in an international setting, as it requires a strong compliance regime that could extend over decades. If 'when flexibility' rests on long commitment periods, it may necessitate decisions on long term goals, which may be premature as uncertainties loom large about future costs and benefits of climate policies. Allocating allowances a long time in advance of commitment periods would create some kind of liability, even if allowances are denied the nature of 'property rights', which may end up difficult to modify at a later stage. However, the other options considered in this section also provide for greater 'when flexibility'.

4.2 Dynamic Targets

One way to get around these difficulties might be to index assigned amounts on actual economic growth. Economic forecast will likely be part of the definition of assigned amounts. Deviation from this forecast could then lead, under 'dynamic targets,' to modifying these assigned amounts, so as to maintain roughly constant the 'gap' between unabated trends and assigned amounts – and the required level of efforts. Dynamic targets need not be 'intensity targets' – a form of target that can be simply expressed in terms of GHG per economic output. Indexation could in fact take a wide variety of forms and be only partial (Ellerman and Wing 2003). One advantage of partial indexation might be to reduce the risk of 'double pain' in case of unexpected economic recession and to drive a greater level of efforts (though allowing greater emission levels than with the original objective) in case of an unexpected economic boom (IEA 2002). One difficulty might be, especially in developing countries, the need to provide accurate measurements of economic variables such as gross domestic product.

Indexing assigned amounts may address the uncertainty arising from uncertain economic forecasting – but its possible effects seem limited, at least in the case of developing countries (Philibert 2005b). Moreover, they would not address other sources of uncertainty on abatement costs arising, say, from the uncertain evolution of availability and costs of various energy sources, and unknown future depth and speed of technical change.

4.3 Price Caps

A more comprehensive way to deal with cost uncertainty might be the introduction of price caps into the international trading regime, as suggested by Pizer (2002) following a concept from Roberts and Spence (1976). This could take the form of making supplementary permits available in unlimited quantities at a fixed price – at the country level (for domestic entities) and/or at the international level (for countries) (IEA 2002). With a price cap, all emission abatement needed to achieve the quantitative commitments would be undertaken as long as the marginal cost of abatement is lower than some agreed price. If abatement costs reach this price, then economic agents and/or countries would be able to cover excess emissions with supplementary permits at the agreed fixed price.

A single international price is necessary for unrestricted global trading. Trading might still be possible, however, albeit with the risk of a loss of cost-effectiveness, if prices vary across countries. One solution to ensuring the integrity of the system is that net sellers do not make 'use' of the price cap (i.e., their actual emissions remain below their assigned amounts). Thus, no Party or entity would 'resell' supplementary permits. However, an agreement on a single price amongst countries of a relatively similar level of development, despite a varying willingness-to-pay, is not necessarily unattainable, as this price cap does not prevent differentiation in respective levels of effort and assigned amounts (Philibert 2005b).

4.4 Non-binding Targets

A similar option for developing countries would be that of non-binding targets. These targets may provide—though emissions trading—an incentive for emission reductions, where sales could occur if (and only if) actual emissions are less than the targets (Philibert 2000). This option may be particularly attractive for developing countries. The existence of such an incentive, however, requires that other countries are potential buyers bound by firm targets.

There are different ways to ensure that countries with non-binding targets only sell emission allowances that exceed the coverage of their actual emissions. The most effective may be to require countries that have over-sold to purchase enough allowances to cover their actual emissions up to the level of the non-binding target—but not beyond. A commitment period reserve, similar to that instituted by the Marrakesh Accords, would also limit inadvertent mistakes.

Non-binding targets are progressively gaining support, or at least interest, from various experts from industrialized countries, newly industrialized ones (e.g., Chan-Woo 2002), or developing countries such as India or China (Chen 2003, Dasgupta and Kelkar 2003). The concept could probably be adjusted so as to accommodate suggestions for defining the 'conditional' targets by Pan (2003) or Viguier (2003). Finally, non-binding targets might be fixed or dynamic, country-wide or sector-wide. Dynamic non-binding targets would offer developing countries a greater chance to participate in international emissions trading, despite possible economic surprises. Sector-wide non-binding targets would likely resemble the concept of sector-wide CDM suggested by various analysts (Samaniego and Figueres 2002, Chung 2003).

5. FLEXIBILITY AND EFFICIENCY

The quest for flexibility has been quintessential to the building of international architecture of commitments against climate change thus far. The current arrangements include what some have termed the 'what flexibility' (flexibility between various greenhouse gases, sources and sinks) and the 'where flexibility' brought by emissions trading and the other well-named 'flexible mechanisms' - though 'where flexibility' could also result from taxes. Tradable permit schemes, however, but not taxes, also offer another type of flexibility, which could be named the 'who flexibility', or the ability to allocate the efforts in a manner felt acceptable by all parties while maintaining cost-effectiveness.

There are, however, other possible dimensions for flexibility. As mentioned earlier the 'when flexibility', which would increase cost-effectiveness in achieving a cumulative reduction of GHG emissions over decades or centuries. All the options considered in the previous section - banking and borrowing, long commitment periods, dynamic targets, price caps, non-binding targets for developing countries - would increase the 'when flexibility', although some may fare better than others in guaranteeing some level of mitigation action.

There is still another desirable dimension of flexibility, however, which could be termed the 'where to' flexibility. The 'where to' flexibility is a way to achieve economic efficiency by making the final result partially dependant on actual costs. The degree of desirable 'where to' flexibility depends on the specifics of the problem at stake; in case of climate change, an important degree seems warranted, given the many uncertainties, on both climate damage and abatement costs. They have prevented any agreement thus far on the precise level and agenda for achieving the ultimate objective of the convention - stabilization of greenhouse gas atmospheric concentrations.

The 'where to' flexibility might result primarily from periodic reassessments of relatively short term objectives in light of past abatement costs, technology prospects and new insights from the climate sciences and assessment of impacts and adaptation possibilities. This necessity may rule out very long commitment periods. Price caps or indexed targets (or taxes), by contrast, are not only compatible with 'where to' flexibility, in fact they would expand it inside the commitment periods. With price caps, likely to facilitate relatively more ambitious targets as is shown below, if abatement costs turn out as expected or lower, the target will be reached. If abatement costs turn out higher than expected, some emissions beyond the target will take place, though limited by the price to pay. Adjustments to actual price would thus be continuous, while adjustments to new scientific assessments would remain periodical.

We now need to consider more closely why price caps, and to some extent indexed targets, would facilitate the adoption of more ambitious objectives. In the face of uncertainties, what concerns decision-makers are the expected benefits and costs, that is, the average of possible outcomes weighted by their probabilities of occurrence. Adding a price cap to a given target reduces its expected costs by 'shaving' the costlier outcomes. It also reduces its expected benefits - if costs reach the level of the price cap, more emissions, and thus more climate damage, will take place than originally sought with the quantitative target.

However, because marginal climate damage (or policy benefits) are roughly constant (over a relatively short period of time), while abatement costs are not, expected benefits are reduced in a much smaller proportion than expected costs. This allows tightening the objective (i.e. reducing the amount of allowed emissions) from the onset. At some point, expected benefits would be the same as originally envisaged but entail much lower expected costs. The target might be tightened again, up to the point where expected costs are the same as with the original target, though providing greater expected benefits. Between these two points there are an infinite number of quantified objectives that, thanks to the price cap, would produce higher expected benefits at lower expected costs than with the original target but no price cap. As a result, the introduction of price caps could allow any agreement to provide greater net expected benefits (Philibert 2006). Wide uncertainties on the policy benefits side probably prevent us from being much more specific on deciding the most efficient target and price cap levels.

The possibility of abrupt climatic changes would modify this analysis, if only we had an idea of the greenhouse gas concentrations most susceptible to trigger off such 'non-linear climate events.' Uncertain as they are, these possibilities do not really modify the rate of change of marginal expected benefits (Pizer 2003).

While some have seen the price cap concept as only a short-term 'fix' to the current difficulties of the Kyoto Protocol (Jacoby and Ellerman 2004), it could be seen instead as a necessary long-lasting element for future agreements dealing with climate change. Rather than being 'inconsistent' with each other, a quantity objective and a price cap would allow a system to spontaneously adjust in real time to the reality of the costs. It would progressively lead us to an efficient level of stabilization, which, given the many uncertainties on both benefit and cost sides, cannot be decided upon today. Decadal revisions of objectives might incorporate new scientific findings on climate change and new assessments of policy benefits, but the process would be too slow to make periodic commitments efficient given uncertain costs.

6. CONCLUSION

The Kyoto Protocol is not the perfect instrument to address climate change. Keeping Kyoto unmodified is likely to provide a partial and weak response to the threat of global climate change. Radical alternatives, however, still have to prove they are negotiable, enforceable, and effective.

Transforming Kyoto may be an efficient way to preserve the achievements of an already long and painful negotiating process. It would keep the advantages of international emissions trading while alleviating the shortcomings of the Kyoto-style fixed and binding targets. A transformation of Kyoto, as illustrated in this chapter, would help make it more cost-effective and more economically efficient. It would provide developing countries with real incentives to participate and comply (finance and technology transfer inflows through emissions trading), as well as reduce the disincentives for industrialized countries to participate and comply.

Whether this should happen by amending the existing Kyoto Protocol or by adopting a new international agreement is a question for climate negotiators. In any case, future climate agreements may differ from the current ones on many other aspects. They will address adaptation, as some climate change will inevitably occur. They may have new dispositions to promote technology transfers. Meanwhile, future climate policies undertaken at country levels will hopefully streamline climate policies in development policies. Energy efficiency improvements, and the development of renewable energy sources, have other merits than contributing to mitigate climate change. Still, in the absence of an international framework the level of action will remain too low - and some indispensable technologies, such as carbon dioxide capture and storage to address emission from coal combustion, may not be implemented on the necessary scale, notably in developing countries. This framework should further elaborate on the existing arrangements and introduce new options to better address abatement cost uncertainties as this may be a key to broadening and deepening the action to mitigate climate change.

References

Aslam, M. 2002. Equal per capita entitlements: A key to global participation on climate change? In: K. Baumert (ed.). Options for Protecting the Climate. World Resource Institute. Washington, D.C., USA.

Asuka, J. and K. Takeuchi. 2004. Additionality reconsidered – Lax criteria may not benefit developing countries. Climate Policy 4: 177-192.

Barrett, S. 2003. Environment and Statecraft. Oxford University Press, Oxford, UK.

Berk, M. and M. den Elzen. 2001. Options for differentiation of future commitments in climate policy: How to realise timely participation to meet stringent climate goals? Climate Policy 1: 465–480.

Bodansky, D. 2004. International Climate Efforts Beyond 2012: A Survey of Approaches. Pew Center on Global Climate Change, Arlington, VA., USA.

Chan-Woo, K. 2002. Negotiations on climate change: Debates on commitment of developing countries and possible responses. East Asian Review 14: 42–60.

Chen, Y. 2003. Chinese perspectives on beyond-2012. Presentation at the open symposium, International Climate Regime beyond 2012: Issues and Challenges, October 7 at Tokyo, Japan.

Chen, Y. and J. Pan. 2003. Equity concerns over climate change mitigation. Mimeo. Global Change and Economic Development Program, Chinese Academy of Social Sciences, Beijing, China.

Chung, R. 2003. CDM linked voluntary dynamic target. Presentation at the CCAP's Dialogues on Future Actions and Clean Development Mechanism, October, Jeju Island, South Korea.

Dasgupta, C. and U. Kelkar. 2003. Indian perspectives on beyond-2012. Presentation at the open symposium, International Climate Regime beyond 2012: Issues and Challenges, October 7. Tokyo, Japan.

Ellerman, A., and I. Wing. 2003. Absolute versus intensity-based emission caps. Climate Policy 3S2: S7–S20.

Grubb, M., C. Hope and R. Fouquet. 2002. Climatic implications of the Kyoto Protocol: The contribution of international spillover. Climatic Change 54: 11–28.

International Energy Agency (IEA). 2002. Beyond Kyoto—Energy dynamics and climate stabilisation. OECD/IEA, Paris, France.

Jacoby, H. and A. Ellerman. 2004. The safety valve and climate policy. Energy Policy 32: 481–491.

Martin, J. 1988. L'intensité énergétique de l'activité économique dans les pays industrialisés: Les évolutions de très longue période livrent-elles des renseignements utiles? Économies et sociétés (4) Avril.

Meyer, A. 2000. Contraction & Convergence. Green Books. Devon, UK.

Moomaw, W. and J. Moreira. 2001. Technological and economic potential of greenhouse gas emissions reduction. In: B. Metz, O. Davidson, R. Swart and J. Pan (eds). Climate Change 2001: Mitigation, Contribution of Working Group III to the Third Assessment Report of the Intergovernmental Panel on Climate Change. Cambridge University Press, Cambridge, UK.

Newell, R. and W. Pizer. 2003. Regulating stock externalities under uncertainty. Journal of Environmental Economics and Management 45: 416–432.

Nordhaus, W. 2002. After Kyoto: Alternative mechanisms to control global warming. Paper prepared for a joint session of the American Economic Association and the Association for Environmental and Resources Economists, Atlanta, GA., USA.

Oppenheimer, M. and A. Petsonk. 2004. Reinvigorating the Kyoto System, and Beyond. Paper presented at L20 and Climate Change Agenda: Council on Foreign Relations, September 20-21.

Pan, J. 2003. Commitment to human development goals with low emissions. Mimeo. Research Centre for Sustainable Development. Chinese Academy of Social Sciences, Beijing, China.

Philibert, C. 2000. How could emissions trading benefit developing countries? Energy Policy 28: 947–956.

Philibert, C. 2003. Technology, innovation, development and diffusion. OECD/IEA Information paper. COM/ENV/EPOC/IEA/SLT(2003)4.

Philibert, C. 2004. International technology cooperation and climate change mitigation. OECD/IEA information paper. COM/ENV/EPOC/IEA/SLT (2004)1.

Philibert, C. 2005a. Approaches to future international co-operation. OECD/IEA information paper. COM/ENV/EPOC/IEA/SLT(2005)6.

Philibert, C. 2005b. New Commitment Options: Compatibility with Emissions Trading. OECD/IEA information paper. COM/ENV/EPOC/IEA/SLT(2005)9.

Philibert, C. 2005c. Climate Mitigation: Integrating Approaches for Future International Cooperation, OECD/IEA information paper. COM/ENV/EPOC/IEA/SLT(2005)10.

Philibert, C. 2006. Certainty *vs.* Ambition-Economic Efficiency in Mitigating Climate Change, IEA Working Paper Series, Paris, France.

Pizer, W. 2002. Combining price and quantity control to mitigate global climate change. Journal of Public Economics 85: 409–434.

Pizer, W. 2003. Climate change catastrophes. Discussion paper 03-31. Resources for the Future. Washington, D.C., USA.

Roberts, M. and M. Spence. 1976. Uncertainty and the choice of pollution control instruments. Journal of Public Economics 5: 193–208.

Samaniego, J. and C. Figueres. 2002. Evolving to a sector-based clean development mechanism. In: K. Baumert (ed). Options for Protecting the Climate. World Resource Institute, Washington, D.C., USA.

Schneider, S.H. and J. Lane. 2006. An Overview of 'Dangerous' Climate Change. In: H.J. Schellnhuber, W. Cramer, N. Nakiæenoviæ, T. Wigley and G. Yohe (eds). Avoiding Dangerous Climate Change. Cambridge University Press, Cambridge, UK.

Socolow, R. 2006. Stabilization Wedges: Mitigation Tools for the Next Half-Century. Presentation at the World Bank's Energy Week, The World Bank, March 6, Washington, D.C., USA.

Viguier, L. 2003. Exploring new tools: The T&B approach. Paper presented at the RFF-IFRI workshop, How to Make Progress Post-Kyoto? March 19. Paris, France.

Winkler, H., R. Spalding-Fecher, S. Mwakasonda and O. Davidson. 2002. Sustainable development policies and measures. In: K. Baumert (ed). Options for Protecting the Climate. World Resource Institute, Washington, D.C., USA.

40

CHAPTER

The 'Action' Approach to Cutting Greenhouse Gases: A Better Model for Addressing Global Warming[1]

Donald M. Goldberg
Executive Director of the Climate Law & Policy Project
Adjunct Professor at American University
Washington College of Law
Washington DC, USA

INTRODUCTION

This chapter discusses a new type of market-based regulation called an 'action approach.' Conventional market mechanisms typically address the problem they are designed to ameliorate by making it more expensive to engage in the activity causing the problem. In the case of global warming, conventional market mechanisms such as an emissions tax or emissions trading (cap-and-trade) make greenhouse gas (GHG) emissions more expensive, creating a financial incentive for emitters to reduce their emissions. But the revenue raised by the tax, or the auction in the case of an auctioned permit trading system, generally is not used to directly achieve additional reductions. Instead, it is redistributed back to the public or used

[1]This paper, and the ideas included herein, were developed and prepared with assistance from David Grossman, Julia Petipas, Tracy de la Mater Craig Hart, and many others.

to finance other governmental activities.[2] In contrast, an action approach dedicates all legally-mandated revenue transfers and expenditures directly to solving the problem at hand – in this case, global warming. It does this principally by using the revenue to achieve additional GHG reductions or acquire them through the carbon market.

Using markets to internalize the costs of environmental damage is still a relatively new concept. In the past, these costs, referred to as "externalities," were borne by actors and non-actors alike – in other words, by society.[3] Where government has seen fit to step in and regulate pollution or other forms of environmental damage, it has done so mainly through 'command-and-control' approaches, which, in essence, tell polluters what they must do and punish them if they fail to do it. It is only in the past few decades that the idea of charging market actors for the costs of avoiding, abating, or repairing environmental damage caused by their activities has really taken hold. Market mechanisms have been shown to be more cost-effective than direct regulation, meaning that they deliver more benefits for the same cost (or the same amount of benefit for less cost), and many applications are just beginning to be tested, with encouraging results.[4]

Despite the significant economic benefits they offer, conventional market mechanisms such as emissions taxes and emissions trading may suffer from a major defect: they generally require the transfer of much more wealth than is actually needed to achieve the environmental objective. For example, in 2005, the United States emitted 1,661 million tons of carbon.[5] Suppose, hypothetically, that an emissions tax of $50 per ton of carbon equivalent (TC_e) would have reduced those emissions by 10%. Assuming that every ton of reduction obtained by the tax cost $50, a 10% reduction would have cost about $8.3 billion.[6] The tax on the remaining 1.5 billion tons of carbon emissions, however, would have produced revenues of $75 billion. In other words, in addition to the actual cost of reductions, an emissions tax on the order of $50 per TC_e would

[2]Some proposals would use a portion of the revenue to smooth the path to abatement in other ways, for example, by helping finance research and development of cleaner technologies or paying to retrain workers that lose their jobs as a result of emissions cutbacks.

[3]For a good discussion of externalities and their internalization, see Charles D. Kolstad, Environmental Economics. Oxford University Press (2000) at 91.

[4]An early example is the sulfur dioxide trading system set up by the Clean Air Act. It is estimated to have reduced overall abatement costs by as much as 17-20%. N.O. Keohane. 2006. Cost Savings from Allowance Trading in the 1990 Clean Air Act: Estimates from a Choice-Based Model. In: C.E. Kolstad and J. Freeman (eds). Moving to Markets in Environmental Regulation: Lessons from Twenty Years of Experience, Oxford University Press, New York.

[5]US Environmental Protection Agency, Inventory of U.S. Greenhouse Gas Emissions and Sinks: 1990-2005 (April 2007), at http://www.epa.gov/climatechange/emissions/downloads06/07ES.pdf

[6]In reality, because marginal abatement costs increase more or less linearly, the cost of the 10% abatement in our hypothetical example would probably be closer to $4 billion.

transfer a huge amount of money from one group - presumably mainly consumers of products that, in their use, manufacture, or both, caused the offending emissions - to the government, which would either redistribute those revenues or use them for other purposes.

An equivalent emissions trading system would produce a similar transfer of wealth, but not necessarily to the government. If some or all of the allowances are allocated to emitters for free, as is currently the case in the European Union emissions trading system (ETS), then these emitters receive a portion of transferred wealth equal to the value of total allowances they are allocated. This large transfer of wealth (or the increase in energy costs that produces it) is a significant impediment to the adoption of stringent controls on GHG emissions.

It is explicitly argued here that action approaches can achieve cost-effectiveness without the enormous transfer of wealth associated with conventional market mechanisms, making them better suited to controlling GHGs. The two action approaches that receive the most attention in this chapter are action targets and action fees, although several other measures that might meet the definition of 'action approach' are briefly discussed as well. As action targets - mainly in the context of new developing country commitments - are discussed at length in Chapter 40, they will not receive a lengthy treatment here. The main focus of this chapter is the action fee, which serves as the basis for most of the examples presented.

This chapter is not intended to comprehensively compare measures to control GHGs, but rather to explain how action approaches work and discuss their main benefits (hopefully without papering over possible drawbacks). Nor are action approaches offered as a 'silver bullet' that will solve all the problems of adopting an effective system for controlling GHGs. It is more likely that a combination of approaches will be required, and that the combination may evolve over time. This chapter is being presented to introduce a new concept that has much to recommend it, in order that it might be studied further and considered alongside other mechanisms already under consideration.

WHAT IS AN ACTION APPROACH?

An action approach, as defined in this chapter, is a market mechanism. While all environmental regulations affect markets, the term "market mechanism" usually is reserved for an approach that utilizes the power of markets to minimize the overall cost of regulation.[7]

[7] According to the Intergovernmental Panel on Climate Change (IPCC), a market mechanism is an environmental measure that, in theory, utilizes the properties of markets to maximize cost effectiveness, i.e., to achieve the environmental aim at the lowest possible cost. *See* Intergovernmental Panel on Climate Change, Climate Change 2001: Working Group III: Mitigation, at §§ 6.2.2.2, 6.2.2.3, and 6.2.2.6.

Action approaches differ from conventional market mechanisms, such as emissions trading systems and taxes on GHGs, in one crucial respect: they use the revenue from any increase in the cost of emissions for the purpose of directly achieving or acquiring additional reductions. As a result, for a given increase in the cost of emissions, an action approach achieves more reductions than a conventional market mechanism, especially in the early stages of implementation.

This chapter focuses on two types of action approaches - action targets and actions fees - that are straightforward and well-suited to illustrating the essential characteristics, including both advantages and disadvantages, of an action approach. Like a conventional emissions tax, an *action fee* levies a charge on GHG emissions. Unlike a conventional emissions tax, however, revenues from an action fee are placed in a fund and used by the administrator of the fund to purchase additional reductions. The term 'fee' rather than 'tax' is used because, in general, tax revenue can be used to fund any legitimate government function, whereas fees must bear a substantial relationship to the cost of providing specified services or allowing specified activities, including the cost of avoiding, mitigating, or repairing environmental damage resulting from GHG emissions.

An *action target* takes a somewhat different approach. Instead of paying a fee to the government based on the quantity of GHGs emitted, an action target requires the regulated entity to achieve a specified percentage of reductions and deliver these to the government. For example, a 20% action target would require the entity to achieve or acquire one-fifth (20%) of a ton of reductions for every ton it emits during a given period.[8]

DIFFICULTIES WITH CONVENTIONAL MARKET MECHANISMS

As noted above, an emissions tax transfers substantial revenues to the government, which can use them as it sees fit (probably to offset other taxes). Emissions trading systems also transfer large amounts of wealth. If allowances are auctioned, the wealth is transferred to the government, just

[8]The choice of period can be critical: Some countries may choose a historical 'base' period, so they know before they commit to a target how many reductions they must achieve or acquire. Other countries may prefer a future base year close to the start of the commitment period, to ensure a close alignment between their economies during the base year and the commitment period. A third possibility would make a country's commitment a percentage of its emissions during the commitment period itself. This approach is discussed in detail in Chapter 40.

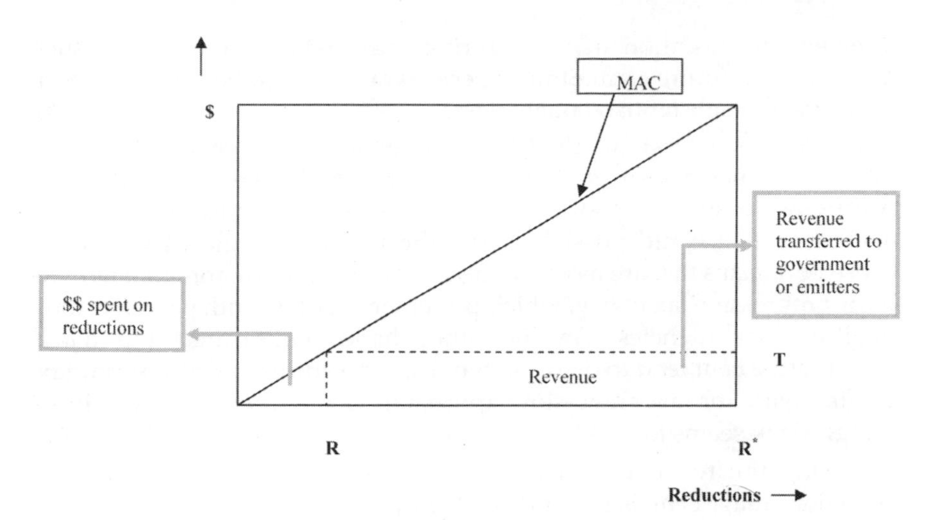

Fig. 1 Transfer of revenue from a conventional market mechanism. **T** is the tax per ton of carbon emitted, **R** is the level of reduction induced by the price effect of the tax. $\mathbf{R}^*$ is 100% reduction (or zero emission). The diagonal line represents the marginal abatement cost (MAC).

as in a conventional emissions tax scheme. If they are allocated for free, then this wealth is transferred to emitters. These wealth transfers are depicted in Figure 1.

Figure 1 shows the flow of revenue from a conventional emissions tax or trading system. Only the money represented by the small triangle to the left of **R** is actually used to reduce emissions. In a pure emissions tax or trading scheme, the revenue in the large rectangle to the right of **R** goes to the government, to be used as it sees fit, or to emitters, to the extent that they are allocated allowances for free.[9]

As discussed below, this initial transfer of wealth - to either the government or the emitter - is highly regressive. The regressivity can be offset by adjusting income taxes or, as some have suggested, simply returning the money directly to citizens on an equal per capita basis. While that would make the system slightly progressive, it is unlikely to happen, given the political power of entrenched industries that profit from the sale or use of fossil fuels. Not surprisingly, these industries strongly favor allocations over auctions. In fact, in the EU, which has the most advanced emissions trading system currently in existence, emitters demanded and

[9]In actuality, most proposals would use some portion of revenues in the box to the right of **R** to help achieve emissions abatement.

received an allocation of 90% during the first commitment period. Allocations in future commitment periods are likely to be somewhat less generous, but substantial, nonetheless.[10]

The large transfers of wealth these taxes or allocations engender and the substantial increase in energy costs that result make it difficult for politicians to support conventional market mechanisms at a level of stringency that could possibly solve the problem of global warming. Those politicians that are most inclined to protect the environment tend to favor progressive tax policy, which puts them at odds with powerful and wealthy constituencies. On the other hand, politicians of a more conservative bent tend to question the need to address global warming in the first place, or favor a go-slow approach. This is a recipe for gridlock, and gridlock seems to be what we have.

Placing the revenue in a fund to be used to purchase additional reductions may seem like a somewhat simplistic solution to this problem. Rather than being revenue neutral, as many have suggested a market mechanism should be, an action fee is tantamount to a tax hike. Theoretically, however, a relatively small action fee could deliver more reductions than a much larger emissions tax (*see* Table 1 below). In fact, this difference is so large that it should leave room for an acceptable compromise. At the very least, it avoids a potentially bruising fight about how to divide up the wealth generated by a conventional market mechanism. And it is simple. There are no winners or losers; every emitter pays the same amount per unit of emissions.

Table 1 Reductions from a conventional tax and an action fee set at the same level.

Conventional tax	5%	10%	15%	20%	25%
Action fee	31%	44%	53%	60%	66%

WHAT IS AN 'ADDITIONAL REDUCTION?'

The astute reader may have noticed that this chapter refers frequently to 'additional reductions' that could be achieved with revenue from an action approach. Chapter 40 extensively discusses what an additional reduction is, so it will receive only a brief explanation here.

What might be termed a 'reduction' depends to a large extent on the overall design of the GHG control system. Generally, there are two types of emissions instruments: allowances and credits, and either of them could

[10]This view is based on conversations the author has had with several EU officials involved with designing and implementing the trading system.

be a reduction for purposes of an action approach. In fact, any instrument that permits an entity to emit a unit of GHG becomes a reduction if it is removed from the system.

Allowances are created by the government. The government (or governments, in an international system) decides how much GHG should be emitted during a given commitment period and allocates or auctions an equivalent number of allowances. In an emissions trading system, these allowances may be bought and sold by emitters during the commitment period, so that emitters with high reduction costs have the opportunity to buy allowances from emitters with lower reduction costs. This ability to trade is what makes the system cost effective. A well-functioning market reveals the marginal cost of reduction for each emitter, giving buyers the ability to identify and purchase the cheapest reductions available.

Not every source of emissions can easily be included in an emissions trading system, however. Such a system lends itself to regulating small groups of large emitters, whose emissions can be monitored by the government. Large groups of small emitters are more difficult to monitor. For this reason, among others, many emitters may not be covered by an emissions trading system. These emitters can participate in a different way. If they can demonstrate that they have taken an action that has resulted in fewer emissions, they may receive credits from a government agency, which can then be sold to emitters operating within the trading system, who can use them in the same way as allowances.[11]

Determining whether an action has actually resulted in fewer emissions is sometimes straightforward and at other times daunting. It usually means demonstrating that the action would not have happened 'anyway.' A huge body of literature has grown up around answering the question of what constitutes an emissions reduction.

As will be shown in the next section, the Kyoto Protocol is a mixed system. It includes an allowance-based system (emissions trading), a credit-based system (the clean development mechanism), and a hybrid approach (joint implementation). Each of these systems creates or transfers some type of legal instrument that allows emissions to occur. For purposes of an action approach, permanent removal from the system of any of these instruments, so that it cannot be used to legally sanction an emission, would constitute an additional reduction.

[11]An important difference between this approach and an action target is that an action target requires every emitter to achieve some reductions (specified by the target) that cannot be sold.

MARKET MECHANISMS AND THE KYOTO PROTOCOL

More than a decade before the Kyoto Protocol was adopted, the international community began a global effort to address the problem of global warming. The outcome of this effort was the United Nations Framework Convention on Climate Change (UNFCCC),[12] which created the outline or "framework" of the approach that was to become the Kyoto Protocol.[13] During the negotiation of the UNFCCC it was concluded, and the document reflects, that market-based approaches would be used to control GHGs.[14] The two market mechanisms initially considered were emission taxes and emissions trading. During the negotiations it soon became clear that the mechanism of choice was emissions trading.

In addition to emissions trading, two other market mechanisms were included in the Protocol: a credit-based approach, known as the clean development mechanism (CDM), and a hybrid approach known as joint implementation (JI). The salient features of the three mechanisms are summarized below:

Emissions Trading (Article 17)

- Provides each Party listed in Annex B of the Protocol (i.e., developed countries) with a fixed number of emissions allowances, known as assigned amount units or AAUs, for the commitment period 2008-2012;
- Allows Annex B parties to trade allowances with each other;
- Requires Annex B Parties to hold allowances for all emissions within their territory of the six GHGs covered by the Protocol.

Clean Development Mechanism (Article 12)

- Allows emission reduction projects to take place in developing countries;
- Requires that each project demonstrate additionality, *i.e.*, that the reductions generated would not have occurred but for the project or, in alternate formulations, the CDM itself;

---

[12]United Nations Framework Convention on Climate Change ("Framework Convention"), 1771 U.N.T.S. 107, reprinted at 31 I.L.M. 849 (1992), *available at* http://unfccc.int/resource/docs/convkp/conveng.pdf.

[13]Kyoto Protocol to the United Nations Framework Convention on Climate Change, *opened for signature* Mar. 16, 1998, U.N. Doc. FCCC/CP/1997/L.7/Add. 1 [hereinafter Kyoto Protocol]. Art. 12.

[14]*See* Framework Convention, at Arts. 3.3, 4.2(a), 4.2(b), 4.2(d), and 4.2(e)(i).

- Creates credits, known as certified emission reductions or CERs, for reductions that have been certified as having met all requirements;
- Requires that each project define a 'business-as-usual' baseline to determine the number of CERs created;
- Allows CERs to be traded and used to offset emissions by Annex B countries.

Joint Implementation (Article 6)

- Allows emission reduction projects to take place in Annex B countries, usually in countries with economies in transition;
- Creates credits, known as emission reductions units or ERUs, for reductions that have met all requirements;
- Allows ERUs to be traded and used to offset emissions by Annex B countries.
- Requires an AAU to be transferred simultaneously with each ERU to avoid double-counting.

Since the Kyoto Protocol includes a system of credit-based reductions (the CDM), it is not possible to know the actual amount of GHGs that will be emitted during the commitment period by countries with reduction commitments. The 'cap' can be exceeded by the number of credits (CERs) generated from projects in countries that did not adopt reduction commitments. In theory, however, this should result in the same amount of emissions reductions as would have occurred had the CDM not been included, but at a lower cost, as many of the cheapest reductions can be obtained in developing countries, where economies are expanding rapidly, and much of the infrastructure remains to be built.

ACTION APPROACH COMPARED TO CONVENTIONAL MARKET MECHANISMS

To demonstrate the difference between the operation of an action approach and a conventional market mechanism, we will contrast an action fee to a conventional emissions tax. First it might be useful to discuss the relationship of an emissions tax to emissions trading (a similar relationship exists between an action fee and an action target).

Emission taxes may look very different from emissions trading to the untutored eye, but in practice, their operation is quite similar. Emissions trading requires the emitter to hold an allowance for each unit of emissions. That allowance has a cash value in the market. In theory, the market-clearing price of an allowance in an emissions trading scheme that would reduce emissions to level X is the same as the tax that would be required to reduce emissions to level X. If the emitter is required to pay for all his allowances (as opposed to receiving allowances for free, as occurs under some systems), the two systems impose the same cost on the emitter, assuming the same level of reduction is achieved.

From a policy perspective, the main difference between an emissions tax and emissions trading is that a tax fixes the emitter's cost per unit of emission (*i.e.*, it aims at price certainty), whereas emissions trading fixes the quantity of GHGs that may be emitted during the commitment period (*i.e.*, it aims at environmental certainty). This relationship is illustrated in Figure 2. The horizontal axis is the amount of reductions, and the vertical axis is the marginal abatement cost (MAC). As can be seen, achieving reductions **R** in an emissions trading system yields a market-clearing price of **T** per allowance. Alternatively, setting an emissions tax at **T** yields reductions **R**.

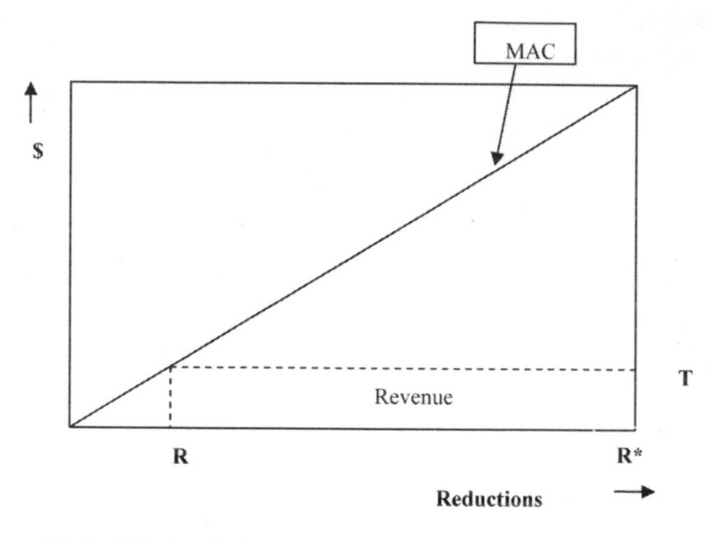

Fig. 2 Relationship between emissions tax and emissions trading

Figure 2 also shows the revenue from a conventional emissions tax.[15] This revenue, in most schemes, would be used mainly to offset other taxes. An action fee instead utilizes the revenue to purchase additional reductions. Assuming a relatively linear MAC, it is possible to calculate

[15]Again, as explained above in Figure 1, revenue is the level of tax **T** multiplied by the emissions the tax is levied upon (which is all of them, or **R***), minus the portion that actually goes to achieving the emissions reductions (the portion of the horizontal axis to the left of **R**).

the approximate amount of additional reductions that would result from a charge treated as a fee as opposed to a tax. To make such an approximation, consider how an action fee would operate in practice. It would involve the following steps:

Step 1: Regulator announces the level of the fee (**F** in Figure 3) per ton of carbon equivalent (TC_e) emitted during the upcoming commitment period.

Step 2: The fee induces emitters that are able to reduce some or all of their emissions for less than **F** per TC_e to reduce their aggregate emissions to point $\mathbf{R_C}$ (so $\mathbf{R_C}$ equals the reductions achieved by the effect of the fee on prices alone).

Step 3: The fund administrator holds a 'reverse auction,' in which companies submit sealed bids to provide additional reductions. The competitive nature of the bidding process should ensure that companies will offer reductions for the lowest price at which they are able to deliver them. If necessary, however, a condition of acceptance of a bid could be a showing by the company that it is offering its reductions at the lowest possible price. In calculating the price at which they could sell more reductions, emitters would factor the avoided fee into their calculations, allowing them to offer some additional reductions for very little cash. For example,

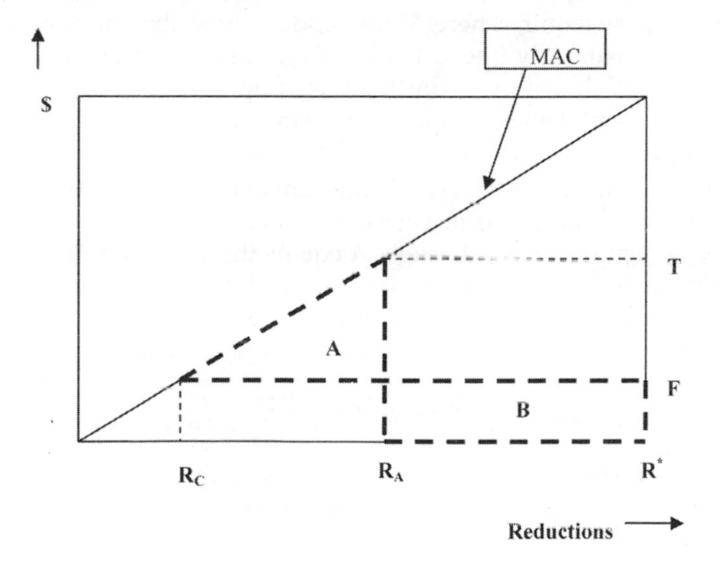

Fig. 3 Comparison of reductions achieved by a conventional emissions tax ($\mathbf{R_C}$) and an action fee ($\mathbf{R_A}$). **F** is the action fee and **T** is the level of tax that would be required to achieve reductions $\mathbf{R_A}$.

if **F** = \$50/TC_e and an emitter can reduce some or all of its emissions for \$60/TC_e, it could offer reductions for \$10/TC_e (\$50 for the foregone fee and \$10 for the additional cost of reduction).

Step 4: After identifying the pool of bids that would meet all government guidelines and produce real, verifiable reductions, the fund administrator accepts bids in order of price, from the lowest to the highest, contracting to purchase as many additional reductions as possible from the revenues to be collected. (No fees are collected on the additional reductions, or, more precisely, the avoided emissions. Fees are collected only on the remaining emissions.) These additional purchased reductions are represented by the distance from $\mathbf{R_C}$ to $\mathbf{R_A}$, and the cost of the additional reduction at each point between $\mathbf{R_C}$ and $\mathbf{R_A}$, by the distance from **F** to the diagonal line representing the MAC. The total cost of these additional reductions is the area of triangle **A**, which should be equal to the proceeds from the fee (**F**) on all remaining emissions (the distance from $\mathbf{R_A}$ to $\mathbf{R^*}$). In other words, the area of triangle **A** and rectangle **B** should be equal, since all revenues are used to achieve additional reductions.

It is possible to approximate the level of reductions achievable by an action fee if one assumes that the marginal abatement cost curve (the diagonal line) is approximately linear. This is a safe assumption for the modest purpose required here. Many models show abatement cost curves that are approximately linear. For example, 14 different models yielded the cost curves shown in Figure 4. At least up to around 25% reduction, they are sufficiently linear to support the simple mathematical model used in this paper.[16]

Assuming a linear cost curve, the amount of reductions achievable with an action fee (as compared to a conventional emissions tax) would be the point $\mathbf{R_A}$ where the area of triangle **A** equals the area of rectangle **B**. That point can be found using the following formula, the derivation of which is explained in the Annex at the end of this chapter:

$$\mathbf{R_A} = \sqrt{2\mathbf{R_C} - \mathbf{R_C^2}}$$

Assuming the abatement cost curve is relatively linear, it is not necessary to know the actual cost of abatement to demonstrate the power

[16]A non-linear cost curve would make mathematical calculations more difficult, but the arguments in favor of action approaches should still be valid.

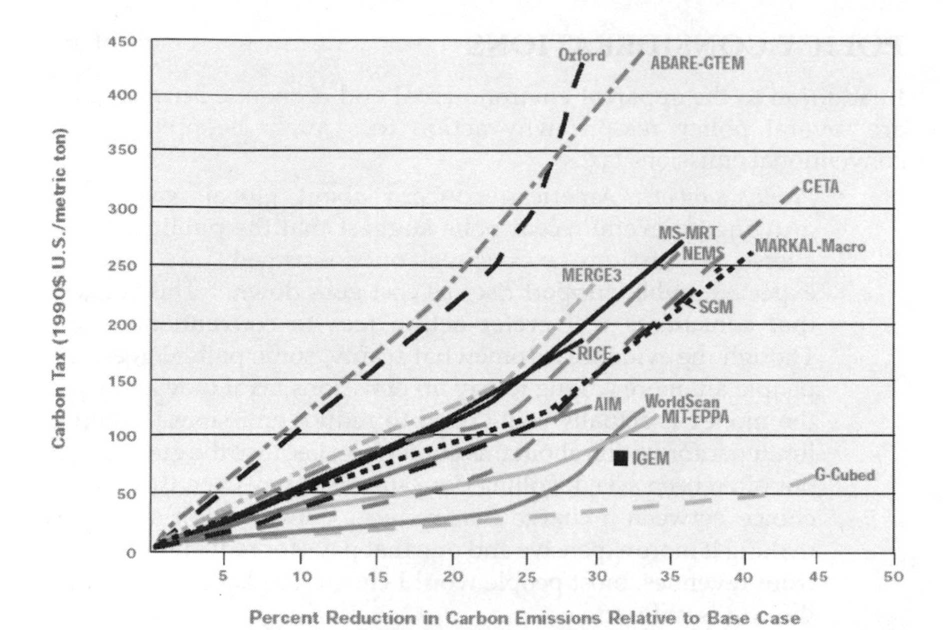

Fig. 4 Compilation of abatement cost curves shows their approximate linearity. *Source:* John P. Weyant, An Introduction to the Economics of Climate Change Policy, Pew Center on Global Climate Change, July 2000.

of action fees.[17] Table 1 compares the levels of reduction achievable with a conventional emissions tax to the levels achievable with an equivalent action fee that uses the revenues to purchase additional reductions.

It must be emphasized that the reductions in Table 1 are highly theoretical and do not take into consideration the cost of implementing an action fee. That cost could be considerable, although, given recent experience with emissions trading in the EU, the same might be said of trading, so it should not be assumed that the cost of implementing an action approach is greater than implementing a conventional market approach. More analysis and empirical data are required to provide the answer. Given the enormous difference in reductions achievable with an action approach, it might prove superior to a conventional market approach even if it turns out to be more expensive to administer.

[17]It should be noted that the abatement costs depicted in Figure 4 do not assume the use of market mechanisms, making the cost estimates considerably higher than they would be if market mechanisms were included in the models.

POLICY CONSIDERATIONS

In addition to the apparent environmental and economic benefits, there are several policy reasons why action fees might be preferable to conventional emissions taxes:

- *Public support*: American concern about global warming is growing.[18] Several recent polls suggest that the public is not as averse to emissions taxes as was once assumed.[19] As would be expected, public support rises as cost goes down.[20] This suggests that consumers will prefer action fees to conventional taxes. Though the evidence is somewhat scanty, some polls suggest that people are more willing to pay an emissions tax if they know that the money is actually being used to reduce emissions.[21] There is limited information about this contention because the question has not often been asked. Polling data suggests, however, that given a choice between a charge that reduces consumption simply by making it more expensive and one that pays for reductions directly from revenues, most people would choose the latter, especially if the cost is reduced.
- *Competitiveness:* One reason both industry and labor have opposed emissions controls is that they fear that increasing the cost of energy will make goods more expensive and, ultimately, drive production to developing countries that do not have emissions controls. Action approaches minimize this threat by minimizing the increase in energy prices for a given level of reductions.
- *Equity*: Environmental taxes and other cost-effective market mechanisms generally tend to be regressive, and hence, it could be argued, less equitable from the standpoint of the consumer.[22]

[18]79% said global warming represented an 'extremely important' (41%) or an 'important' (38%) threat to the US in the next ten years. German Marshall Fund, 2005; a third consider global warming as the most important environmental problem, and 7 out of 10 want the federal government to take more action, Washington Post/ABC News, Apr 5-10, 2007.

[19]Seventy three percent of Americans back a $10-per-month charge to underwrite renewable energy production. Stanford University, Resources for the Future, New Scientist Magazine, June, 2007.

[20]*Id.*, p.t 13.

[21]When asked: "What if the increased tax on gasoline would cut down on energy consumption and reduce global warming, then would you favor or oppose an increased federal tax on gasoline?" Fifty nine% of respondents said they would favor such a tax, while 34% opposed it. The New York Times/CBS News Poll, Feb. 22-26, 2006; three-quarters of Americans polled said they would be willing to pay more for electricity generated by renewable sources like solar or wind energy. The New York Times/CBS News Poll, Apr. 07, 2007.

[22]See Costanza, R. et al. 1997. An Introduction to Environmental Economics, p. 216.

Companies, of course, may have a different view of what is equitable and insist on keeping a portion of the taxes to compensate for reductions in sales. This would be comparable to allocating free allowances to companies in an emissions trading system. While it would still be regressive in character, judged from the standpoint of the consumer, an action fee would be more equitable than a conventional emissions tax because it achieves a comparable level of abatement at considerably lower cost. The significance of this is made clear when one considers that, compared to high-income households, low-income households expend substantially more of their disposable income on energy. For example, in 2001, U.S. households with income less than $10,000 spent more than 10% of their income on energy while households above $75,000 spent less than 2.7% of their income on energy.[23]

- *Tax policy*: While there are legitimate arguments for using the proceeds of environmental taxes to offset other types of taxes (the main argument being improved economic performance), there are also risks to this approach. One is that politicians will become 'addicted' to the revenue stream from the product they are seeking to eliminate. If proceeds from an emissions tax constitute a significant portion of total tax revenue, politicians may find them hard to give up. Thus, rather than being an incentive to reduce consumption of fossil fuels, the tax could become the reverse, locking us into a level of consumption from which it would be difficult to reduce. The 'break even' point (depending on the shape of the marginal abatement cost curve) comes when approximately 50% of emissions have been eliminated. Beyond that point, consumption decreases faster than the tax increases, so overall revenue declines. Action fees avoid the risk of tax addiction by ensuring that the revenue stream is dedicated to only one purpose: reducing GHG emissions.

SIMILAR PROGRAMS AND PLANS

Several countries, states, and municipalities are already experimenting with approaches that are similar to or have elements in common with action approaches. It is still too early to assess their effectiveness, but they

[23]Supplemental Measures of Material Well-Being: Basic Needs, Consumer Durables, Energy, and Poverty, 1981 to 2002, U.S. Census Bureau, Dec. 2005, at 16.

will be watched closely and widely copied if they prove to be effective. A few prototype programs are described below:

- *Swiss Climate Cent* - Put forward in 2001 and operating since October 2005, the Swiss Climate Cent may be the first government-level proposal for an action-type mechanism.[24] It levies a charge of 1.5 Swiss cents per liter on all gasoline and diesel imports in Switzerland. The proceeds, estimated at about EUR 60 million annually, will be spent on measures to reduce GHG emissions both domestically and abroad, with the majority of the revenue to be spent domestically. Unlike an action fee, the Swiss Climate Cent is voluntary. The plan will be assessed in 2007 and may be made mandatory if the voluntary approach does not achieve its objectives.[25]
- *Colorado Climate Action Plan Tax* - In November 2006 the city of Boulder, Colorado voted 58% to 41% to adopt an energy tax to directly combat global warming. The tax will be collected by the local electric utility company based on consumed electricity, most of which comes from the burning of coal. The tax will be used to fund the city's Climate Action Plan approved by the city council in June 2006. It is estimated the average household will pay $1.33 per month and an average business will pay $3.80 per month. The tax will generate about $1 million annually through 2012 when the tax is set to expire. The Climate Action Plan aims to reduce Boulder's emissions by 24% by 2012. Estimated energy cost savings from implementing the Climate Action Plan are $63 million over the long term.[26]
- *Canadian Climate Fund* - In October 2005 the government of Canada established a fund to purchase 75-115 Mt of reduction credits a year, up to 40 percent of the total reduction needed in 2008-2012. Priority would be given to domestic reductions from farmers, forestry companies, municipalities, and other sources, including 'Large Final Emitters' that do better than their targets. Purchases would be made on a competitive basis. Reductions also would be purchased through the Kyoto mechanisms, with

[24]The Swiss Climate Cent was the brain-child of Ann Arquit Niederberger, at the time the Deputy Head of Climate Change Affairs at the Swiss Agency for Environment.

[25]Swiss Federal Office of Energy, at http://www.bfe.admin.ch/energie/00572/00575/index.html?lang=en.

[26]City of Boulder Climate Action Plan, http://www.bouldercolorado.gov/files/Environmental%20Affairs/climate%20and%20energy/cap_final_25sept06.pdf.

safeguards against the purchase of so-called "hot air."[27] The government agreed to allocate from general revenue CAD$1 billion per year over the next 5 years and projected funding of $4-5 billion 2008-2012.[28] Due to the change in government from liberal to conservative that occurred in 2006, the future of the Climate Fund is uncertain.[29]

- *EU white certificates* - Beginning January 1, 2005, several EU countries have adopted schemes to create offsets, mainly aimed at energy efficiency. Unlike a pure action approach, a white certificate involves the purchase of reductions that are additional to those that are created by the primary policy instrument. This has led to a degree of complication, such as potential double-counting, that would not be expected in a pure action system. In the EU the primary instrument is not a tax but an emissions trading system. White certificates also differ from action fees in their method of acquisition, in that reduction certificates are obtained directly by regulated firms, not by governments. In this respect, they are more like action targets, discussed below (and in greater detail in Chapter 40). Placing an obligation on firms to obtain reductions, as is the case with action targets, could eliminate the need for the government to levy a tax or create a fund dedicated to purchasing additional reductions. Blended systems such as the EU's, in which White Certificates are separate from the ETS, can create complications, but these may be resolvable, and much research is currently aimed at understanding how different systems interact and how potential conflicts can be reconciled.[30]

OTHER POSSIBLE ACTION APPROACHES

As defined above, an action approach is a market mechanism that avoids transferring wealth for any purpose other than to achieve reductions.

[27] In the Kyoto Protocol some countries, notably Russia and Ukraine, received more allowances than they were expected to need during the commitment period, as an enticement to join the Protocol. The excess allowances are frequently referred to as hot air.

[28] This program appears to have been scaled back considerably by the current government. *See* Canada International Development Agency, Canada Climate Change Development Fund, http://www.acdi-cida.gc.ca/CIDAWEB/acdicida.nsf/En/JUD-4189500-J8U. For a brief description of the 2005 fund, see World Energy Outlook, Canada Climate Fund, http://www.iea.org/textbase/pamsdb/detail.aspx?mode=weo&id=2307.

[29] *See, e.g.,* The Honourable Stéphane Dion' s remarks to the House on Climate Change, February 1, 2007, at http://www.liberal.ca/story_12428_e.aspx

[30] *See, e.g.,* European Commission Directorate-General Environment, Interactions of the EU ETS with Green and White Certificate Schemes: Summary Report for Policy Makers, 17 November 2005.

Funds generated by an action approach are used solely for the purpose of averting global warming. In contrast to a conventional carbon tax, or its cap-and-trade equivalent, both of which create price signals that discourage activities that result in greenhouse gas emissions, an action approach relies mainly on subsidies or similar incentives to induce reductions. Several other action-type approaches are described below:

- *Action targets* – As opposed to capping emissions, an action target sets an amount of reductions to be achieved by a country (or company) during a commitment period. It is a market mechanism because, like cap-and-trade, it creates a market that reveals the price of reductions, making it possible to purchase the cheapest reductions available. But unlike cap-and-trade, the government does not issue allowances, so there is no transfer of wealth to the government or emitters. Money is expended only for the reductions themselves plus the cost of administering the system and any other expenses directly related to averting global warming. The target could be set in a number of ways: as a percentage of past emissions; as a percentage of future emissions prior to the commitment period; or as a percentage of emissions during the commitment period itself.

  One formula that could be used for the third option is:

  AT = X% of domestic (or company-wide) emissions during the commitment period.

  This formula gives more weight to domestic reductions than to reductions achieved overseas, hence it might appeal to countries that prefer to maximize domestic actions, rather than rely on actions taken abroad.

  A second formula that would weigh all reductions equally is:

  AT = X% of (domestic emissions + units purchased from other countries or firms).[31]

As explained in Chapter 40, an important advantage of action targets is their predictability. Compared to other market mechanisms, they provide emitters with more assurance that the level of effort required to meet their commitment will not be much different than expected. If they work as predicted, action targets could be particularly beneficial for developing countries, which often face large uncertainties about current inventories and future economic growth. Given all the other important demands on their limited resources – such as poverty reduction and improved health

[31] At the company level: AT = X% of (company-wide emissions + units purchased from outside sources).

care – developing countries need to be confident that the level of effort they will actually expend to meet their commitments will closely match the level they intended to make.

Several other action approaches are theoretically possible. Indeed, the variations on the theme may be limitless. One approach that might merit further study is an emissions trading system that auctions all allowances. Proceeds from the auction could be used to buy back and retire some of the allowances previously auctioned. Such a system would need to be carefully constructed to avoid artificially inflating the price of allowances. For example, if the government uses auction proceeds to retire some allowances, the market clearing price presumably will rise to the level that would have occurred if those retired allowances had never been issued. This could make future trades very expensive. This problem might be avoided by allowing companies to purchase additional allowances at a lower price, determined prior to the commencement of the commitment period. This is known as a 'safety valve' because it prevents the price of allowances from rising above a predetermined level. In effect, if allowance prices reach the safety valve level, the system converts to an emissions tax. Of course, emitters cannot be allowed to purchase additional allowances only to sell them back to the government at a higher price. This problem might be avoided by restricting purchases of additional allowances to sectors or countries that have no caps, or to countries that do not have a safety valve or similar mechanism. As noted above, such blended systems are inherently more complex than 'pure' ones and require further study.

CONCLUSION

As defined above, an action approach is a market mechanism that transfers wealth only for the purpose of achieving reductions in order to avert global warming. While the main objective should be to achieve additional reductions, the fee could be increased to cover other related costs, e.g., worker retraining, public education, or even research and development of new technologies. The essential point is that all funds generated by an action approach should be used for the purpose of reducing emissions to avert global warming.

The goal is to maximize reductions per dollar of additional cost to the consumer. The government could charge the highest fee the public would accept, in order to maximize reductions, or the lowest fee required to achieve the level of reductions science demands. Political negotiation likely would result in something between these two extremes.

For any given marginal reduction cost, the ability of an action approach to achieve more reductions than a conventional market mechanism assumes that the costs of implementing an action approach are not so great that any environmental and economic benefits would be wiped out. It is too soon to make such an assessment. The cost of implementing the EU ETS has been higher than expected, suggesting that the costs associated with implementing such systems cannot be known with certainty until they are up and running. It may turn out that action approaches are no more expensive, or are even cheaper, to implement than an effective carbon tax or cap-and-trade system.

One final point must be emphasized. The ultimate test of any system is its environmental efficacy. Thus, any system - conventional market-based approach, action approach, or any other approach - must undergo constant monitoring and assessment to ensure that it is meeting its objectives. If it is not, it must be strengthened. No matter what system of GHG regulation is adopted, emissions targets that are set must be met. If the adopted approach does not produce the necessary downward trend in emissions demanded by science, it must be tightened or abandoned. This holds true for action approaches as well as the more conventional approaches currently under consideration.

ANNEX

Derivation of the equation: $\mathbf{R_A} = \sqrt{2\mathbf{R_C} - \mathbf{R_C^2}}$

Given the percentage of total emissions that would be reduced by a conventional emissions tax, it is possible to find the percentage of reductions that can be achieved if the tax is converted to a fee, that is, if all revenues from the tax are used to purchase additional reductions.

As in the body of the paper, this highly theoretical model assumes a perfectly competitive market, equivalent transaction costs, and a linear MAC.

In Figure A1, the marginal abatement cost of reductions (MAC) is represented by the diagonal line

F is the action fee.

$\mathbf{R_0}$ is 0, the point where no reductions occur (*i.e.* emissions in the absence of any effort to reduce, which we will refer to as business-as-usual (BAU) emissions).

$\mathbf{R_C}$ is the amount of reduction, as a percentage of BAU emissions, induced solely by the price effect (the increased cost of emitting) of action fee **F**.

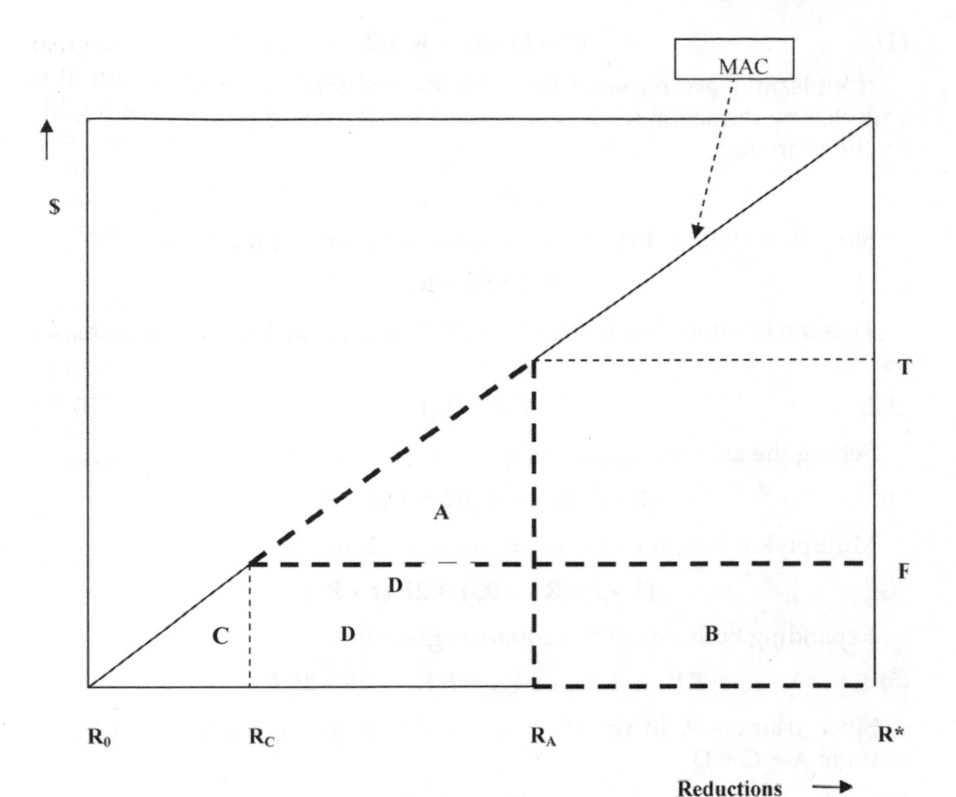

Fig. A1 Reductions induced by a conventional emissions tax compared to reductions achievable by equivalent action fee.

$\mathbf{R_A}$ is the amount of reduction that can be achieved if, in addition to reductions induced by the price effect of the fee, additional reductions are purchased using the revenues from the fee.

T is the level at which a conventional emissions tax would achieve reduction $\mathbf{R_A}$.

$\mathbf{R^*}$ is the 100% reduction, or zero emission, point.

Emitters pay no fee on abated emissions, so the cost of abatement to emitters with reduction opportunities to the right of $\mathbf{R_C}$ is the horizontal distance between **T** and the diagonal MAC, and the aggregate cost of reducing from $\mathbf{R_1}$ to $\mathbf{R_2}$ is represented by the area of triangle **A**, given by the formula

(1) $$(T - F)(R_A - R_C)/2$$

If emissions are reduced to point R_2, the total amount of revenue collected by the action fee is represented by the area of rectangle **B**, given by the formula

(2) $$F(R^* - R_A)$$

Since R^* = 100% reduction, we can write the area of **B** as

(2.1) $$F(100\% - R_A)$$

This can be simplified further by substituting 1 for 100% and writing **B** as

(2.2) $$F(1 - R_A)$$

Setting the area of triangle **A** equal to the area of rectangle **B** gives us

(3) $$(T - F)(R_A - R_C)/2 = F(1 - R_A)$$

Multiplying both sides of the equation by **2**, we have

(4) $$(T - F)(R_A - R_C) = 2F(1 - R_A)$$

Expanding both side of the equation gives

(5) $$T R_A - T R_C - F R_A + F R_C = 2F - 2F R_A$$

Since triangle **A** in the diagram is similar (*i.e.* equiangular) to the triangle **A** + **C** + **D**

(6) $$F + R_C = T + R_A$$

Multiplying both sides by R_A gives

(7) $$T = F R_A + R_C$$

Now we can eliminate **T** from *equation (5) by substituting the right side of equation (7)*

(8) $$(F R_A + R_C) R_A - (F R_A + R_C) R_C - F R_A + F R_C = 2F - 2F R_A$$

Dividing both sides by **F**, we get

(9) $$(R_A^2 + R_C) - R_A - R_A + R_C = 2 - 2R_A$$

Simplifying gives us

(10) $$(R_A^2 + R_C) + R_C = 2$$

Rearranging and multiplying both sides by R_C gives

(10) $$R_A^2 = 2 R_C + R_C^2$$

Finally, taking the square roots of both side yields

(11) $$R_A = 2R_C + R_C^2$$

41

CHAPTER

Action Targets: A New Approach to International Greenhouse Gas Controls

Kevin A. Baumert[1*] and Donald M. Goldberg[2]
[1]Former Senior Associate, World Resources Institute, 9106 Woodland Drive Silver Spring, MD 20910 USA
[2]Executive Director of the Climate Law & Policy Project and An Adjunct Professor of Law at the American University Washington College of Law

1. INTRODUCTION

For over a decade, governments and observers have struggled with establishing an appropriate form of participation in the international climate regime for developing countries. Industrialized countries, for their part, have largely acceded to a system of fixed emission limits, coupled with market-based trading mechanisms, through the Kyoto Protocol to the UN Climate Convention. For a variety of reasons, developing countries have shown little inclination to join such a system. In particular, GHG targets - Kyoto-style fixed targets and even some alternative target formulations - seem to be viewed by many countries as a threat to development aspirations.

Much greater support has been evidenced for the Kyoto Protocol's clean development mechanism (CDM) which, for its part, has an explicit

*Corresponding author: E-mail: kbaumert@wri.org
Reproduced by permission from Climate Policy 5 (2006) 565-580.

sustainable development purpose. As a market-based mechanism, however, CDM project proponents tend to seek out the cheapest emission reductions, not the most robust development opportunities. Furthermore, the CDM operates only at a project level, suggesting that it is unlikely to drive the large-scale energy and development transformations needed to achieve the Climate Convention's objective.

The Climate Convention urges all countries to integrate GHG considerations into development planning.[1] Yet there seems to be no effective mechanism through which the climate regime promotes such a climate – development integration in developing countries. Considering that developing country engagement is essential to long-term climate protection, and also that economic and social development are the foremost considerations for poorer countries, creating new mechanisms that integrate these vital interests ought to be a major priority. Through several means outlined in this article, action targets attempt to promote such an integration, and thereby transform the notion of a target from threat to opportunity (Goldberg and Baumert 2004).

Section 2 of this Chapter explains the basic mechanics of an action target. Section 3 illustrates a particularly important aspect of action targets – the reduction of uncertainty over abatement efforts required by a given target. To do this, emissions and economic projections are used to simulate uncertainties inherent in three forms of hypothetical targets: fixed, intensity, and action targets. Section 4 examines the kind of GHG accounting system that would be needed to enable a system of action targets to operate effectively. To date, the only internationally agreed system for GHG accounting is through the CDM. To make more room for sustainable development, action targets would build upon and expand this project-based system. Section 5 addresses emissions trading under an action targets system, and how the attendant environmental risks can be understood and managed. Sections 6 and 7 explore the sustainable development dimension of action targets, including how 'actions' with mutually beneficial climate and development outcomes might be more effectively financed. Section 8 outlines how countries would plan for and assess compliance with action targets.

[1]See United Nations Framework Convention on Climate Change, 1771 U.N.T.S. 107, reprinted at 31 I.L.M. 849 (1992), art. 4, 1 (b), 1 (f) and art. 3, 4 [*hereinafter* UNFCCC] (urging Parties to be 'guided' by the principle that '[p]olicies and measures to protect the climate system ... should be ... *integrated with national development programmes*').

2. MECHANICS OF AN ACTION TARGET

An action target is an obligation to achieve or acquire an agreed amount of GHG emission reductions. The amount of reductions required by the action target is expressed as a percentage of the country's actual emissions during the compliance period. For example, if a country adopted an action target of 2% for the period 2013–2017, it would need to demonstrate emission reductions equal to 2% of its actual emissions during this period. In this way, an action target defines the amount of abatement to be achieved during a commitment period. This differs from Kyoto-style or dynamic targets, which define a level of *emissions* (or *emissions per unit of GDP*) to be achieved during a particular period.

Mathematically, an action target can be illustrated as:

$$RR = AT \times E \tag{1}$$

where required reductions (RR) is the number of reductions a country must achieve, the action target (AT) is the percentage by which the country has agreed to reduce its emissions, and E is the country's emissions during a given compliance period. Required reductions (RR) is equal to the action target (AT) multiplied by the country's emissions (E). To illustrate, suppose Country A agrees to an action target (AT) of 5% for the year 2015. If Country A's emissions (E) in that year are 100 million tons of carbon (MtC), then the required amount of reductions is 5 MtC. According to Equation 1:

$$RR = AT \times E \tag{1}$$

$$RR = 5\% \times 100 \text{ MtC}$$

$$RR = 5 \text{ MtC}$$

This illustration demonstrates that action targets would have the effect of bending the emissions trajectory of a country downward. It follows that, if emissions are *actually* 100 MtC during the compliance year *and* the country has demonstrated 5 MtC of domestic reductions, then emissions *would have been* 105 MtC in the absence of any actions taken to reach the target.[2]

Because the required emission reduction is a function of the actual emissions during the commitment period (100 MtC, see above), large

[2]This dynamic holds true only if the target is achieved through domestic actions, since making international purchases to achieve compliance will not reduce domestic emissions. Because of this asymmetry, the mechanics of an action target actually favour, albeit slightly, compliance through domestic action rather than through international purchases. This effect is relatively small and could be eliminated mathematically if desired by policymakers. These calculations are available from the authors.

fluctuations in economic and emission levels have only moderate effects on the level of abatement required. In the example above, suppose that Country *A*'s economy grew faster than expected, causing emissions to rise to 120 MtC during the commitment period. In this case, Country *A* would need to demonstrate 6 MtC of reductions (5% of 120), either domestically or through international purchases. Conversely, economic stagnation would have the opposite effect. If emissions turn out to be only 80 MtC during the commitment period, Country *A*'s required reductions drop to 4 MtC (5% of 80). Thus, extremely large emission fluctuations, on the order of 40 MtC, have the effect of altering this particular target by only 2 MtC.

This contrasts with Kyoto-style fixed targets, which are formulated as a percentage change in emissions relative to a fixed base year. If Country *A*, in the example above, had agreed to a fixed target of 100 MtC, then this target could turn out to be extremely onerous (e.g. if Country *A* ended up on an emissions path of 120 MtC) or require no effort at all (e.g. if economic stagnation put Country *A* on a path toward 80 MtC) resulting in a windfall of excess emission allowances.

As the name implies, some amount of 'action' – in the form of domestic reductions or international purchases – is required to meet any target. This is true for very small targets (e.g. 0.5%) or more ambitious action targets (e.g. 10%). The amount of action can be tailored to a relatively high level of certainty. As the above example illustrates, a country could adopt an action target and be relatively certain, even a decade in advance, of the level of effort (i.e. emission reductions) that will be required to meet that target. This relative certainty regarding level of effort is shown in more detail in Section 3.

3. UNCERTAINTY IN LEVELS OF EFFORT: COMPARING FIXED, INTENSITY, AND ACTION TARGETS

The presumptive approach to target setting, employed in Kyoto, is to set a fixed level of emissions that will be achieved at some point in the future. This can be a technically challenging task, given that business-as-usual (BAU, or 'baseline') scenarios – which are necessary to gauge the stringency and economic acceptability of a particular emission target – are often highly uncertain. Achieving a fixed level of emissions at some future year might be very easy under conditions of low economic growth and industrial stagnation but exceedingly difficult if economic growth were instead robust. Thus, fixed emission targets can entail widely varying levels of effort, depending on underlying socioeconomic conditions (especially GDP growth) present in the country. This problem is especially acute in developing and transition countries, where economies may be more volatile and affected by external conditions.

This uncertainty presents serious technical and political difficulties. If the target is set too stringently, it may constrain economic development (or lead to non-compliance). Given their aversion to risk, governments, especially in developing countries, might avoid emission targets that have the *potential* to adversely affect economic growth, even if that potential is remote. Yet, if the target is set too loosely it will create surplus emission allowances (i.e. 'hot air') which, when traded, will effectively weaken the targets of other countries.

Accordingly, a number of alternative forms of GHG targets have been proposed to try to address the problem of uncertainty. Included among these are dynamic targets, which allow the amount of emissions for a country to adjust according to a variable, presumably GDP. These kinds of targets can take the simple form of 'intensity' targets, which typically frame the commitment in terms of a ratio (e.g. emissions *per unit of GDP*), although other possibilities also exist (Baumert et al. 1999, Philibert 2002). Dynamic targets tend to reduce the economic uncertainty associated with taking a particular target by adjusting that target to economic reality, that is, by allowing faster-growing economies more emissions and contracting economies fewer emissions.[3] While helping to reduce uncertainty, dynamic targets also introduce additional complexity into both target setting and the interplay between targets and market mechanisms, such as emissions trading. In addition, substantial uncertainties may remain, especially if emissions of non-CO_2 gases and sinks are factored into targets (Kim and Baumert 2002).

To illustrate the different levels of uncertainty, we compare a modest 2% target in 2015, using three different forms of international target – fixed, intensity, and action – in five large developing countries where emissions are expected to grow significantly.

The fixed target is set at 2% below the US Energy Information Administration (EIA) 'reference case' emission scenario for each country. Similarly, for each country, the intensity target is set at 2% below the 'reference case' intensity (emissions per unit GDP) scenario. Using EIA's 'High GDP' and 'Low GDP' scenarios, we then evaluated the uncertainty in the level of abatement effort that inheres in a target set at 2% below the reference case. In other words, what would be the required amount of

[3]Yet another approach is *dual* targets, involving two national targets with differing legal characters: one non-binding (selling target) and another which could be binding (compliance target) (Q4 Philibert and Pershing 2001). This approach can also reduce uncertainty and the risk of hot air. Non-binding and dual targets, which can also be deployed with dynamic targets (Q5 Kim and Baumert 2002), are not explored further here.

abatement in 2015 if emission or intensity levels turned out to follow a High or Low GDP growth pattern, rather than the reference case?

The results, shown in Table 1, demonstrate the well-understood shortcoming of fixed targets when applied to developing countries, where significant growth is expected and uncertainties are high. A fixed target set at 2% below BAU levels (i.e. reference case) could entail, in China for example, either large reductions in emissions (10%, if GDP growth is High) or significant amounts of surplus emission allowances (i.e. 22% 'hot air', if GDP growth is Low). The results are similar for the other countries shown, although the uncertainties tend to be smaller than for China. In every case examined, higher-than-expected GDP growth results in potentially burdensome reductions (–9% to –13%), whereas lower-than-expected GDP growth results in hot air (+3% to +22%).

For intensity targets, there is less uncertainty in the level of abatement effort required to reach a target.[4] In the scenarios examined for a 2% reduction in intensity, almost all require *some* level of reductions, although several are close to zero. The overall level of abatement effort ranges from zero (Brazil and S. Korea, High GDP scenario) to a 7% reduction (Brazil, Low GDP scenario). One factor that is potentially troubling is that *higher* levels of effort tend to be needed when GDP is *lower* than expected (i.e. targets are most stringent in the Low GDP scenarios). This is the case for Brazil (–7% in the Low GDP case), India (–5%), China (–4%) and Korea (–3%). This could be problematic, as economic stagnation will reduce the capacity of countries to take actions on climate, as other social and economic issues rise in priority. It is possible that, in some cases, this dynamic can be remedied mathematically, as the target proposed by Argentina in 1999 attempts to do (Bouille and Girardin 2002). However, such refinements would be achieved at the expense of added complexity and less transparency in the climate negotiations, and some amount of continuing uncertainty over the abatement effort implied by a given target (Kim and Baumert 2002).

For action targets, the level of abatement effort varies rather little between scenarios. This is due to the fact that the reduction requirement is based on actual rather than projected emissions. The nature of action targets ensures that the level of abatement effort remains at the agreed target, 2% in this case. If GDP (and consequently emissions) growth levels are lower than expected, then slightly fewer tons of reductions will be needed. Conversely, if growth levels are higher than expected, slightly more emission reductions are required. In China, for example, due to the large uncertainties in future emissions, a 2% action target would entail

[4]The conclusions of this relatively simplistic analysis are confirmed by more complex modelling efforts (see Jotzo and Pezzey Q6 2005).

Table 1 Comparison of uncertainty in level of effort: fixed, intensity, and action targets

Country (2001 emissions)	GDP scenario	Projected emissions in 2015 (MtC)	–2% fixed targets			–2% intensity targets			–2% action targets		
			2015 emissions (MtC)	Required reductions (MtC)	Required reductions (%)	2015 intensity (t/1000$)	Required reductions (MtC)	Required reductions (%)	2015 emissions (MtC)	Required reductions (MtC)	Required reductions (%)
	Low	145	149	4	3%	0.105	–10	–7%	142	–3	–2%
Brazil	Reference	152	149	–3	–2%	0.105	–3	–2%	149	–3	–2%
(95 MtC)	High	164	149	–15	–9%	0.105	0	0%	161	–3	–2%
	Low	1063	1293	229	22%	0.438	–45	–4%	1042	–21	–2%
China	Reference	1319	1293	–26	–2%	0.438	–26	–2%	1293	–26	–2%
(832 MtC)	High	1435	1293	–142	10%	0.438	–16	–1%	1406	–29	–2%
	Low	354	368	14	4%	0.341	–19	–5%	347	–7	–2%
India	Reference	375	368	–8	–2%	0.341	–7	–2%	368	–8	–2%
(250 MtC)	High	423	368	–55	13%	0.341	–19	–4%	414	–8	–2%
	Low	163	174	11	7%	0.155	–5	–3%	160	–3	–2%
S. Korea	Reference	178	174	–4	–2%	0.155	–4	–2%	174	–4	–2%
(121 MtC)	High	193	174	–18	–9%	0.155	–1	0%	189	–4	–2%
	Low	157	171	14	9%	0.176	–2	–1%	154	–3	–2%
Mexico	Reference	174	171	–3	–2%	0.176	–3	–2%	171	–3	–2%
(96 MtC)	High	191	171	–21	11%	0.176	–4	–2%	187	–4	–2%

Notes: Projected emission in 2015 are from EIA (2003) and include CO_2 from fossil fuels only. "MtC" is millions of tons of carbon. Fixed targets are a 2% reduction below the EIA reference case scenario. Intensity targets (emissions per unit GDP) are a 2% reduction below the projected EIA reference case intensity level (not shown). Action targets are, by definition, a 2% reduction below actual emissions in 2015 (see Section 1).

emissions abatement of between 21 and 29 MtC, depending on the economic scenario that actually unfolds.

Because action targets eliminate much of the uncertainty in target setting - at least at the national level - they might make it more likely that countries would participate. Certainly, other factors also determine whether a government chooses to adopt a target. But compared to other target forms, action targets may better enable governments to tailor a target that matches a level of effort at which they are politically ready to commit. Hypothetically, a target level of 0.1% or less would still require *some* level of demonstrated action.

The international political ramifications of broader participation should not be overlooked. Agreeing to an action target, however modest, could reduce the perception in industrialized countries - especially the USA, but also elsewhere - that developing countries are not contributing to global climate protection efforts. Several studies since Kyoto have illustrated that developing countries are indeed taking action to bend the trajectory of their emissions downward (Goldemberg and Reid 1999, Chandler et al. 2002). However, they are not getting sufficient recognition for climate-friendly actions, and genuine efforts being made in the developing world remain largely invisible to politicians in some wealthier countries, who point toward inaction in the developing world as part of a justification for their own lack of effort.

4. ACCOUNTING FOR EMISSION REDUCTIONS

While action targets substantially reduce the uncertainties associated with setting the target, they may introduce uncertainty as to what constitutes an 'emission reduction' that could be recognized in pursuance of that target. Devising accounting standards to quantify emission reductions with reasonable accuracy and simplicity is perhaps the most significant challenge to the viability of action targets. Much progress has already been made on defining emissions reductions for CDM purposes, but action targets, as discussed below, may require a different, more expansive, approach. It is here, at the level of an accounting system, that uncertainty can be reasonably managed and reduced. This requires, first and foremost, that GHG accounting principles, definitions and rules be agreed ahead of time so as to guide the subsequent behaviour of governments, the private sector and relevant international organizations.

To date, the most prominent GHG accounting system is the one underpinning the CDM. This may provide a useful starting point from which to build. Defining the precise contours of an appropriate accounting framework for action targets is beyond the scope of this

Chapter. However, three desirable features of such a framework merit initial discussion here.

First, to promote sustainable development and maximize GHG abatement, the accounting system would need to have broader coverage than merely projects. More specifically, an accounting system for action targets should be able to accommodate *policies* and even private-sector-led initiatives that have a sectoral or national reach. As discussed in Section 6, these could include policies such as renewable energy portfolio standards, vehicle efficiency standards, and appliance efficiency standards, among others. In addition to promoting policy change, this could reduce the high transaction costs associated with project-by-project assessments. Some observers have suggested expanding the scope of the CDM to cover entire national sectors or geographical areas and encompass policy changes (Samaniego and Figueres 2002, Schmidt et al. 2004). Indeed, the Protocol Parties have already begun to expand the CDM accounting system by allowing multiple activities - undertaken collectively to implement a policy or standard - to be registered as a single CDM project (see UNFCCC, 2005).[5]

Second, broadening the scope of the accounting system would require altering additionality rules. While challenging for the CDM, additionality assessments could be virtually impossible in the context of multidimensional government policy making. A more promising approach might be to define a set of activities or policies - such as those mentioned above - that are unquestionably climate-friendly and therefore *a priori* eligible for crediting, regardless of the motivation for enactment. In other words, the system would recognize such actions, even if they were adopted primarily for oil security, air pollution or other non-climate reasons. Accounting standards, based on such a set of activities and policies, would then need to be developed to enable emission reduction determinations in a manner that is reasonably simple and transparent, but also in a manner that strives to avoid emission reductions accruing from normal, business-as-usual investments.[6] This might be done through a system of performance benchmarks or rate-based emission baselines (e.g. CO_2 per unit of output), probably on a sector or subsector level.

[5]This decision (UNFCCC, 2005) establishes that 'a local/regional/national policy or standard cannot be considered as a clean development mechanism project activity, but that project activities under a programme of activities can be registered as a single clean development mechanism project activity' provided that CDM methodological requirements are met. It remains to be seen how this language will be interpreted.

[6]See the GHG Protocol Initiative (http://www.ghgprotocol.org), convened by the World Resources Institute and World Business Council for Sustainable Development, for an example of such accounting standards at the corporate and project level.

Third, it is important for negotiators to agree on an accounting system – at least the main contours of one – *prior to* adopting action targets. In doing so, governments may avoid the approach taken under Kyoto, which turned negotiations on CDM project eligibility, additionality methodologies and other issues into *de facto* re-negotiations of national targets.[7] To the extent possible, an accounting system should be developed through broad stakeholder participation (given the inevitable policy issues that will arise) coupled with the input of technical competence and expertise.[8] Furthermore, as noted above, failure to agree on accounting matters *ex ante* would undermine the uncertainty-reduction benefits of action targets discussed in Section 3. In short, countries would not know what kind of actions would be required to meet a target.

Collectively, these characteristics of an accounting system make it apparent that its overriding purpose is not achieving absolute quantitative accuracy, which no system can deliver. The accounting system should be shaped, instead, with an eye towards promoting the kinds of *actions* that are needed to achieve the Climate Convention's objective, including those actions taken mainly for economic, social or other purposes. Motive, in other words, should be irrelevant, as is the case with projects and policies that help Annex I Parties achieve their Kyoto commitments.

5. EMISSIONS TRADING AND ENVIRONMENTAL PERFORMANCE

Action targets could operate in a manner that is complementary and consistent with the prevailing Kyoto system of fixed targets, emissions trading and the CDM. Like countries with emissions targets, a country adopting an action target could comply with its obligation by purchasing Kyoto-compliant emission allowances or credits in lieu of (or in concert with) taking domestic action. Likewise, countries could be permitted to sell allowances if they over-comply with their action targets. This would provide a potentially strong incentive for vigorous domestic implementation of action targets, as deeper reductions would generate financial flows.

[7]The adoption of expansive project eligibility and additionality rules that would have granted credits for projects that countries were likely to have undertaken anyway had the potential to significantly reduce the stringency of national targets. Likewise, extremely onerous requirements that would have denied credits for even the most uncontroversial projects held the potential to make targets more stringent than some Parties had expected.

[8]The GHG Protocol may be a useful multi-stakeholder model for developing such standard (see note 6).

On the other hand, with a more expansive GHG accounting system – outlined in the preceding section – trading could introduce significant new environmental risk. Environmental risk is affected by the three policy variables shown in Figure 1: the GHG accounting rules, target size, and quantity of allowed trading. If the GHG accounting system is designed to be expansive (i.e. many types of 'reductions' can be recognized), country targets are very small (e.g. 0.1%), and if trading is unrestricted, then environmental risks may rise to unacceptable levels, as countries might be able to transfer large amounts of credits that entailed little or no new efforts. In short, the system could create excessive environmental risk if the rules of each policy variable are aligned fully on the left, as in Figure 1.

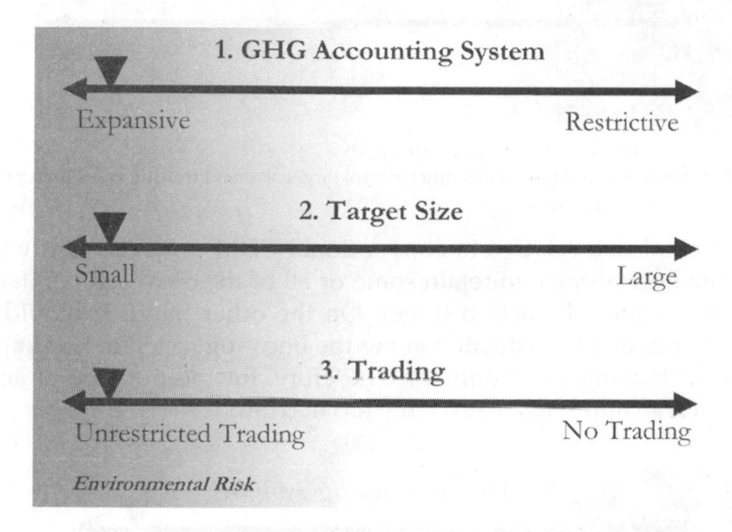

Fig. 1 Risk factors: greenhouse gas (GHG) accounting system, target size and trading rules.

For comparison purposes, the CDM can be viewed as an action target of 0%, as it does not require retention of any credits for commitment purposes, and it places no restrictions on trading of credits once they are generated. To avoid excess environmental risk, the CDM has therefore placed a heavy emphasis on project accounting rules: additionality, monitoring, verification, etc. Variable 1, in other words, is shifted to the right side, while variables 2 and 3 are shifted to the left, as depicted in Figure 2.

A simple way to introduce action targets into the Kyoto system would be to use the same accounting rules currently employed by the CDM for defining and measuring reductions as well as monitoring and verifying projects. As depicted in Figure 3, this should have the effect of reducing

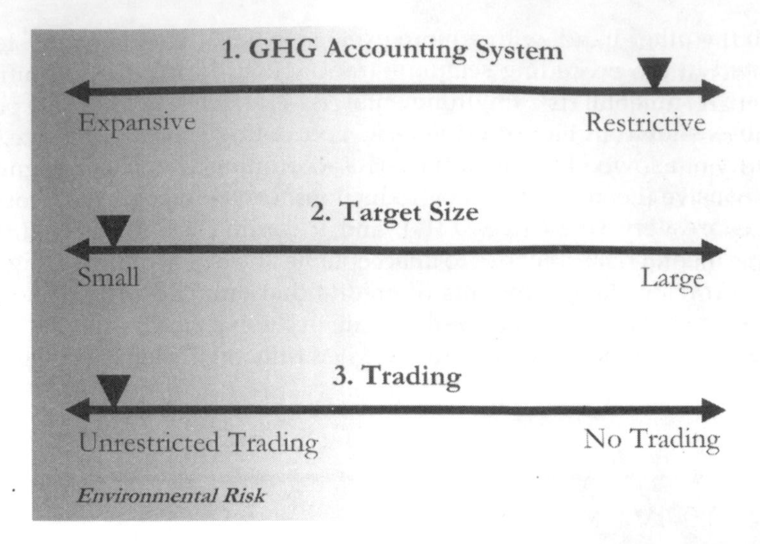

Fig. 2 Greenhouse gas (GHG) accounting system, target size and trading rules under CDM.

environmental risk relative to conventional CDM projects, as it would require the host country to retain some or all of the credits generated by the project to meet its action target. On the other hand, it would not expand the notion of a reduction along the lines suggested in Section 4 to promote sustainable development. Therefore, for the purpose of action targets, such an approach is probably too narrow.

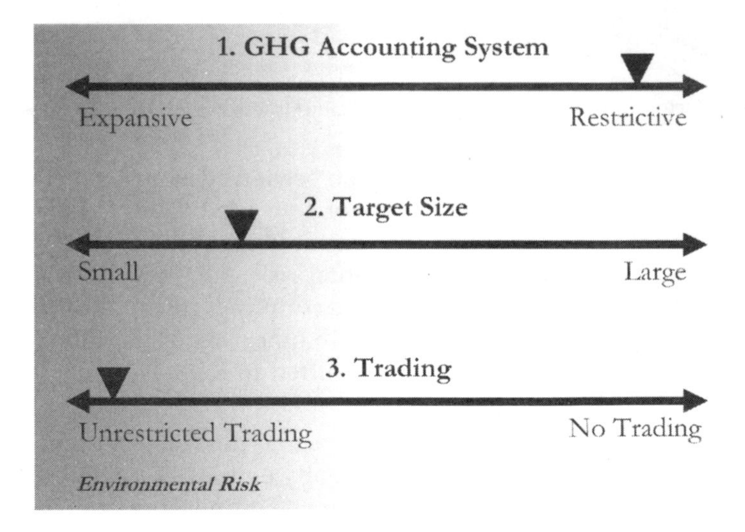

Fig. 3 Greenhouse gas (GHG) accounting system, target size and trading rules under CDM-based action targets.

If the definition of an emissions reduction is to be expanded, consideration should be given to the other two risk variables: size of the target and availability of trading. First, as to target size, it is likely that in many cases developing countries will already be taking some actions that could qualify to earn future emission reductions. Accordingly, an appraisal of the expected abatement quantities generated by existing actions might constitute a useful starting point for setting an action target. Establishing an action target that is equal to, or higher than, the emission reductions expected under current policies would, for example, give recognition to past actions taken while at the same time avoiding the creation of surplus emission reduction credits. Second, with respect to trading rules, it might be that all surplus reductions (i.e. in excess of the target requirement) are tradable. Another possibility, however, would be to limit the amount of trading to only a portion of the surplus reductions generated. Figure 4 depicts one such set of trade-offs: a more expansive accounting system (relative to the CDM), modest targets, and modest discounting or other restrictions on trade. Finding the optimal target size and trading rules is a subject for further analysis (and eventual negotiation). Key factors influencing the optimal mix include the parameters of the GHG accounting system that are agreed to *ex ante*, as well as the relative stringency of industrialized country commitments, which in large part will determine whether surplus reductions have a market value. If the value is small, trading becomes a less important driver of actions.

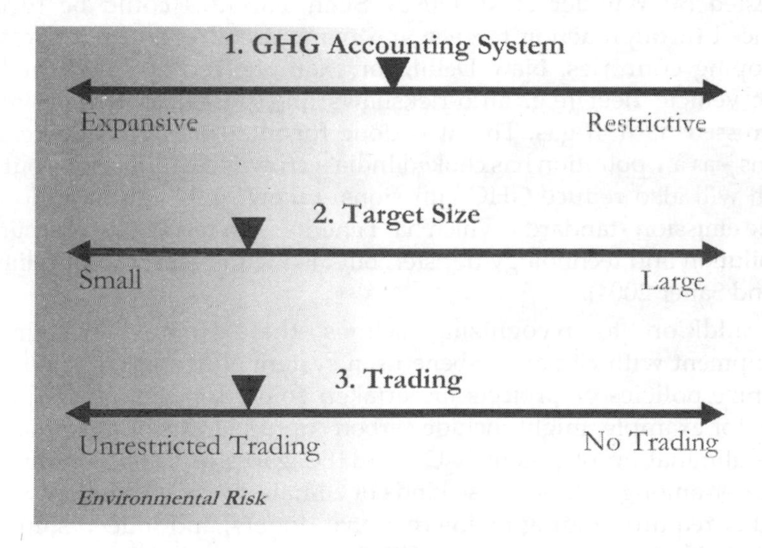

Fig. 4 Greenhouse gas (GHG) accounting system, target size and trading rules under action targets.

If trading is restricted under action targets, an additional consideration would be how action targets interact with the already existing CDM which, as noted, includes no trading restrictions. One approach would be to leave the CDM intact, so that developing countries would have available to them two means of generating reductions: the CDM and action targets. CDM accounting and trading would remain as is, while action targets might have a different set of accounting and trading rules that would promote a broader, but complementary, set of actions not recognized under the CDM.

6. PROMOTING SUSTAINABLE DEVELOPMENT

An important purpose of action targets, noted above, is to improve the prospects of integrating climate protection measures into Parties' development objectives, along the lines urged by the UNFCCC (Art. 3(4)). Action targets provide a mechanism through which countries - alone or in cooperation - can undertake development initiatives in a manner that delivers some tangible climate benefit, even if reduction of GHG emissions is not the primary purpose.

Renewable energy initiatives, energy efficiency standards, forest conservation programmes and biofuels programmes are examples of actions that could be pledged. These kinds of policies and measures reinforce the important priorities of developing countries, and have tangible GHG benefits that could be recognized and captured, as suggested by Winkler et al. (2002). Such activities could be further advanced through action targets, and many are already under way in developing countries. New Delhi, for example, recently switched its public vehicle fleet (e.g. auto-rickshaws and buses) to less-polluting compressed natural gas. This was done for vital public-health-related reasons - as air pollution has choked India's crowded capital city - but the switch will also reduce GHG emissions. Likewise, China has adopted vehicle emission standards, which has benefits with respect to oil security, air pollution and technology transfer, but also with respect to the climate (An and Sauer 2004).

In addition to recognizing actions that promote sustainable development with climate co-benefits, a system of action targets would recognize policies or projects undertaken solely for climate purposes. These, for example, might include carbon capture and storage initiatives or the elimination of potent N_2O and HFC gases in various industrial processes, among others. These kinds of climate-specific activities would probably require funding by international donors, and indeed some are already being funded through the CDM at the project level.

The CDM itself, however, has little capacity to assist developing countries in 'achieving sustainable development' - one of its stated purposes. A genuine altering of development paths is likely to require *policy* interventions of the kind not yet recognized by CDM project rules. A comprehensive assessment of the CDM in Latin America, for example, found that the sustainable development component of the CDM basically amounts to ensuring that 'the GHG mitigation project is congruent with the nation's existing environmental policies', rather than actually precipitating policy changes in a manner that promotes cleaner development (Figueres 2004). Furthermore, even at the project level, development-related benefits are likely to be secondary to climate benefits. The fact that a majority of CDM credits are expected to come from projects generating low-cost reductions of non-CO_2 gases, such as elimination of N_2O or HFC, suggests that a mechanism based mainly on market principles encourages project proponents to seek out the cheapest emission reductions, not the most robust development benefits.

A true 'sustainable development' instrument under the Convention or Protocol would need to promote larger scale, transformative initiatives - for example, providing access to electric power in Africa or southern Asia - in a manner that delivers tangible climate benefits. With an appropriate accounting framework, discussed in Section 4, climate-friendly elements of such large-scale initiatives could be developed, formally recognized, and quantified under an action target. Given the decisions at the first meeting of the Protocol Parties in December 2005,[9] the CDM Executive Board may begin to shape a more expansive accounting framework that could accommodate some climate-friendly sustainable development policies within the existing CDM.

7. FINANCING ELIGIBLE ACTIVITIES

The UNFCCC establishes a framework for financing GHG mitigation in developing countries. In short, developing countries are required to formulate and implement national GHG mitigation 'programmes',[10] and industrialized countries are obligated to provide the finance and technology to meet the 'agreed full incremental costs' of implementing these programmes ([Q2]UNFCCC, Art. 4(3)) Financial resources can be

[9]See UNFCCC (2005), and also note 5.

[10]See [Q2] UNFCCC, Art. 4, para 1(b) ([f]ormulate, implement, publish and regularly update national ... programmes containing measures to mitigate climate change ....'). Although Art. 4, para 1(b) constitutes a mitigation obligation applicable to all Parties, it is invoked primarily in the context of developing countries, since Annex I Parties are subject to additional obligations under the UNFCCC and Kyoto Protocol.

provided through the Global Environment Facility[11] (GEF) or through bilateral, regional, or multilateral channels ([Q2]UNFCCC, Art. 11).

The present model for funding mitigation in developing countries has had only limited success, perhaps because it is so vague and indefinite. There are no definitions, guidelines or requirements as to what constitutes a national mitigation 'programme'. There is no systematic accounting of funding provided (aside from the GEF), nor of the resulting emission reductions.[12] Both the mitigation programmes (in developing countries) and the associated financing and technology transfer (from developed countries) are viewed as more hortatory than mandatory.

A system of action targets could improve the situation in at least three ways. First, action targets provide a tangible commitment toward which financial resources can meaningfully be directed. The successful financial mechanism of the Montreal Protocol on Substances that Deplete the Ozone Layer (together with bilateral assistance) finances the phase-out *commitments* agreed to by developing countries. If the GEF were similarly geared towards assisting the implementing of action targets, then developing countries would be able to negotiate additional funding, and all stakeholders could monitor progress.

Second, because action targets incorporate the concept of sustainable development into their basic operation, they could help tap and eventually influence the much larger 'non-climate' funding sources. This might increase the overall funding that mutually supports climate protection and sustainable development. Funding could come from any source: bilateral aid agencies, the GEF, multilateral development banks, export credit agencies, the private sector, the host government (federal and perhaps state/local), state and local communities, or others. Some funders - host governments, development banks, and aid agencies - would be primarily concerned with alleviating poverty or otherwise boosting economic development. Other funders, such as the GEF, would invest because of the explicit climate benefit. Still others, such as private banks or corporations, would have commercial purposes, or finance the GHG component of a policy or project in order to acquire resulting emission reductions. The intent is to align and strengthen the linkages

[11]During the year 2003–2004, the GEF (as the financial mechanism of the Convention) contributed about $217 million to climate change activities, about $150 million of which was targeted at GHG mitigation efforts related to wind power, energy efficiency and other areas (UNFCCC, 2004a).

[12]See UNFCCC (2004b). The most recent estimates of bilateral assistance are from 1998–2000, when the OECD estimated 'climate-change-related aid[0]' (broadly defined) at about $2.7 billion per year (OECD 2002). Multilateral funding through the World Bank, UNDP and others for the support of Convention implementation is significant, but not presently known.

between the relevant financial institutions in a manner that maximizes resource and technology flows to development initiatives that deliver climate benefits.[13] In some cases, public–private partnerships might come together to forge large, transformative strategies that offer both development and climate benefits. In other cases - such as energy efficiency - measures may be sufficiently attractive on non-climate grounds that they would not require international assistance.

Third, action targets could integrate financial flows associated with 'carbon financing' with the other financial flows mentioned above. (These flows are explicitly disconnected under the CDM.[14]) As discussed in Section 5, a system of action targets could allow for the transfer of surplus emission reductions to industrialized countries that are covered by emission caps. Accordingly, should there be a reasonable price of carbon in the future, carbon finance could provide a further tangible boost to pledged actions. While carbon finance could play an important role, however, action targets differ from the CDM or proposals to expand the CDM in that they are not predicated on the existence of a market for emission reductions.

8. ASSESSING COMPLIANCE

Compliance assessments under action targets would entail two basic steps. First, a determination of required reductions would need to be made at the end of the commitment period (or, during a 'trueup' period following the commitment period). To do this, according to Equation 1 (see Section 2), a country's action target would simply be multiplied by its actual emissions during the commitment period. This is not to suggest that countries should wait until the end of the commitment period to determine what actions are needed to meet their action targets. Just as fixed targets require countries to look ahead to determine the actions they will need to take during, or even preceding, the commitment period, action targets require countries to assess the number of reductions they are likely to need to meet their target (as demonstrated in Sections 2 and 3, this assessment cannot be far off the mark) and to have a plan in place to achieve the amount of required reductions.

[13]For an excellent discussion of this concept, see Heller and Shukla (2003, p. 132) (referring to 'programmatic climate cooperation').

[14]CDM project participants must provide an 'affirmation' in the registered project design document that any public 'funding does not result in a diversion of official development assistance and is *separate from and is not counted towards the financial obligations of those Parties'* (UNFCCC 2001, Appendix B) (emphasis added).

To undertake this first step, a national GHG emissions inventory would be needed. However, the degree of accuracy and international oversight such inventories would require is less under action targets than under fixed or dynamic targets. This is because measurement inaccuracies have a relatively small effect on the required reductions (*RR*) under an action target. Repeating the Country *A* example used above: if the action target (*AT*) is 5%, and the emissions (*E*) inventory during the commitment period is understated by 10% (90, instead of 100 MtC), then the required reduction will be 4.5 MtC (5% of 90). Similarly, a 10% overstatement in emissions during the commitment period would increase the reduction requirement to 5.5 MtC (5% of 110). Thus, the same dynamics that reduce uncertainty in target setting also help to offset the potentially deleterious effects of inaccurate national inventories.

By contrast, under a system of fixed or dynamic targets, a bias of a few percentage points might substantially alter the level of effort needed to achieve compliance. Accordingly, inventories must be prepared to a higher degree of quality and are subject to rigorous international standards and oversight procedures. Were developing countries to adopt such targets, achieving high quality inventories would entail major financial and institutional capacities, which might otherwise be directed toward substantive action. Indeed, almost all developing countries have reported difficulty in compiling their emissions inventories under the Climate Convention.[15]

The second step in a compliance assessment is determining the amount of reductions a country has generated domestically and transacted internationally (purchases and sales).[16] Thus, for action targets, the compliance assessment would need to be directed primarily at assessing the efficacy of pledged *actions*, rather than a Kyoto-style assessment of actual *emissions*. This assessment would need to proceed in accordance with the accounting standards that were adopted (see Section 4 for discussion). This kind of process - examining actions, or the lack thereof - might help accelerate learning in climate protection efforts and help build capacity to take further actions. Emissions inventories may tell policy makers whether emissions have gone up or down, but they do not explain the reasons for those changes. In contrast, the information required to assess compliance with action targets should enhance the ability of regulators and stakeholders to distinguish between actions that were effective from those that failed to produce desired reductions.

[15]See UNFCCC (2002, para 161). Problems reported include lack of quality data, lack of technical and institutional capacity, and problems related to methodologies.

[16]This could be accomplished through the same registry system as developed under the Kyoto Protocol.

If a process to deal with instances of non-compliance is needed to protect the integrity of the trading system,[17] it should be facilitative. First, non-compliance may not be deliberate; rather it may be the result of lack of capacity, or even the failure of industrialized countries to deliver on the promised financial assistance needed to achieve these reductions. Thus, a facilitative process might improve the prospects of future compliance and better North–South cooperation. Second, a facilitative process is in step with the Convention principles, which grants transition economies, such as Russia and the Ukraine, a 'certain degree of flexibility' in achieving commitments. Accordingly, it seems appropriate that developing countries be likewise offered flexibility in meeting action targets.

CONCLUSION

This chapter does not answer all of the questions surrounding action targets and their implementation. Indeed, there are significant challenges associated with an action target approach, and further study is needed. How can we be assured that countries would adopt reasonably stringent targets, especially in light of the USA's non-participation in Kyoto? Can a suitable accounting system be developed? How would action targets be implemented at the domestic level? How would action targets (governmental obligation) best avoid conflicts with pre-existing CDM projects (private sector-led)? Should industrialized countries be able to pursue action targets?

There is no silver bullet for protecting the climate system. The approach advanced in this chapter is a modest one that, if viable, would be only one part of a broad and ambitious effort to protect the world from dangerous climate change. Such an effort would no doubt include deeper reductions from industrialized countries, perhaps through fixed targets or even a hybrid approach involving a combination of fixed and action targets. It must also include measures to help vulnerable countries adapt to impacts of unavoidable climate change, provisions for technology development and transfer, and perhaps greater clarity as to the regime's long-term objective.

In one respect, however, the approach advanced here could significantly alter the way we think about and implement our response to climate change. Adopting action targets would shift the focus of climate protection somewhat away from short-term fluctuations in emissions and

[17] Even if the target were made non-binding, the system would need to ensure that a country is not a net seller at the end of the commitment period if it cannot comply with its target (Philibert 2005).

toward the actions that give rise to those fluctuations, but without abandoning quantitative commitments. Of course, any system to address climate change must keep a continuous eye on greenhouse gas emissions and be prepared to make corrections as new information about emissions, atmospheric concentrations and the response of the climate system comes to light.

Ultimately, governments will need to decide whether targets and trading, as conventionally understood, are workable for developing countries. Existing evidence increasingly suggests that this presumptive path is fraught with difficulties. While our preliminary analysis suggests that action targets could ameliorate some of these difficulties, much work remains to be done before the workability of action targets can be reliably assessed. It is our hope that this article will stimulate sufficient interest in this new approach to motivate additional research and analysis, so that such an assessment soon can be made with confidence.

References

An, F. and A. Sauer. 2004. Comparison of Passenger Vehicle Fuel Economy Standards and Greenhouse Gas Emission Standards Around the World. Pew Center on Global Climate Change, Washington, DC.

Baumert, K.A., R. Bhandari and N. Kete. 1999. What Might A Developing Country Climate Commitment Look Like? World Resources Institute, Washington, DC.

Bouille, D. and O. Girardin. 2002. Learning from the Argentine voluntary commitment. In: K.A. Baumert, O. Blanchard, S. Llosa and J. Perkaus (eds), Building on the Kyoto Protocol: Options for Protecting the Climate. World Resources Institute, Washington, DC.

Chandler, W., R. Schaeffer, D. Zhou, P.R. Shukla, F. Tudela, O. Davidson, S. Alpan-Atamer (eds). 2002. Climate Change Mitigation in Developing Countries: Brazil, China, India, Mexico, South Africa, and Turkey. Pew Center on Global Climate Change, Washington, DC.

EIA [Energy Information Administration]. 2003. International Energy Outlook 2003. US Department of Energy, Washington, DC.

Figueres, C. 2004. Institutional Capacity to Integrate Economic Development and Climate Change Considerations: An Assessment of DNAs in Latin America and the Caribbean. Report to the Inter-American Development Bank, Washington, DC.

Goldberg, D. and K. Baumert. 2004. Action targets: a new form of GHG commitment. Joint Implementation Quarterly (Oct), 8–9.

Goldemberg, J. and W. Reid (eds). 1999. Promoting Development while Limiting Greenhouse Gas Emissions: Trends and Baselines. UNDP and World Resources Institute, New York.

Heller, T. and P.R. Shukla. 2003. Development and climate: engaging developing countries. In: Beyond Kyoto: Advancing the International Effort Against Climate Change. Pew Center on Global Climate Change, Washington, DC.

[Q7] Jotzo, F. and J.C.V. Pezzey. 2005. Optimal Intensity Targets for Emissions Trading under Uncertainty. Program on Energy and Sustainable Development Working Paper 41. Center for Environmental Science and Policy, Stanford University, CA [available at http://pesd.stanford.edu/publications/20915/].

Kim, Y.-G. and K. Baumert. 2002. Reducing uncertainty through dual-intensity targets. In: K.A. Baumert, O. Blanchard, S. Llosa and J. Perkaus (eds), Building on the Kyoto Protocol: Options for Protecting the Climate. World Resources Institute, Washington, DC.

OECD, 2002. Aid Targeting the Objectives of the Rio Conventions 1998–2000. OECD, Paris.

Philibert, C. 2002. Evolution of Mitigation Commitments: Fixed Targets Versus More Flexible Architectures. Annex I Experts Group on the UNFCCC, OECD and IEA, Paris.

Philibert, C. 2005. New Commitment Options: Compatibility with Emissions Trading. IEA and OECD Information Paper, Paris.

Samaniego, J. and C. Figueres. 2002. A sector-based clean development mechanism. In: K.A. Baumert, O. Blanchard, S. Llosa and J. Perkaus (eds), Building on the Kyoto Protocol: Options for Protecting the Climate. World Resources Institute, Washington, DC.

Schmidt, J., K. Lawson and J. Lee. 2004. Sector-Based Greenhouse Gas Emissions Reduction Approach for Developing Countries: Some Options. Working Paper. Center for Clean Air Policy, Washington, DC.

UNFCCC, 2001. Report of the Conference of the Parties on its Seventh Session, held at Marrakesh from 29 October to 10 November 2001: Decision 18/CP.7. UNFCCC Document FCCC/CP/2001/13/Add. 2.

UNFCCC, 2002. National Communications from Parties Not Included in Annex I to the Convention. Note by the Secretariat. Document FCCC/SBI/2002/16.

UNFCCC, 2004a. Report of the Global Environment Facility to the Conference of the Parties. Note by the Secretariat. Document FCCC/CP/2004/6.

UNFCCC, 2004b. Implementation of decisions 12/CP.2 and 12/CP.3: determination of funding for the implementation of the Convention. Note by the Secretariat. Document FCCC/SBI/2004/6.

UNFCCC, 2005. Further Guidance Relating to the Clean Development Mechanism. Decision -/CMP.1 (advance unedited version).

Winkler, H., R. Spalding-Fecher, S. Mwakasonda and O. Davidson. 2002. Sustainable development policies and measures for sustainable development: starting from development to tackle climate change. In: K.A. Baumert, O. Blanchard, S. Llosa and J. Perkaus (eds), Building on the Kyoto Protocol: Options for Protecting the Climate. World Resources Institute, Washington, DC.

42

CHAPTER

Towards Diffused Climate Change Governance – A Possible Path to Proceed After 2012

Norichika Kanie
Associate Professor, Department of Value and Decision Science,
Graduate School of Decision Science and Technology
Tokyo Institute of Technology
Tokyo, Japan

INTRODUCTION

An international institutional framework beyond 2012 does not mean that we should aim at a completely different approach from the Kyoto Protocol. If we return back to the fundamental principle of multilateral agreements, there is nothing more difficult and costly than to dissolve agreements and policies that have been created by international consensus. The only way to create an institutional framework that will surpass the Protocol is to build it upon the institutions that have already been established under the Kyoto Protocol along with strengthening its measures to mitigate and adapt to climate change. In the debate on future international institutions on climate change, however, some have suggested to revoke the Protocol due mainly to the US non-ratification to the Protocol as the largest emitter of the GHGs (Greenhouse Gases). Some others accuse the Kyoto Protocol for not posing any targets to developing countries, which are expected to be the larger emitters by absolute numbers in a few decades. The debate gets even more complicated in a country like Japan, whose largest export and import trading partners, the US and China, do not have a binding target under the current international framework. Under such circumstances, it may appear for those who are concerned about trading that Japan is the only country that must 'suffer' from reduction commitments out of its major trading partners. Although there are such voices and proposals, it is highly unlikely that a 'revolutionary' approach will be created to replace the existing Kyoto Protocol. Considering the socio-economic costs and implications of establishing a new framework completely from scratch, it is best to take an incremental approach to create an international framework going beyond the Kyoto Protocol.

However, politics on climate change is not as straightforward as to make it possible to agree on something based on the Kyoto Protocol for post-2012. On the one hand, the European Union has defined clear rules for cap and trade measures that follow the provisions of the Protocol. It has also set its ultimate emission control target for the overall rise of Earth's surface atmospheric mean temperature within 2°C compared to pre-industrial level. The US, on the other hand, has excluded itself from the confines of the Protocol, at least for the time being, and has created its own measures as if dreaming for the revival of a post war American hegemonic regime. The US climate change policy is centered primarily on technological developments and deployments, and it is committed to bear this cost to a certain degree.[1] Meanwhile, there is no sign of greenhouse gas (GHG) emissions slowing down in the US, although the recent surge in oil prices may reduce GHGs emissions temporarily.[2] It is, however, important to acknowledge that both of these policy factors are not conflicting against each other. Under these circumstances, what international institutions would be the best possible solution to move forward?

Another important factor that is related to climate change politics is the variety of actors that have been, and are likely to continue to be, related to and affected by climate-change policies. Although the impact of global warming extend to various areas in society, it is the (less wealthy) citizens, who are in the most vulnerable position, that are affected the most. This standpoint is especially important when considering adaptation to climate change. Businesses and industries will also be affected by climate change policy measures and will particularly play a key role in mitigating GHGs. This aspect of climate change politics is especially important in the context

[1]http://www.state.gov/documents/organization/75455.pdf.

[2]The US 2004 GHG emissions levels show an increase of 15.8% compared with 1990 levels. *Source:* USEPA, Inventory of U.S. Greenhouse Gas Emissions And Sinks. 1990-2004" April 2006, USEPA #430-R-06-002.

of post 2012 international negotiation on climate change, because what is happening now is the 'implementation phase' of international agreement under the Kyoto Protocol and other frameworks is taking place simultaneous to the 'negotiation phase' on the future international institution. Unlike former negotiations, success and failure of implementation would matter in the dynamics of international negotiation.

Multi-stakeholder involvement for sustainable institution is also in line with the current study on complexity. For 'complex issues' such as environmental ones - in which many social factors and natural phenomena are interlinked - involving various stakeholders and creating dense network of agents would work best in solving problems. In this way autonomous dispersed cooperation can be created by each agent with some space for innovation, and information can be distributed through a network of agents. After all, distributive governance, rather than a centralized structure, may be the way to proceed in climate governance. One big difference that makes climate change (and other sustainable development issues) peculiar from other complexity issues (such as computer networking etc.) is that climate change needs to the change the direction of a guiding principle of the society, which requires a huge amount of driving force. In other words, this change entails a change in a value set. In order to avoid a high risk of a dangerous level of climate change, it is the role of international agreement that would give guidance to such a directional change - an important role for an international regime. The UN Framework Convention on Climate Change (UNFCCC) shall be the basis for formulating such cooperation and norms, but it is more important to keep a clear numerical target to reduce GHGs. Unless clear target is set, strong motivation to reduce GHGs may not emerge as history has shown us. When there is an upper limit of GHG emissions on the Earth, as most scientists argue, then that should be acknowledged in an international agreement in a clear manner. Of course, we must also recognize the importance of various partnership initiatives for the management of complex issues. However, it must be reminded that it is the states that are the fundamental units of contemporary international relations. Therefore, ultimately it should be the states, and not NGOs or businesses, which are responsible for ensuring to minimize the risks to climate change. Partnership initiatives, including sectoral approach, could be complementary, but not replacement for international agreements.

Therefore, keeping the Kyoto Protocol (or improved Kyoto Protocol) is a prerequisite for post-2012 institution. Under such a prerequisite, institutions must be created where the various functions of governance are directed to the most appropriate agents. The U.S.A.'s withdrawal from the Protocol has led to create or enhance a variety of international frameworks that are related to climate change but not necessarily directly addressing climate change. To date, this is perhaps their greatest side effect of abandoning Kyoto by the U.S.A.

MAGNITUDE OF THE CHALLENGE

Before proceeding with this discussion, first a review of the magnitude and seriousness of the situation on climate change should be described. Article 2 of the Convention describes the objective of the Convention that is used frequently for an ultimate goal of climate change policy. Article 2 reads the objective of climate change policy as "stabilization of greenhouse gas concentrations in the atmosphere at a level that would prevent dangerous anthropogenic interference with the climate system". In other words, it aims to prevent a dangerous level of climate change to humans and the environment. As 192 countries, including the US, have ratified the treaty, we can say that Article 2 has reached an international consensus and is applied by most signatories at present.[3] The next step shall attempt to clarify a question on "what is the definition of dangerous level of climate change?" The European Union (European Community, then) designated their definition of 'dangerous level ' back in 1996, which was to maintain the increase in earth's mean temperature increase within 2°C compared with pre-industrial levels.[4] In Japan, the special committee to the central environmental commission started discussions in 2005 by setting the '2°C ' target as a point of departure for the discussion of the dangerous level for the following reasons.. If the target was set at 3°C, the latest scientific findings estimated that the damages would outweigh the benefits. For example, the agricultural sector in various parts of the world would be afflicted, and vulnerable areas would particularly see an alarming rise in the number of human deaths. Furthermore, although uncertainties surround the science of global warming, evidence has shown that the probability has increased for irreversible damage and instability inflicted on the earth. If we regard placing minimum risk on the human environment as most important, the magnitude of risk the world would face at the 3°C temperature increase is unacceptable. At the same time, the recommendations of specialists on the actual cost of emissions reductions have shown that the costs are enormous to limit the temperature increase by 1°C - an amount that cannot possibly be met by many countries.

[3]Source UNFCCC homepage as of 8 June 2008.

[4]Refer to the following for more information on the EU 2°C target issue. Y. Matsumoto, H. Ohta, N. Kanie. 2005. *Environmental Research*, No. 138, pp. 93-101 (in Japanese).

Furthermore, if we consider the uncertainties on our current knowledge of the actual effects of global warming, it is best to examine the effects of GHGs calculated at 1°C intervals. Because differentiating the effects between 1.5 to 2°C, 2 to 2.5°C is difficult for the current development of scientific knowledge, the '2°C' target would serve as a point of departure.

In sum, since this debate has not reached a consensus, the definition of dangerous level of climate change is still unclear. From a scientific and political standpoint, however, the '2°C' target would be used as the starting point of our discussions.

In order to achieve the '2°C' target, how much GHGs must be reduced, and by when? According to our research project applying an energy/ economy model called AIM/Impact[Policy], we calculated that in order to maintain the '2°C' target GHG stabilization level should be at 475ppm.[5] To realize such a path requires reduction of GHGs by approximately half by 2050 on the Earth as a whole. For each national level reduction amount which is necessary to achieve such a target, differentiation of GHG reduction amount among nations is necessary.

Many people in developing countries still live on less than US$1 per day and lack the basic necessities, such as electricity and energy supplies. It is physically impossible to demand these countries to reduce emissions from the current level. In light of these differences, if we calculate the percentage of emissions reductions needed per country, we found that an industrialized country like Japan needs to reduce its emissions by at least 60 percent by 2050.[6] Existing studies have also shown similar figures.[7] Furthermore, when looking at the mid-to-long term political aspiration goals that have been established already at national and regional levels, many have set their reduction targets ranging from around 60 to 80 percent (from 1990 levels). Even in the US, the state of California has set a CO_2 reduction target by 80 percent by 2050.

Aside from the small differences in these figures, what is important essentially is to recognize the reality of the seriousness and magnitude of the challenge we are facing in tackling with climate change. There is, of course, some scientific uncertainties since these figures have been

[5]The views stated in this paper include research findings from the Research Project on Establishing a Methodology to Evaluate Mid-Long term Environmental Policy Options towards a Low Carbon Society in Japan (Japan Low Carbon Society Scenarios toward 2050). For AIM Impact [policy], refer – Stabilizing Levels of Green House Gas to stop Global Warming. Environmental Research Quarterly, 2005/No. 138 (2005), pp. 67-76.

[6]Refer to the pervious project report for calculation results.

[7]N. Kanie. 2005 Mid-Long term Target Setting and Challenges for its Internationalization: Reduction of Global GHG emissions and Japan's Targets. Environmental Research Quarterly. No. 138, pp. 84-92.

Table 1 Mid-Long term targets per country

Country/Date	*Organizations setting targets and reports*	*Long-term targets*	*Mid-term targets*
Germany (Oct/2003)	German Advisory Council on Global Change (WBGU)	• Overall rise in air temperature compared with pre-industrial levels: 2 degrees maximum. The rise in air temperature for the next 10 years: within 0.2°C • CO_2 concentration levels: under 450 ppm	45-60% reduction in CO_2 levels by 2050 compared with 1990 levels
Britain (Feb/2003)	Energy White Paper	CO_2 concentration levels within the atmosphere: under 550 ppm	60% reduction of CO_2 emissions by 2050
France (Mar/2004)	The Government's Climate Change Commission	CO_2 concentration level stabilized under 450 ppm	• Reduce CO_2 emissions per person to 0.5 tC by 2050 • Reduce 3 billion tC of emissions per annum for the entire planet by 2050
Sweden (Nov/2002)	Swedish Environmental Protection Agency	Stabilize concentration of all GHGs defined under the Kyoto Protocol: within 550 ppm (CO_2 concentration within 500 ppm)	Reduce CO_2 and GHG emissions per person to 4.5 tC by 2050 in all industrialized nations and a continued reduction of emissions as becomes necessary (emissions are 8.3tC at present)
European Union (Mar/2005)	EU Environmental Commission	To maintain the overall rise in air temperature within 2 degrees, the concentration of GHGs to be stabilized under 550 ppm	15-30% reduction by 2020 compared with 1990 levels in industrialized nations, and a 60-80% reduction by 2050

Source: Chart created by author based on various reports and policy papers

calculated based on a series of supposition. Furthermore, there is still much to be debated concerning the '2°C' target itself. Nevertheless, considering international consensus on the Rio Principle 15 (the precautionary approach), it is clear that substantial reductions with a level of 60 to 80 percent from 1990 level are necessary for industrialized countries by 2050. We must realize the gravity of the situation we are facing. Moreover, even with such efforts we must prepare ourselves to adapt to a society which is 2°C warmer in average than now. In other words, when considering climate change policy measures and institutions, we must look at both mitigation measures as well as adaptation measures.

NETWORKED DIFFUSED GOVERNANCE

Existing research on institutions provide a guidance in designing a next phase of the international institutions on climate change as a long-term and global issue. Current research has shown that the best institutional design for managing complex problems such as global environment is a loose, decentralized and dense network of institutions and actors that are able to relay information and provide sufficient redundancies in the performance of functions so that inactivity of one institution does not jeopardize the entire system.[8] In such an institutional setting, key stakeholders such as the states, NGOs and businesses share information and carry out the functions necessary for effective governance. In other words, when solving issues that are essentially complex, it may seem at a first glance as though the institutional frameworks and various agents involved in the problem-solving process are independent and complete in themselves. However, as the problems are inherently complex and interdependent, networks of agents and institutional inter-linkages would be emerged throughout various (seemingly independent) frameworks. In some cases agents and problem-solving functions could even overlap at times. The most effective way to solve complex issues is, therefore, to create institutions designed to foster and functionalize the loose but dense

[8]V.K. Aggarwal, 1998. Institutional Designs for a Complex World. Cornell University Press. E. Ostrom. 2001. Decentralization and Development: The New Panacea. K. Dowding, J. Hughes and H. Margetts. 1999. Challenges to Democracy: Ideas, Involvement and Institution, Palgrave Publishers. pp. 237-2560. C.K. Ansell and S. Weber. 1999 Organizing International Politics. International Political Science Review, January. P.M. Haas, N. Kanie and C.N. Murphy. 2004 Conclusion: Institutional design and institutional reform for sustainable development. N. Kanie and P.M. Haas (eds). Emerging Forces in Environmental Governance, UNU Press.

networks between these stakeholders.[9] Such a structure would, furthermore, help ensure that the problem-solving function's survival. For example, even if one institution became de-functionalized or ineffective for some reason, other networked institution would back it up so that it would not cause the entire system failure.

Climate change governance should also be considered in such a context. In the first place, climate issues are complex, in which various issues are interlinked: mitigating climate change is an issue which involves the increased use of renewable energy such as wind power, solar power, and biomass, as well as improving energy efficiency. Renewable energy also involves the development, diffusion and transfer of technology. Climate change is also linked to issues of deforestation, forest use, desertification, and biodiversity. Adaptation to climate change also concerns sustainable development issues in developing countries, and so on.

Such complex issues will involve various stakeholders in its problem solving. In order to create effective networked diffused climate governance, network of agents that can accommodate the optimum governance functions and actor relations that would create synergies should be developed. Under the current international agreements, for example, a nation state is ultimately responsible for consensus building and implementation of agreements, whereas industries are not the ones that should be responsible instead of a nation state. Although some agreements established by the industrial sector may have a complementary function to international agreements, agreements created by the interest-led business community do not have a role to secure global public goods that would override international agreements. On the other hand, when looking at the financial mechanisms for climate change measures, private monetary flows and investments supplied by the business sector have a strong influence along with the public flows from national and international organizations. It is, therefore, important to create an institutional framework, including such formal and informal

[9]In reality, there are over 200 Multilateral Environment Agreements (MEA). There is, however, debate that many of these agreements overlap and that they should be coordinated. This is one of the key factors towards reforming the international system in the environment arena. Interlinkages: Synergies and Co-ordination between MEAs. United Nations University: Tokyo, pp. 31. 1999. D.C. Esty and M. Ivanova (eds). 2002. Global Environmental Governance: Options & Opportunities. Yale School of Forestry and Environmental Studies. N. Kanie and P.M. Haas. 2004. A. Rechkemmer (ed). 2005. UNEO–Towards an International Environment Organization. N.W. Bradnee Chambers and J.F. Green. 2005. Reforming International Environmental Governance: From Institutional Limits to Innovative Reforms. UNU Press.

monetary flows, in which governance functions for solving climate issues can be optimized in a synergistic manner.[10]

Furthermore, taking the recent developments in the theory of "constructivism" into account, "common knowledge" and "consensual knowledge" would be developed though networked distributive governance as a process of knowledge diffusion.[11] Loose but dense network of stakeholders may create a chain reaction of information diffusion that would diffuse knowledge and norms on climate change measures even further. As a result, this will create a dynamic form of diffused governance. In other words, the structure of diffused governance may play a key role in distributing the norms related to climate change. This is an important aspect particularly for climate change measures in the long run.

Looking at the recent development of international relations on climate change from this perspective, it is not necessarily all pessimistic, as the (former) British Prime Minister Tony Blair said in his speech in early 2006.[12] The USA's withdrawal from the Protocol, along with implementation of its own climate measures, may be seen as a process where the various initiatives that are related to climate change, but not necessarily addressing directly climate change, are permeating and progressing into various sectors. What is important is not an outright 'Yes or No' to Kyoto, but is to maintain the Protocol while developing other initiatives to find a win-win solution mitigation. This is an obvious choice given the history of how climate change measures have been developed and also since the current international system holds the state responsible for the consensus building and implementation of international

[10]For views on the functions of governance and on actors, refer to N. Kanie. 2003 and the previously stated book by N. Kanie and P.M. Haas.

[11]The term 'common knowledge' refers to Wendt's theory: "the faith of the actors concerning the situation in other countries, as well as the rationality, strategies, and preferences of the mutual actors. 'Consensual knowledge' refers to the following kind of scientific knowledge within causal relationships "(1) where various important political issues are reassembled in an accessible and adaptable way (2) common knowledge by both scientific and political experts". A. Wendt. 1999. Social Theory of International Relations, Cambridge University Press. E.B. Haas. 1990. When Knowledge Is Power: Three Models of Change in the International Organization, University of California Press. A. Wendt. 1999. Social Theory of International Relations. Cambridge University Press. E.B. Haas. 1990. When Knowledge Is Power: Three Models of Change in International Organization. University of California Press. Yamada. 2004. Multiple Governance and Changes in Global Public Order – from the Perspective of Evolutionary Constructivism in International Politics, No. 137.

[12]Statement by Tony Blair on March 28th 2006, Source: ABN-AMRO, Pacific Hydro, Australian Business Council for Sustainable Energy, Phillips Fox. Show me the money, May 2006, p. 8. Source:ÿABN-AMRO, Pacific Hydro, Australian Business Council for Sustainable Energy, Phillips Fox. Show me the money, May 2006, p. 8).

agreements and securing global public goods. A totally new framework for Kyoto may be considered only if we find an alternative to nations to be responsible for securing emissions reductions in a measurable way. At present, however, such an alternative is yet to be found. Hence, the best way to realize a reduction in emissions is to implement Kyoto along with other initiatives, and to create synergies among them.

UNFCCC/KYOTO PROCESS AND OTHER CLIMATE-RELATED INITIATIVES

UNFCCC/Kyoto Process

The international institutions that deal with the climate- change issue per se are organized under the United Nations Framework on Climate Change (UNFCCC) and the Kyoto Protocol. Taking into consideration the state centered rule-making process, a key here for the debate over post-2012 institutions is the future developments in the USA – a countries that has not yet ratified the Protocol, although they participated in the negotiations. One of the biggest challenges that lie ahead will be to explore ways to bring the U.S.A into the future climate change framework discussions while trying to achieve the goal of the Convention.

Article 3.9 of the Protocol defines that parties must start consideration of the commitment after the first commitment period at least seven years before the first commitment period expires. As the first commitment period is between 2008 and 2012 this means that the negotiations must start in 2005 at the latest. As such is the case, it was anticipated that the Conference of the Parties to the United Nations Framework Convention on Climate Change (COP) would hold active, full-fledged discussions to explore future commitment issues which were initially discussed at COP 10 held in Buenos Aeries, Argentina in December 2004. At the COP10, a 'seminar' on future developments of the framework was proposed by the President of the COP in order to gather ideas and to prepare for future discussions, as such seminar could function as 'position mapping' through dialogue. Although the Articles that mention commitment issues are defined only within the Protocol, negotiations were held in Buenos Aires without limiting the discussions to Kyoto parties, so that the US and Australia who had not ratified the Protocol would be able to participate in the dialogue process. However, such dialogues, that might function as 'pre-negotiations', were strongly opposed by the US and Saudi Arabia. As a result, a Seminar of Government Experts (SOGE) was convened in May 2005. The seminar was held in a way that would not lead to any future negotiations, commitments, processes, frameworks and mandates under the Convention and the Kyoto Protocol.[13] Twenty six countries gave presentations in which many elaborated on the issues of the future of the framework in the end.

The Kyoto Protocol (KP) came into effect on February 16, 2005, which would work as the backbone of multi-level networked diffused climate change governance. To be more specific, the Kyoto regime became nested in the UNFCCC regime. The Protocol depicts concrete international measures such as monitoring the progress of industrialized countries in achieving GHG emissions reduction targets which should essentially lead to realizing the ultimate targets. KP also defines various functions to promote climate change governance including capacity building, financial mechanisms and monitoring. These functions are to be undertaken by the appropriate and suitable agents (ideally) such as: corporate business initiatives through CDM, awareness-raising activities by civil society organizations, reporting by the signatory countries and screening and assessment by international organizations. Like it or not, the international system of the KP has started.

With the KP's entry into force, the Meeting of the Parties to the Protocol (MOP) was held alongside the COP meeting in November and December 2005, also referred to as the COP/MOP. With the clock ticking in face of a deadline in late 2005, the two week long negotiations centered on how to create an environment that would set the course for future action on climate change. Two agreements were made as a result of the marathon negotiations and consultations which ran from morning to morning. Under Article 3.9 (future commitments) of the Protocol, Parties agreed to establish an 'Ad hoc Working Group (AWG)' as a process to discuss the future development of the KP. On the other hand, discussions on a future framework which include countries that have not ratified the Protocol were decided to be undertaken as "the dialogue on long-term cooperative action to address climate change by enhancing implementation of the Convention (the Dialogue)" in order that the dialogue would be carried out without prejudices to any future negotiations, similar to the conditions to SOGE. The workshops are planned to be convened four times during 2006 to 2007.[14]

As such, consensus building on the future institutions on climate change by the states, which is ultimately responsible for consensus building and implementation of climate change policies, are decided to be

[13]Refer to the following websites for further information:0 http://unfccc.int/meetings/seminar/items/3410.php, http://www.iisd.ca/climate/sb22/

[14]Refer to the following websites for further information:0 http://unfccc.int/meetings/cop_11/items/3394.php, http://www.iisd.ca/climate/cop11/

promoted both under the KP – the core of the climate change regime that is composed of multi-layers, and the Convention, as the meta-regime that ranks highest within hierarchical structure of climate change regime. In other words, international negotiations on how to 'go beyond Kyoto' within the framework of the KP were separated from discussions of a generic, long-term climate change policy efforts that would involve the US. This line of thought was set even clearer in the Bali Action Plan decided at CoP13.

Other Climate-related Initiatives

As the Convention and Protocol process continued to show developments, the preparatory processes of the Gleneagles G8 Summit and the Fourth Assessment Report of the Intergovernmental Panel on Climate Change (IPCC AR4) confirmed findings on the actual progress of climate change. Furthermore, the effectiveness of climate change norms has brought on a situation where the international community can no longer just continue to wait for change of policy directions of countries that did not ratify the KP such as the US. In addition, there have been efforts to try to somehow get these two countries on board, which these two countries - that are responsible for around one-fourth of the world's CO_2 emissions - would not be exempted from the responsibilities of tackling climate change as a global effort. Ever since the US's withdrawal from Kyoto, there has been a continuous debate on how to bring the US and Australia into the post 2012 negotiations. As a result, various climate change initiatives have been established. These initiatives are not bound to the multilateral UN framework. Many of these initiatives take form of a partnership by/among various entities including governments and businesses and industries, and are called 'Type II partnerships' which have been registered at the UN Commission for Sustainable Development (CSD) since the Johannesburg Summit. Many of these partnerships have been established under an international cooperation framework of voluntary consensus that does not go through the process of international negotiations.

As a follow-up to the Johannesburg Summit, the Commission on Sustainable Development (CSD) reviews and monitors the progress and implementation on sustainable development areas. The focal areas during the period of 2006 and 2007 include energy and climate change. During a review session, the CSD found that out of the 319 partnerships, 25% focused on climate change.[15] Initiatives such as the Asia-Pacific Partnership on Clean Development and Climate (APP), Renewable Energy Policy Network for the 21st Century (REN21), the Renewable

[15]The number of partnerships as of Feb. 24 2006, E/CN.17/2006/6.

Energy and Energy Efficiency Partnership (REEEP), Methane to Markets (aims to reduce methane emissions) fall within this framework.

There are also other initiatives such as the Carbon Sequestration Leadership Forum (CSLF) and an international partnership for a hydrogen economy society that are not registered at CSD.The reality is, however, that many of the initiatives that fall outside of the Convention and the Kyoto Protocol process take the form of partnerships.

Although these initiatives do not solely focus on climate change issues, what they have in common is that they take on climate change measures as one of its main themes. In other words, while these initiatives focus on other issues related to climate change, they also focus on climate change within its range at the same time. Many of these initiatives have not yet started and their future developments are unclear. However, if these initiatives are carried out in full-scale towards the respective objectives, the interdependence of the various actors will be facilitated within the related diverse frameworks. Furthermore, as the diverse framework and the actors build a network among each other through interdependent relationships, it is anticipated that the problem solving structure for climate change-related challenges will permeate into various sectors, and take a form akin to the structure of diffused governance. In order to realizing a more effective form of networked diffused climate governance, however, the relationship between governance functions and actors/ agents required for tackling climate-change issues must be organized, and an international system that encourages the optimization of this structure must be established. It should be noted that a moderate overlap is important as it provides sound competition among the initiatives. For example, the Asia Pacific Partnership on Clean Development and Climate covers eight taskforce including the development of new technology for renewable energy. It aims to promote information exchange within the private sector as well as to establish procedures to identify, evaluate, and to provide solutions for the challenges and obstacles faced in technology development.[16] These procedures are difficult to undertake within the framework of conventions and protocols. It is, however, possible for the actors that are implementing the CDM projects within the KP to also be actors of the partnership. Therefore, as different actors repeatedly work together interdependently in various frameworks, new information, knowledge and awareness is created, and interdependency and networks are further enhanced. It can be noted that such elements may, in the long run, be a short-cut to achieving the ultimate objectives underlined in Article 2 of the Convention.

[16]The first ministerial meeting of the Asia-Pacific Partnership on Clean Development and Climate took place on January 12th, 2006 in Sydney, Australia – summary and evaluation.

What is important is that these partnerships are not alternatives to the Convention or to the Protocol process. Only when the Convention and Protocol process - particularly the international systems defined in KP - specifically sets down the responsibility of the states for climate change, will a dispersed governance system structure be realized. It is similar to the Johannesburg Summit Partnership documents that have a complementary function to the political declaration and the Plan of Implementation, and do not solely represent a comprehensive agreement. According to contemporary international relations, even if governments become a partner, the collection of independent initiatives alone does not guarantee the realization of achieving targets.

CONCLUSION: GOING BEYOND THE KYOTO PROTOCOL

In the international arena surrounding climate change, the perception is that the two bipolar big coalitions will continue for the time being – the EU-centered block which aims to establish measures that respect the Kyoto Protocol, and the US-centered block which created its own initiative. The important point, however, is that both parties are not exclusive of each other. In the first place, when solving complex issues such as climate change, a co-existing framework of convention, protocol and other initiatives may stimulate structural innovation (or synergy effects) based on the reinforced characteristics of the latter system (other initiatives) which are constituted by 'partnerships'. From the perspective that the climate change issue is long term in nature, it is recommended to establish a security system by creating a mechanism for an autonomous dispersive cooperation which permeates into various sectors.

In this case, however, what is important when dealing with climate issues is the existence of the Convention and the KP.[17] As mentioned earlier, if there are other means for the respective countries to seriously deal with climate-change measures, then there is no reason to dwell on the Protocol. However, until the actions aimed at combatting climate change take firm root, the Protocol which sets clear goals and objectives will continue to play an important role. This sounds more convincing when looking back the history under the UNFCCC of non-compliance towards non-binding targets, and when looking at the evidence of the situation in the US in 2004 which has shown an increase of green-house gas emissions

[17]Network theory states that in a network that does not generally have a vertical command and control system, the hub which acts as the core body plays an important role. This hub coordinates the relationship among various actors. W.R. Scott. 1997. *Organizations: rational, natural, and open systems*, Prentice-Hall, Inc.

by 15.8 percent compared to 1990 levels and that there is no sign of putting a stop to the increase in emissions.[18] Emphasis has been, furthermore, placed on the fact that the Kyoto Regime is now actually in operation after 16 February 2005.[19] That is, if the current mechanism happens to fail, the international community will not only be pressured to change future measures, but will also be faced with a decline in investment willingness under the current system. For instance, even in Brazil, also known as the 'CDM world leader', which accounts for nearly 15 percent of the total registered CDM projects world wide, it is said that due to the uncertainty of the future framework debate, there are now signs of reluctance towards CDM investments.[20]

In this light, from the perspective of systems, history and incentives, it is important to maintain momentum towards continuing beyond the first commitment period of the Protocol as part of the next step towards future actions to climate change. As stated in the early sections of this chapter, it is too costly to start the Kyoto process again from scratch by a revolutionary approach. In addition, based on the role of the state within the structure of distributive governance, it is important to strategize on how to bring the US back into the negotiations. For instance, this could be an intensity-based target, or enhanced capacity development measures in developing countries, especially the larger ones, through initiatives outside the Convention and Protocol process. If the US is severely impacted by the effects of climate change, there may be room for large developments. Before that happens, however, the US should be called for to propose a blueprint for taking responsibility and action on climate change, as a responsible unit in the internationally system. Only then, aside from the Convention, the Kyoto process and other initiatives, respecting the Rio Principle 7 of 'common but differentiated responsibilities', will the discussion on the role of developing countries become clearer. According to the calculations of the aforementioned Japan Low Carbon Society 2050 project, it would be impossible to reduce GHG emissions globally by half by 2050 with measures taken only by industrialized countries. In the near future developing countries will also need to take actions in order to minimize the risks to the dangerous climate change.

---

[18]As stated earlier in the USEPA document.

[19]According to Hovi, Skodvin and Andersen (2003), once a system is created, there are bureaucratic interests and opportunities created within the market. A system change is highly unlikely if there is no large political change. J. Hovi, T. Skodvin and S. Andersen. 2003. The Persistence of the Kyoto Protocol: Why Other Annex I Countries Move on without the United States. Global Environmental Politics 3:4, November, pp. 1-23.

[20]CDM figures refer to March 2, 2006, http://cdm.unfcc.int
Interview to a World Bank officer in Brazilia, 06 March 2006.

In reality, many international regimes have been formed that do not include the US, similar to the situation with the Kyoto Protocol. To mention a few: the International Court of Justice (2002), the Convention on the Rights of Child (1989), Optional Protocol to the Convention on the Rights of the Child on the involvement of children in armed conflicts (2002), and the Convention on the Prohibition of Anti-Personnel Mines (1999). Despite the non-participation by the US in these treaties, there have been developments towards reaching a consensus. One main factor in solving the issue of how to bring the US in these negotiations lies in the hopes for the diffusion of norms and creation of networked distributive governance structure, but also for diplomacy.

In the era of networked diffused governance, transparent and accountable diplomacy will be a key factor.[21] In order to reflect the negotiators' contributions and devotion working day in and day out in creating the international agreement, the negotiation process should not resort only to a diplomatic approach in a narrow sense. Diplomacy should be maximized on the comparative advantages of the diverse actors and utilize their dispersed and extensive multi-channel networks. These methods will be essential for establishing a system as well as for realizing effective diplomacy. Based on these factors, furthermore, there is a greater possibility for creating effective synergies. In addition, in an era where information holds value in governance issues, the factor of rapid information flow is the foundation for effectiveness of diplomacy that could create comparative advantage in negotiation settings. Information sharing through communication and dialogue should be promoted among various stakeholders, avoiding confrontation at all costs. What is important is to explore a roadmap for effective diffused governance institutions that promotes networks and enhanced governance functions through diffusion of information.

[21]Governance is defined as not only the implementation of government rules and laws, but includes an unofficial non-government mechanism that works towards a common goal based on consensus and self-motivation. J.N. Rosenau and E.-O. Czempiel. (eds). 1992. Governance without Government: Order and Change in World Politics, Cambridge University Press, Cambridge, U.K.

43

CHAPTER

Background on CDM and Carbon Trading

Graham Erion
512 Whitmore Ave.
Toronto, ON, M6E 2N8
Canada
Email: graham@erion.ca

A BRIEF HISTORY OF CARBON TRADING

The intellectual origins of carbon trading can be traced back to a small publication in 1968 titled, 'Pollution, Property, and Prices' by the Canadian economist John Dales. Like Garrett Hardin who penned his famous essay, 'The Tragedy of the Common' in the same year, Dale believed that natural resources in their unrestricted common property form would face tragic overexploitation by people acting in their rational self-interest.[1] Yet Dales went much further than Hardin in his solution to this problem. Dales proposed to control water pollution by setting a total quota of allowable waste for each waterway and then set up a 'market' in equivalent 'pollution rights' to firms to discharge pollutants up to this level.[2] These rights, referred to as "transferable property rights…for the disposal of wastes" would be sold to firms and then they could trade them amongst themselves.[3] The more efficient firms would make the largest pollution reductions and sell their credits to less efficient firms, thereby guaranteeing a reduction of pollution at the lowest social cost.

[1]Hardin, Garrett. 1968. The Tragedy of the Commons 162 Science 1243
[2]John Dales. 1968. Pollution, Property and Prices: An Essay in Policy-Making and Economics. University of Toronto Press. Toronto, Canada. p. 81
[3]*Ibid*. p. 85.

Though Dale's proposal took a backseat to the command and control approach of the environmental policy during the 1970s, his idea would resurface in the following decades. Proponents of pollution trading – typically a mix of industry groups and self-described 'free-market environmentalists' – echoed Dales' logic about greater efficiency, and added claims of lower administrative costs and greater incentives for innovation. After a series of proposals and pilot projects by the Environmental Protection Agency, the United State Congress amended the *Clean Air Act* in 1990 to create a national emissions trading (ET) scheme in sulphur dioxide, the main pollutant behind acid rain. Up until 1997, the United States was the only country in the world with any significant pollution trading scheme. This of course would change following the Kyoto Protocol.

Though carbon trading was initially met with hostility from some European countries and environmental non-government organizations (ENGOs) during the third Conference of the Parties to the UNFCCC in Kyoto, it was eventually adopted and appears in three separate articles of the final text of the Protocol. Article 17 of the Protocol establishes a system of 'Emissions Trading' whereby Annex 1 countries (e.g. developed countries that have accepted binding emissions reductions targets) can trade emissions credits amongst themselves if they overshoot their targets. This aspect of trading can be controversial, especially when applied to the Eastern Bloc countries such as Russia and the Ukraine. The collapse of the former Soviet economy around 1990, the same base year for Kyoto, has meant that these countries get a 'free pass' on trying to reduce their emissions as their contracted economies have already reduced gross emissions by nearly 40%.[4] For this reason the unfavourable label 'hot air' has been widely applied to this form of trading since it has nothing to do with deliberate efforts to reduce emissions and everything to do with economic collapse. However, Europe's larger considerations about energy security and access to natural gas may still benefit Russia and the Ukraine in this market and increase the likelihood of future trading under Article 17.[5]

The second type of carbon trading is Joint Implementation (JI) – Article 4 – whereby Annex 1 countries can invest in projects other Annex 1 countries to reduce emissions with the investing country receiving credit for the host country's reductions. Like Emissions Trading, JI has thus far

[4]BBC News: "Q & A on the Kyoto Protocol online: http://news.bbc.co.uk/1/hi/sci/tech/4269921.stm

[5]Prototype Carbon Fund. 2005. 'Carbon Market Trends 2006', World Bank Group, Washington, DC., USA. p. 45.

not played a significant role in the international carbon market.[6] According to the World Bank's Prototype Carbon Fund, this is due to a lack of investor confidence and institutional set up in JI countries.[7]

With Emissions Trading and Joint Implementation playing minimal roles, the global carbon market is at present almost entirely made up of transactions under Article 12 of Kyoto, the Clean Development Mechanism. The CDM provides an opportunity for Annex 1 countries to receive emission reductions credits to use against their own targets by investing in projects to reduce or sequester GHG emissions in non-Annex 1 countries (i.e. developing countries.) One of the most controversial aspects of Article 12 is that it requires projects to show "Reductions in emissions that are additional to any that would occur in the absence of the certified project activity."[8] This requirement has become known as "additionality" and is intended to ensure there is a net emissions reduction.[9] Another controversial aspect of the CDM is the requirement that projects must also help developing countries in "achieving sustainable development."[10] The sustainable development requirement represented a hard fought victory by many of the countries and ENGOs that were initially against the CDM. However, in subsequent meetings of the Conference of the Parties (COP) to the UN Framework Convention on Climate Change (UNFCCC) countries have been allowed to set their own definition of sustainable development and judge whether a project meets these criteria, rather than adopt a universal definition that could better ensure the accountability of those authorities overseeing project approval.

A number of domestic and international governance structures have been set up to oversee CDM projects. There are three key institutions governing CDM projects through their validation. The first of these is each host country's Designated National Authority (DNA.) The DNA is the first institution to review a project's documents, namely the Project Design

[6]As of September 2006 there were only 126 Joint Implementation projects in various stages of validation compared with 1150 projects in the CDM pipeline. (Source: Jørgen Fenhann, UNEP Risø Centre, 'CDM Project Pipeline ' updated 14-09-06, online: www.cd4cdm.org)

[7]*Supra* note 6, p. 26.

[8]The Kyoto Protocol, Article 12, paragraph 5 (2).

[9]Since at its root, carbon trading is about Northern countries 'offsetting ' their pollution by reductions in Southern countries, if these reductions were to occur without the intervention of the carbon market - and thus not additional to the status quo - there is no net benefit for the climate.

[10]*Supra* note 9 at Article 12, paragraph 2.

Document that lays out all the relevant information about the project.[11] Assuming everything is in order, the DNA will write a letter of approval saying that all participants are voluntary and that the sustainable development criteria have been met. As to the actual makeup of the DNAs, they will often be housed in government departments and staffed with public sector employees, such as in South Africa where the DNA is in the Department of Minerals and Energy. However, in other cases, such as Cambodia, the DNA is contracted out to private consultancies.

Once the letter of approval has been issued by the DNA, the PDD is then assessed by a Designated Operational Entity. Unlike the DNAs, the DOEs are all private sector entities. To date 12 companies have been accredited as DOEs, though not all of them can accredit every single methodology. To validate a project the DOE will review the PDD to consider whether the project's methodology is in line with approved methodologies, the claimed emissions reductions and baseline scenarios are accurate, and the project is 'additional. ' In making its determination, the DOE will also post the PDD on the internet for a 30-day public comment period.

With the approval of the DOE and the DNA, the final stage in project validation is the CDM Executive Board (EB/CDM) whereby the findings of the DOE and DNA are reviewed and a final decision is made whether to allow the project to start generating Certified Emissions Reductions (CER).[12] There is also a final 30-day public comment period while the project is at the EB/CDM With only 12 members on the EB/CDM they do not have the resources to closely scrutinize every project that comes across their desk. As such they rely heavily on decisions of DOEs. According to Eric Haites, a private sector consultant in the carbon market, "the vast majority of validation and certification decisions by DOEs expected to be final; the Executive Board only deals with the problem cases."[13]

[11]In addition to the PDD, in some countries, project developers can submit a Project Identification Note (PIN.) A PIN tells the DNA what the project plans to do but need not include all the details required in a formal PDD. The purpose of this stage is to allow a project developer to get a sense of how the project will be viewed by the DNA. If the DNA has some initial concerns, these can be addressed prior to the submission of the PDD to save both time and money. If the DNA has no concerns about the PIN then the project developer can ask for a letter of 'no objection' that will expedite the process later on. While this is the case in South Africa, it is important to remember that every DNA can establish its own procedures around reviewing PINs and what is applicable in one jurisdiction may not be the case in another.

[12]Certified Emissions Reductions are the currency of the carbon market: once a project starts reducing emissions that are verified by a DOE, they pass on certificates for the reductions in carbon dioxide equivalent (CO_2e) to the country investing in the project, which is then used against the Annex 1 country's Kyoto targets.

[13]Haites, E. and M. Consulting. 2006. Presentation to York University's Colloquium on the Global South. 25 January.

CARBON MARKET TRENDS

With the process of validation now established and some of the relevant institutions explained, let us turn our attention to how the global carbon market has developed since Kyoto. The first thing to note is the large role played by Northern firms and consultants - such as Ecosecurities - who are able to provide a certain level of capacity and expertize that might not be as readily accessible in Southern countries. Another example of this has been the prominence of the World Bank's Prototype Carbon Fund (PCF.) In partnership with 6 governments and 17 companies plus a budget of US $180 million, the PCF describes itself as "a leader in the creation of a carbon market to help deal with the threat posed by climate change."[14] As the single largest purchaser of CERs, as of September 2006 the PCF had 32 projects in development with a total CER value potential of US $165 million.[15]

A second noteworthy trend is that the market is heavily concentrated in large middle income countries led by India, China, and Brazil. The PCF admits that "this concentration of CDM flows towards large middle-income countries is consistent with the current direction of Foreign Direct Investment."[16] By contrast poorer countries, especially in Africa, have almost entirely been left behind. As of September 2006, South Africa and Morocco were the only countries on the continent to have validated a CDM project. According to the PCF, "This under-representation of Africa raises deep concerns about the overall equity of the distribution of the CDM market, as the vast majority of African countries have not, for the moment, been able to pick up even one first deal."[17] This seems to dispel the notion that the CDM would help uplift the world's poorest countries to a cleaner path of development.

The other major trend in the carbon market has been the enormous profitability of non-carbon related projects. While renewable energy projects (which offset CO_2 emissions) make up nearly 58% of the total number of projects, they account for only 15% of the total number of CERs that have been issued.[18] By contrast, projects abating nitrogen (N_2O) and

---

[14]Prototype Carbon Fund. 2004. 'PCF Annual Report' World Bank Group, Washington, D.C., Usa. p. 7

[15]*Ibid.* p.7

[16]*Ibid.* p. 5

[17]Prototype Carbon Fund. 2005. 'Carbon Market Trends 2005', World Bank Group, Washington, D.C. USA. p. 25.

[18]Fenhann, J. UNEP Risø Centre, 'CDM Project Pipeline' updated 14-09-06, online: www.cd4cdm.org

hydroflorocarbons (HFC23) are less than 2% of the overall *number* of projects, yet make up 74% of the CERs issued to date by project sector.[19] These projects are known as 'low-hanging fruit' since their high returns mean they are the first to be picked by investors. The reason is that HFC23 has 11,700 times the potency of CO_2 and since credits are in CO_2 equivalent (CO_2e) a relatively small capture of HFC23 can bring an enormous windfall of credits. According to the PCF, the large amount of non-CO_2 projects in the carbon market has meant that "traditional energy efficiency or fuel switching projects, which were initially expected to represent the bulk of the CDM, account for less than 5% [of it now.]"[20] How these trends affect the legitimacy of the carbon market and to whose benefit will be central questions in future climate debates.

[19] *Supra* note 12.

[20] *supra* note 12, p. 5.

44

CHAPTER

Land-use and Climate Change in China with a Focus on the Shaanxi Province in the Chinese Loess Plateau – Lessons for Future Climate Politics

Madelene Ostwald[1,2] and Deliang Chen[1]
[1]Earth Science Centre, Göteborg University
P.O. Box 460, 405 30 Göteborg, Sweden
[2]Centre for Climate Science and Policy Research
Linköping University, Norrköping, Sweden
E-mails: madelene.ostwald@gvc.gu.se deliang@gvc.gu.se

LAND-USE AND CLIMATE IN THE CHINESE LOESS PLATEAU

Land-Use Change

The processes associated with climate change are very relevant for China since the country is highly dependent on climate and is susceptible to climate change (Smit and Cai 1996). One area is agriculture and hence land- use. The impact of climate change on agriculture has been part of a national key project (National Climate Centre 2000, Song et al. 2006). Apart from the impact related to a changing climate, China's economic success has draw attention to the changes in the environment, including land-use, agriculture and hence food production (Zhao et al. 2005). During the last decades, several large scale land-use policies have been implemented causing changes in land-use and hence, land-cover (Hu 1997, McElroy et al. 1998, Skinner et al. 2001). The latest of these reforms, Slope Land Conversation Programme, started in full force in 2000 and Shaanxi (Fig. 1) has been the province converting the greatest amount of land area as a result of this policy (Xu et al. 2006).

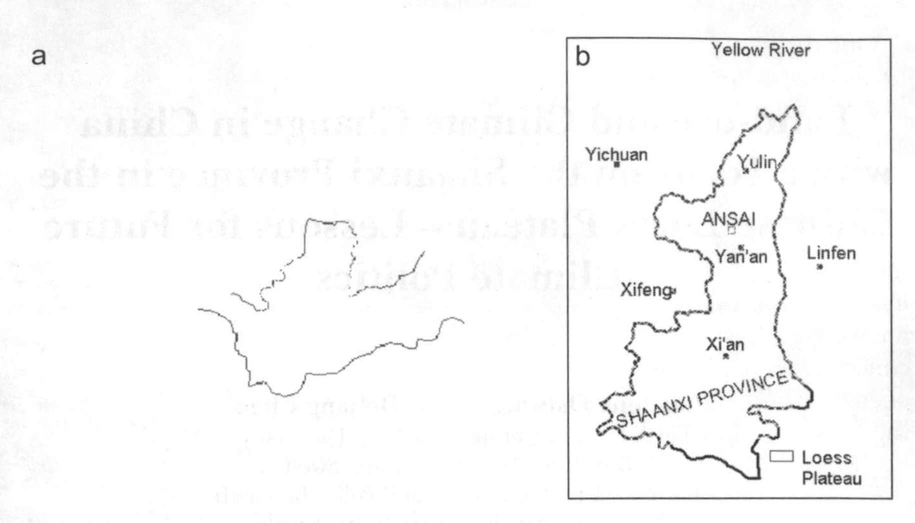

Fig. 1 a) The area of Shaanxi Province and the Loess Plateau within China. b) Shaanxi province with county seat Town Ansai indicated[1]. Modified from Hageback et al. (2005).

The Shaanxi province covers 205 600 km^2 in northern inland China between 31°45′-39°35′N and 105°29′-111°15′E, approximately 2% of the total country. Climate and topography divide the province into three distinct regions: the semi-arid Loess Plateau in the north, the warm-temperate central plain and the sub-tropical region south of the Qinling Mountains (Zhao 1986); which influences land-use. The main land-use in the north is characterized by grassland, rainfed-summer crops and sparse woods. Moving to the central area, irrigated croplands, mainly winter wheat and maize, are dominating. In the south, the main crops are maize, wheat, and rice in combination with economic forest, sparse woods and mixed needle- and deciduous broad-leaved types (Zhao 1986, IIASA 2001). The province is inhabited by 37 million unevenly distributed people, with the greatest density in the central most industrialized part

[1]Research regarding impact of climate change on small scale land-users and agricultural production and impact of land-use policies has been the authors focus with northern Shaanxi and the county of Ansai as a central node. Several of the findings presented in this chapter are based on result from this area.

(Knutsson 2005). Approximately 20% of the provincial area is used for agriculture, while half of Shaanxi was forested in 2004. The area used for agriculture decreased to 10% from 2000 to 2004, and for forests to 11% from 2003 to 2004 (China Statistical Yearbook 2005). Between 1980 and 2000 the cultivated land per farmer changed from 0.164 ha to 0.307 ha despite a population increase (Liu and Chen 2005).

One threshold for the land-use changes in Shaanxi can be traced to the 1978 economic reform that opened up China to the outside market-world in a controlled way, a change forced by years of economic stagnation (Lu and Wang 2002). This reform, the Open-Door Policy, involved China in increased globalization (Yeh and Li 1999, Bao et al. 2002) with expanding trade of agricultural products. The change was further accentuated with the land reform of 1982, the Household Responsibility System, which resolved the communes. As a result, land-use rights were distributed to individual farmers with leases of 15 years. The effects of these two reforms have had a clear impact on land-use. During the last decade, a rising concern about the environmental effects of these reforms and policies has been highlighted (McElroy et al. 1998, Skinner et al. 2001), particularly in the area of the Loess Plateau in Shaanxi, which is focussed here (Hu 1997, Liu 1999, Skinner et al. 2001).

Climate Change

Climate change and variability have significant effects on land-use, including agriculture and forestry. These changes occur both on the large scale, related for example to the East Asian monsoon system (Tao et al. 2004, Gordon et al. 2005) and on a regional scale related to human impact (Menon et al. 2002). In the northern part of the Shaanxi province, regional and local climate has shown a change over the last 50 years (Hageback et al. 2005, Ostwald et al. 2004). Hageback et al. (2005) showed an increase in annual mean temperature of 0.9°C for the period 1955-2004. Furthermore, the number of days with daily mean temperature above 0 and 5°C increased to around 20 and 15 days respectively in 40 years (1955-2004), indicating a longer potential growing season.

Precipitation is generally erratic and 40-70% falls between May and September. The total annual amounts therefore show decadal rather than linear trends (Song et al. 2005, Zhao et al. 2005). Looking at the monthly levels between 1956 and 2004 (Table 1), the most significant seasonal trends occurred during spring (March-May) and winter (December-February) for both mean temperature, which increased between 1.2 and 1.9°C/49 yr. The winter precipitation increased significantly with 20 mm/

Table 1 Mean, variability and linear trends of monthly temperature and precipitation during 1975-2004 on the annual and seasonal basis. Pearson's correlation coefficients were tested (one-tailed) with Pearson's test for significance using SPSS with ** indicating the significance level of 0.01 and * 0.05.

Shaanxi					
Period (1956-2004)	*Yearly*	*Spring*	*Summer*	*Autumn*	*Winter*
Mean temperature (°C)	17.8	19.2	29.8	17.5	4.7
Standard deviation of temperature (°C)	0.7	1.0	0.9	1.0	1.4
Linear trend (°C/yr)	0.02**	0.02**	0.00	0.02*	0.04**
Linear trend (°C/49 yrs)	1.0	1.2	0.1	1.0	1.9
Mean precipitation (mm)	558.2	110.6	286.6	151.3	9.7
Percentage of the annual precipitation (%)		20	51	27	2
Standard deviation of precipitation (mm)	92.3	36.2	63.1	54.2	9.0
Linear Trend (mm/yr)	-0.8	-0.4	-0.4	-0.4	0.4**
Linear trend (mm/49 yrs)	-38.2	-18.3	-20.7	-20.6	20.1
Relative changes of the trend (%/49 yrs)	-7.1	-18.0	-7.5	-14.7	100.6

49 yr, and decreased in the other seasons. The total annual precipitation went down by 38 mm during the 49-year period.

The response of these climatic changes and variabilities has been studied at a local level in Ansai County (Fig. 1) to see how land users perceive climate and hence react and adapt to it (Hageback et al. 2005, Ostwald and Chen 2006). The interactive and participatory tool, the 'climate game' was used to describe temperature and precipitation in the past by land users.

The farmers' recollections of the climate in the past are consistent with the instrumental climate data from the last 40 years. Generally there is no dispute regarding temperature, since all agree that it has constantly increased since the late 70s. Precipitation, on the other hand, caused long discussions and disagreements. Hence, the farmers are well aware of the changing climate.

Another approach to describe climate variability is by noting changes in an extreme event. Farmers were asked to choose which disaster had occurred most frequently and which caused the most damage to their agriculture in the 20 years and in recent years (Fig. 2). According to the farmers there was large variation in the types of disasters as well as their damages in the 80s. More recently, 82% of the responded farmers agreed that drought had occurred most frequently in the recent years alongwith with associated damages.

Very few direct farm practices have been modified due to the effects of change in climate. More drought-resistant crops have been disregarded (e.g. different types of millet) for more sensitive crops (e.g. potatoes).

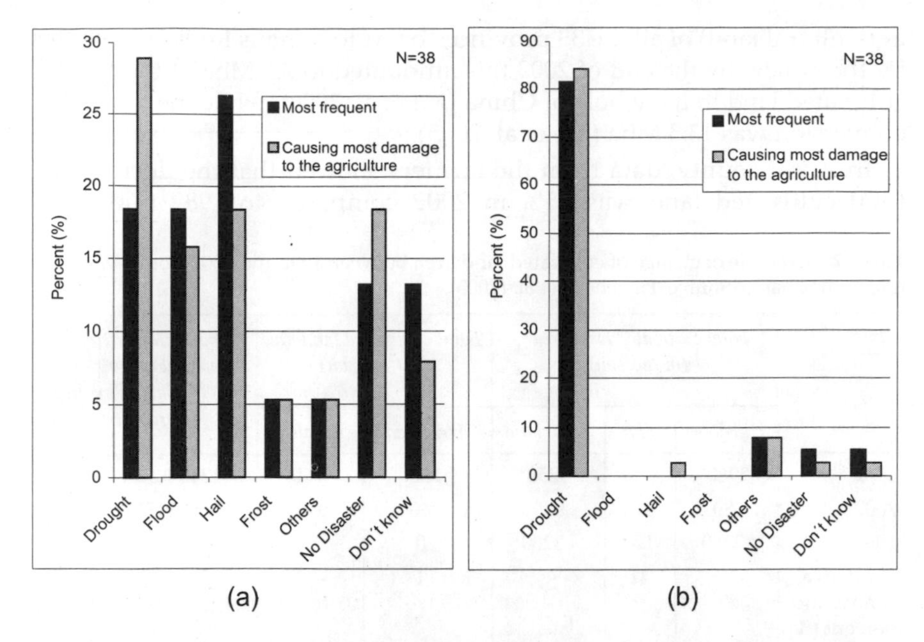

Fig. 2 Most frequent disaster and disaster causing the most damage according to the asked farmers (%) in (a) 80s and (b) recent years. 1980s refers to the period after the introduction of the Household Responsibility System in the area (1982). Source: Hageback et al. (2005).

Economic values, including better prices and higher yields were stated as the main reason for this crop change.

THE SLOPING LAND CONVERSION PROGRAMME

Changing the Land Use

In 1999 the Chinese government introduced the Slope Land Conversion Programme at the national level, also known as the Grain for Green Policy (Feng et al. 2004), later called Cropland Conversion Program (CCP) at the provincial level in Shaanxi. The massive policy had a budget of over US dollars 40 billion (Xu et al. 2004) and can be seen as one of the most ambitious environmental initiatives globally. The objective was to halt the erosion, and indirectly to solve the problem of sedimentation in the big rivers, by encouraging farmers to exchange crop cultivation on slopes for tree- and grass plants, hence reduce the arable land. This programme provides compensation in terms of money and food grains for up to eight years. The land targetted for the Shaanxi province has slopes exceeding 25° and includes the plantation of locust (*Robinia pseudoacacia* L.) (Rui et al. 2001). Shaanxi has converted most of the land (0.82 Mha of former agricultural land) of all the 33 provinces or Autonomous Regions covered by the policy. By the end of 2003 this amounted to 7.2 Mha of the former cultivated land in the whole of China (Xu et al. 2006), while the total land conversion was 13.3 Mha (Xie et al. 2005).

In Ansai county, data from the farmers showed that the decrease in total cultivated land was 75% in 2002 compared to 1982 (Table 2).

Table 2 Example of change of cultivated land area between 1982 and 2002 from a valley in northern Shaanxi. Source: Hageback et al. (2005).

	Total cultivated land area (m^2/person)*			*Total cultivated land area (m^2/person)*		*Mean total cultivated land area(m^2/person), with variations in brackets*
	Hillslope	*Flat*	*Total land*	*Irrigated*	*Rainfed*	
1982	3960	290	4250	180	4060	(4240-4250) ~ 4245
2002	580	410	990	200	950	(990-1150) ~ 1070
1982-2002	-3380	120	-3260	20	-3110	-3175
Land area change in percent (%)	-85	41		11	-77	-75

Cultivation in the hill slopes has decreased to 85%, while flat land and irrigated cultivation have increased to 41 and 11% respectively. Needless to say, the changes in land use and farming practices have changed greatly in recent years.

Environmental Impact

The overall aim of the land-use policy is to combat the environmental degradation such as water and soil erosion and to increase China's forest cover (Xu et al. 2004). One way to examine the possible vegetation cover change is to use remote sensing data. One index commonly used to do this is the Normalized Difference Vegetation Index (NDVI) that uses the characteristics which exists at different wavelengths. NDVI is often used

Table 3 Change in NDIV from 2000 to 2004. Values are taken from August since it is the peak for vegetation in Shaanxi.

Shaanxi	*% decreased NDVI*	*% unchanged NDVI*	*% increased NDVI*
North	11	10	79
Central	33	15	52
South	31	18	51
Mean for province	25	14	61

as an index to reveal seasonal and/or inter-annual change in vegetation cover. MODIS data produced into NDVI have been used to detect temporal and frequency in vegetation change in Shaanxi (Ostwald and Chen 2006, Ostwald et al. 2007). Table 3 shows the changes in NDVI for different regions in Shaanxi. The north, characterized by hilly landscape, has therefore converted more land under CCP than the more irrigated agricultural central part or the sub-tropical area in the south. Seventy-nine percent of the area in the north showed an increase in NDVI values from 2000 to 2004, but only an 11% decrease. When correlating the NDVI data with climate data in different pre-seasonal and simultaneous analysis, little correlation was found. Hence, increased vegetation is not explained by climate. The differences between the regions also suggest that CCP might have had a positive impact on the overall vegetation in Shaanxi.

Income Strategies and Impact

The effects of the present situation, characterized by the land-use policy and subsidies to the farmers, are several in terms of income. A large part of the subsidies that the farmers receive are in grains and a minor part in cash. This means that households are given food which before the policy, they had to produce themselves or buy. The main indirect impact of the policy among the farmers is the time saved by not having to farm the inaccessible hillslopes. This has opened up several possibilities.

First, off-farm work is expanding as an income among farmers, which is made available by the increasing demand for daily labour. This effect is considered important among farmers, since it generates cash. Three fourths, or 73%, of the total income was received by off-farm work in 2003 (Ostwald et al. 2004).

Second, the time is used to increase returns from flatland. This is shown in Table 4, where the variety of crops, particularly vegetables, has increased due to market value. Small businesses, such as noodle shops and arts and crafts productions, are also expanding, especially among the women.

IMPLICATIONS FOR THE FUTURE

Policy and Livelihood – A Feed-back Analysis

One important factor regarding the present shift from farming to other types of income, due to the subsidies from the land-use policy is that the trend is very welcomed by the farmers. The majority of farmers participating in our studies in the northern part of Shaanxi are eager to

Table 4 Data from farmers in northern Shaanxi on crops planted in flat lands and hill slope over three different time periods from 1980s to 2003. Source: Ostwald and Chen (2006).

	Flat land	*Hill slope*
Early 1980	Corn Foxtail millet Pearl millet Tobacco Potatoes	Foxtail millet Pearl millet Buck wheat Wheat Potatoes Beans
Mid-90s	Corn Foxtail millet Pearl millet Tobacco Foxtail millet Pearl millet Buckwheat Potatoes Beans	
2003	Corn Pearl millet Potatoes Tree plantations Vegetables Green houses Tree nursery	Trees Grass

have their children take on other occupations. This is shown in the education expenditure, which is 10-15% of the households' income in demand.

The difference between low, middle and high income households are many. In terms of total gross income, the differences are 6,000, 14,500 and 20,000 yuan for the low, middle and high income groups respectively. Further, low income groups have less land meaning less land converted under CCP and hence less subsidies. The land left for cultivation is also used differently depending on the income group. The low income groups use the land extensively, while high income groups find the time more valuable for doing off-farm work.

By seeing the situation in Shaanxi it can be assumed that the resources provided to the low income groups and their land would alleviate their poverty. It has been suggested that this would be counterproductive since the processes of value for different capitals are not taken into account (Knutsson and Ostwald 2006). This means that the value of land is at present decreasing in value and so are traditionally produced goods. For investments in areas showing increasing value, greater network building leading to off-farm work opportunities generating more cash (flexibility) would be more feasible.

One important conclusion which can be drawn is that the farmers having access to other types of income will not return to farming. This is of crucial importance with regards to the time after 2008 when land-use policy subsides will end. Again, the impact will differ depending on wealth and opportunities.

Future Climate and Its Impacts

In order to look at the future climate change, present and future climate simulations by General Circulation Model (GCM) have been used. Future projections of IPCC scenarios of two GCMs, ECHAM4 and HadCM3, are derived from the IPCC data distribution center (http://ipcc-ddc.cru.uea.ac.uk/). Annual mean temperature and precipitation of one grid (109.7E, 37.7N for ECHAM4 and 108.7E, 37.5N for HadCM3) were chosen to represent climate in Ansai and part of the Shaanxi province. For each model three runs were selected, i.e. a control run for the present climate and two future scenarios based on IPCC A2 and B2 Emission Scenarios, referred to as SRES (IPCC 2000). The A2 is characterized by a very rapid increase in population, mostly regional economic growth and relatively slow implementation of new technology. The B2 scenario is mainly focussed on sustainable development of the economy as well as social and environmental issues on a local scale. The GCM-runs refer to different time-periods with ECHAM4 covering the following periods: control run 1860-2099, A2 and B2 1990-2100, and HadCM3: control run 1858-2097, A2 and B2 1950-2099.

Figure 3 shows the differences in the temperature between the scenarios A2 and B2 and the mean of the control run. Although there are some differences between the two models and the scenarios, it is obvious that the future temperature may increase with 2-3°C during the next 50-year period and 3-6°C during the next 100-year period. The increase would increase the length of the growing season in the area. However, higher temperature in summer would enhance the heat stress for vegetation which is already a problem in the area.

Figure 4 shows results for precipitation in the same way as Figure 3. Interestingly, the simulations between 1950 and 2000 show a slightly decreasing trend which is consistent with the observation. There is a general increasing trend for future precipitation in the area. Due to dryness in the area, this increase should have a positive effect on the vegetation in the area.

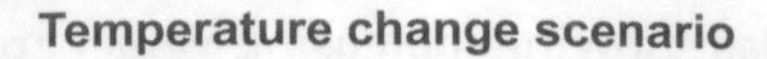

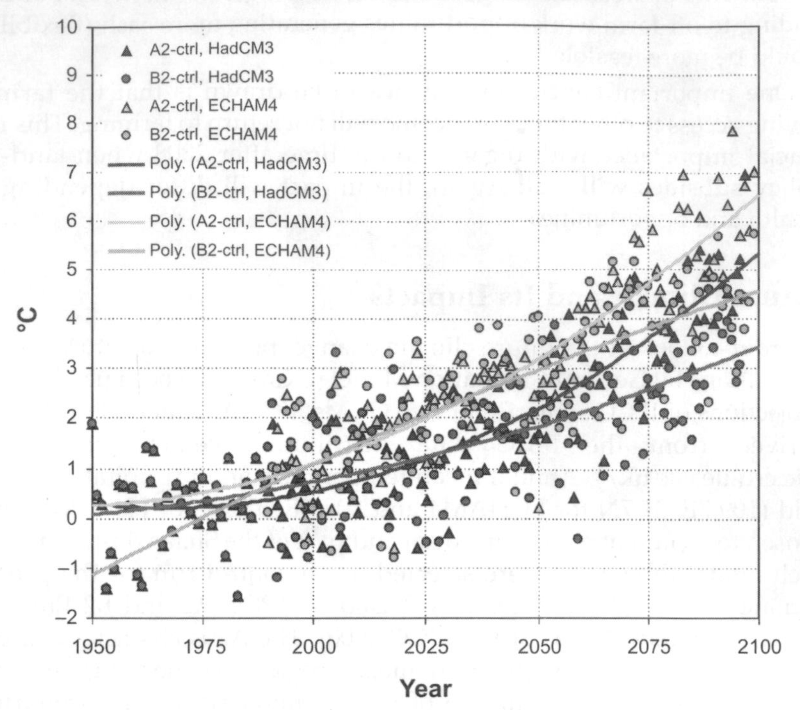

Fig. 3 Shaanxi/Ansai GCM-grid with simulated temperature change by year presented as difference from control scenario. Polynomial regression lines are calculated to indicate trends.

Colour image of this figure appears in the colour plate section at the end of the book.

The Lesson Learned for Future Climate Policies in Developing Environments

There are several issues that are worth pointing out from this case, in terms of impact of policies in the future Kyoto regime. We will focus on three areas.

The impact of slow moving climate change and variability is evident and recognized in a large part of the developing world, as exemplified by results from Shaanxi. But climate change is seldom at top of the agenda due to other direct and more livelihood-based needs. This means that for large a part of the developing world, people need direct returns from an action. This implies that even though they are aware of the long term negative impact of an action, the short term benefit is often chosen. The introduction of the land-use policy was not met with resistance, partly due

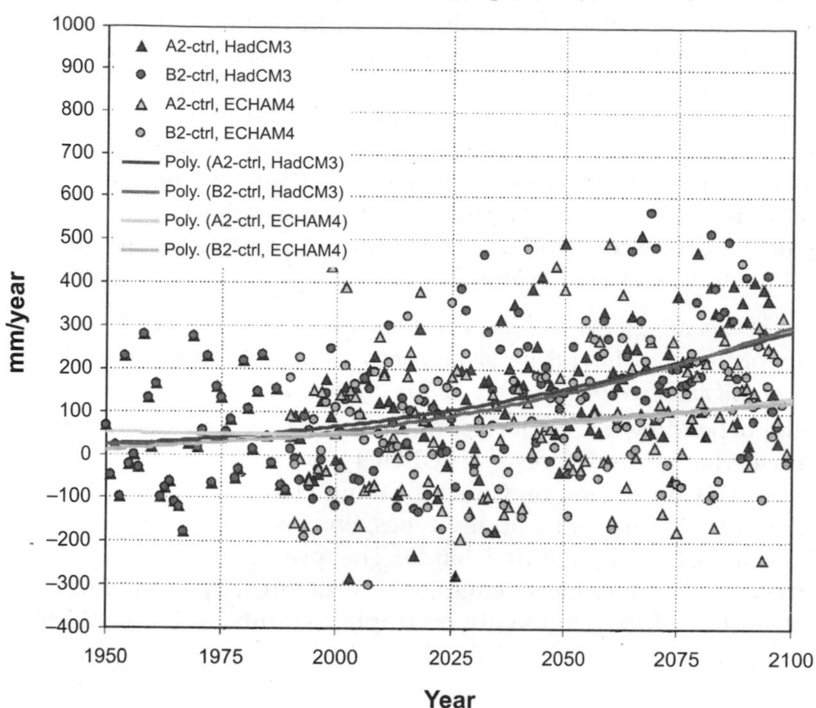

Fig. 4 Shaanxi/Ansai GCM-grid with simulated precipitation change by year presented as difference from control scenario. Polynomial regression lines are calculated to indicate trends.

Colour image of this figure appears in the colour plate section at the end of the book.

to its welcomed effects of less farming on inaccessible lands causing a lot of time consumption, but mainly due to the direct return of subsidies making livelihood better for many farmers.

Lesson 1: Direct positive effect, mainly in livelihood improvements, are needed for successful implementation of policies.

Inherent in the meaning of developing is also the complex processes of several changes, of which climate change is only one. As has been exemplified here, there are environmental changes (addressed on a national level) and economical changes (addressed on a household level), making climate change less important.

Lesson 2: Climate change is not isolated from other changes, particularly for the developing world. Incorporating these changes in policies is needed for successful implementation.

Climate change interacts with other environment changes and economical settings, which implies that feedback should be taken into account to predict the final result of climate impact. For example, while climate change affects vegetation, changed vegetation can also have an impact on local and even regional climate. Therefore, the projected temperatures with the GCM which do not take the interaction account may not be realistic. As for local economical changes, the value of land will decrease with the increase of off-farming income. This means that for the poor, with very little assets except land (meaning little network for off-farming work etc.), the future looks harsh after 2008.

Lesson 3: Climate change and its impact are dynamic processes. The dynamics, exemplified here as increasing or decreasing values of capitals, needs to be incorporated and accounted for in policy development.

Finally, depending on the development of the next generation land- use, land-use change and forestry (LULUCF) within the United Nations Framework Convention on Climate Change (UNFCCC), including the baseline year of different land-use types, Shaanxi might benefit from the 0.82 Mha former agricultural lands. The present debate on avoided deforestation and reduced emission for deforestation in developing countries have focussed mainly on tropical countries, but could in the future apply to other regions as well.

Acknowledgement

Elisabeth Simelton is acknowledged for processing the climate data. Funding for this project was given by Knut och Alice Wallenberg Foundation, Swedish International Development and Cooperation Agency and STINT.

References

Bao, S., G. Chang, J, Sachs and W. Woo. 2002. Geographic factors and China's regional development under market reforms, 1978–1998. China Economic Review 13: 89–111.

China Statistical Yearbook. 2005. National Bureau of Statistics. www.stats.gov.cn downloaded 2006-08-01.

Feng, Z., Y. Yang, Y. Zhang, P. Zhang and Y. Li. 2004. Grain-for-green policy and its impacts on grain supply in West China. Land Use Policy 22: 301-312.

Gordon, L., W. Steffen, B.F. Jönsson, C. Folke, M. Falkenmark and Å. Johannessen. 2005. Human modification of global water vapour flows from the land surface. PNAS 102: 7612-7617.

Hageback, J., J. Sundberg, M. Ostwald, D. Chen, Y. Xie and P. Knutsson. 2005. Climate variations and land use in Danangou watershed, China – Examples of small-scale farmers' adaptation. Climatic Change 72: 189-212.

Hu, W. 1997. Household land tenure reform in China: its impact on farming land and agroenvironment. Land Use Policy 14: 175-186.

IPCC. 2000. IPCC Special Report on Emission Scenarios Summary for Policy Makers. Knutsson, P. 2005. Interdisciplinary knowledge integration and the sustainable livelihoods approach. PhD Thesis School of Global Studies, Human Ecology Section, Göteborg University.

Knutsson, P. and M. Ostwald. 2006. A process-oriented sustainable Livelihoods Approach—a tool for increased understanding of vulnerability, adaptation and resilience. Mitigation and Adaptation strategies for Global change DO1 10.1007/s11027-006-4421-9.

Liu, G. 1999. Soil conservation and sustainable agriculture on the Loess Plateau: challenges and prospects. Ambio 28: 663-668.

Liu, X. and B. Chen. 2005. Efficiency and sustainability analysis of grain production in Jiangsu and Shaanxi Provinces of China. Journal of Cleaner Production (on line)

Lu, M. and E. Wang. 2002. Forging ahead and falling behind: Changing regional inequalities in post-reform China. Growth and Change 33: 42-71.

McElroy, M.B., C.P. Nielsen and P. Lydon. 1998. Energizing China Reconciling Environmental Protection and Economic Growth. Harvard University Press, USA.

Menon, S., J. Hansen, L. Nazarenko and Y. Luo. 2002. Climate Effects of Black Carbon Aerosols in China and India. Science 297: 2250-2253.

National Climate Centre. 2000. An assessment system for impact of climate anomaly on social-economic aspect. National key project (1996-200): Studies on short-tem climate prediction in China, subproject 3, Beijing, China.

Ostwald, M., D. Chen, Y. Xie, P. Knutsson, S. Brogaard, K. Borne, and Y. Chen. 2004. Impact of climate change and variability on local-scale land use - Shaanxi Province, China. Research report Earth Science Centre Series C61, Göteborg University.

Ostwald, M. and D. Chen. 2006. Land-use change: Impact of climate variations and policies among small-scale farmers in the Loess Plateau, China. Land Use Policy 23: 361-371.

Ostwald, M., E. Simelton, D. Chen and A. Liu. 2007. Relation between vegetation changes, climate variables and land-use policy in Shaanxi Province, China. Geografiska Annaler 89: 223-236.

Rui, L., W. Zhongming, W. Fei, W. Yue and Z. Mingliang. 2001. Case study on conversion of farmland to forest and grassland in Ansai County, Shaanxi Province. Prepared for the CCICED Western China Forest Grassland Task Force.

Skinner, M.W., R. G. Kuhn and A.E. Joseph. 2001. Agricultural land protection in China: a case study of local governance in Zhejiang Province. Land Use Policy 18: 329-340.

Smit, B. and Y. Cai. 1996. Climate change and agriculture in China. Global Environmental Change 6: 205-214.

Song, Y., E. Simelton, D. Chen and W. Dong. 2005. Influence of climate change on winter wheat growth in north China during 1950-2000. ACTA Meteorologica Sinica 19: 501-510.

Song, Y., D. Chen and W. Dong. 2006. Influence of climate on winter wheat productivity in different climate regions of China during 1961-2000. Climate Research (in press).

Tao, F., M. Yokozawa, Z. Zhang, Y. Hayashi, H. Grassl and C. Fu. 2004. Variability in climatology and agricultural production in China in association with the East Asian summer monsoon and El Niño Southern Oscillation. Climate Research 28: 23-30.

Uchida, E., J. Xu and A. Rozelle. 2005. Grain for green: Cost-effectiveness and sustainability of China's conservation set-aside program. Land Economics 81: 247-264.

Xie, C., J. Zhao, D. Liang, J. Bennet, L. Zhang, G. Dai and X. Wang. 2005. Livelihood impacts of the conversion of cropland to forest and grassland program. Australian Centre for International Agricultural Research (ACIAR) Project. Research Report No. 3.

Xu, Z., M.T. Bennet, R. Tao and J. Xu. 2004. China's sloping land conversion programme four year on: current situation, pending issues. International Forestry Review 6: 317-326.

Xu, A., J. Xu, X. Deng, J. Huang, E. Uchida and S. Rozelle. 2006. Grain for green versus grain: conflict between food security and conservation set-aside in China. World Development 34: 130-148.

Yeh, A.G. and X. Li. 1999. Economic development and agricultural land loss in the Pearl River delta, China. Habitat International 23: 373-390.

Zhao, S. 1986. Physical geography of China. Science Press. Beijing.

Zhao, W.Z., H.L. Xiao, Z.M. Liu and J.Li. 2005. Soil degradation and restoration as affected by land use change in the semiarid Bashang area, northern China, Catena 59: 173-186.

Section X

Kyoto Protocol: Bali and Beyond

45

CHAPTER

Climate Governance Post Bali: Signs of Hope

Joyeeta Gupta[1]
De Boelelaan 1087
1081 HV Amsterdam
E-mail: joyeeta.gupta@ivm.vu.nl

1. INTRODUCTION

The nature of global governance on climate change is unprecedented. The number of actors engaged, consistently, over long periods of time, in the domestic and international context is rapidly multiplying and substantial. Climate change governance has also evolved incredibly rapidly from a historic point of view. With five yearly assessment reports provided by the Intergovernmental Panel on Climate Change (IPCC) since 1990[2], the adoption of the United Nations Framework Convention on Climate Convention[3] in 1992 and the Kyoto Protocol[4] in 1997, and annual meetings of the Parties to these treaties, to culminate year long work undertaken by

[1] The author is professor in climate change policy and law at the Institute for Environmental Studies of the Vrije Universiteit Amsterdam, and professor on policy and law on water resources and the environment at UNESCO-IHE Institute for Water Education in Delft.

[2] For an overview of the history of IPCC and its publications, visit www.ipcc.org.

[3] United Nations Framework Convention on Climate Change, (New York) 9 May 1992, in force 24 March 1994; 31 I.L.M. 1992.

[4] Protocol to the Framework Convention on Climate Change (Kyoto), 37 ILM (1998) 22, in force 16 February 2005.

the secretariat and the subsidiary bodies, the density of rule making is immense. The nature of the rules is unprecedented. Highly formalised market-based instruments are being institutionalised in a way that stock markets take note of what is happening within the context of climate change.

The question is: Does all this activity get us any closer to the ultimate goal of keeping the climate change problem within manageable limits? This essay briefly recalls the expectations from Bali, discusses the outcomes of Bali, and examines the changing political context of the post 2007 period, before finally drawing some conclusions.

2. EXPECTATIONS FROM BALI

2.1 Introduction

The key question is: Why was the meeting at Bali critical? The failure of the meeting in Bali would only have meant postponement of policymaking by one additional year - so why was it seen as so critical? Although the Climate Change Convention and even the Kyoto Protocol have a life beyond 2012, the key elements of mandatory greenhouse gas emission targets and timetables for industrialised countries expire in 2012. This has two implications. For those making long-term infrastructural investments, the lack of a post-2012 set of targets and timetables implies lack of clear information that can be taken into account in current investments. For those engaged in market mechanisms established in the climate regime - Joint Implementation, the Clean Development Mechanism (CDM) and emissions trading - the lack of targets beyond 2012 reduces the incentive to trade; and since these mechanisms are not, in general, spot transactions but have long lead-times, there is need for information to facilitate planning and to ensure economic viability.

2.2 Scientific Expectations

The publication of the Fourth Assessment Report of the Intergovernmental Panel on Climate Change (IPCC)[5] shows that the global concentrations of greenhouse gas emissions have increased from 280 ppmv in pre-industrial times to 379 ppmv in 2005. While most greenhouse gases have been emitted in the course of using fossil fuels, land use activities have also been a major contributor. Methane contributions are higher than they have been in the last 650,000 years and

[5] IPCC-1 AR4 2007. Summary for Policymakers, in *Climate Change 2007: The Physical Science Basis*, Cambridge University Press.

are mostly emitted from the energy and agricultural sectors. There is a significant rise in nitrous oxide emissions mostly from the agricultural sector. The rise in greenhouse gases is expected to lead to enhanced warming despite the impacts of temporary cooling effects of aerosol emissions. The IPCC reports confirm that sea level rise over the last forty years (1961-2003) is in the order of 1.8 mm per year, but that in the last decade the annual rise has been 3.1 mm per year.[6]

The IPCC's Policy Response[7] report submits that greenhouse gas emissions are expected to grow substantially in the future. The energy sector is projected to emit from 40-110% more by 2030 in relation to 2000. At the same time, the report states that the literature shows that a large number of measures which have zero to negative costs exist that can enable societies to reduce emissions by 6 GT CO_2-eq/year by 2030. The key question is finding appropriate policy measures that will ensure that societies will be able to find and utilize these reduction options. The report calls not only for technological measures but also changes in lifestyle. It argues that any efforts to keep the global average temperature rise below 2 degrees with a considerable degree of certainty implies that we need to stabilize concentrations of greenhouse gases at 445-490 ppmv. This means that global emissions must peak in 2015 and emissions in 2050 should be substantially lower than in 2000. This calls for urgent action. Of course, if we wish to accept a temperature rise above 2 degrees, the urgency is lower.[8] The Stern Report of 2006 assessed that without taking action, we would face costs and risks "equivalent to losing 5% of global GDP each year, now and forever" and that the estimates would be higher if a larger set of risks were taken into account. However, if action were taken, that would cost about 1% of global GDP annually.[9]

2.3 Thus the Recent Scientific Reports Highlight the Need for Urgent Action if we Wish to keep the Worst Effects under Control. Political Needs

The urgency of the scientific material was stressed by several political actors. Over 2007, the political and public pressure began to build up. Al Gore went around the world with his slide show on climate change and

[6] See Summary for Policymakers, above at note 5.

[7] IPCC-3 AR4 2007. Summary for Policymakers, in *Climate Change 2007: Mitigation of Climate Change*, Cambridge University Press.

[8] Supra note 7 above.

[9] Stern, N.H. 2006. *Stern Review on the Economics of Climate Change*. Available at: http://www.hm-treasury.gov.uk/independent_reviews/stern_review_economics_climate_change/stern_review_report.cfm (accessed 30 September 2007).

his documentary - An Inconvenient Truth - was given much publicity.[10] A number of recent films on climate change, music events like Live Earth, and Bill Clinton's own global climate initiative created large-scale public news and awareness on climate change. Religious organizations like the World Council of Churches also started to promote their perspectives on the issue. The press attention was not just focused on climate change as an uncontroversial problem, but it also looked at the controversies and presented climate change as a contested problem and often created more confusion as well. The recognition of climate change as a serious problem affecting global security reached a high point when the Nobel Prize for Peace was awarded jointly to the IPCC and Al Gore. The political expectations world wide were thus reaching a high point prior to Bali, where more than ten thousand participants gathered to negotiate and/or to put pressure on the negotiators to take action.

2.4 Formal Expectations

The Bali meeting was significant since it presented the absolute last moment that a process could be set in motion to negotiate a follow-up agreement. A two year process could lead to a new Protocol or amendments in the existing legal texts which could be adopted at the Conference of the Parties in 2009 at Copenhagen at the very latest; thus providing just about enough time for preparations to reduce emissions in the post-2012 period and to try and peak global emissions, if at all by 2015. This process under the Climate Change Convention requires constructive participation by the US. This process is over and above the existing process established under the Kyoto Protocol. This latter process was established at the 11th meeting of the Conference of the Parties in Montreal in 2005, under Article 3.9 of the Kyoto Protocol which would ensure that an Ad Hoc Working Group (AWG) would be established to ensure that there would be no gap between the first commitment period mentioned in the Kyoto Protocol and the next commitment period to be identified. There is less active participation of the US in this group since the US is not a Party to the Kyoto Protocol.

2.5 Needs as Expressed by Other Actors

Within the scientific and policy community, a large number of proposals have been made about the possible follow-up action to Kyoto. There have been heated discussions about whether the follow-up should take place within the UN multilateral framework or outside of it. Many who wished

[10] Gore, A. 2006. *An Inconvenient Truth – A documentary*, and A. Gore 1992. *Earth in the Balance*, Plume Books.

to keep the US on board the discussion talked in terms of follow-up outside the UN framework. The huge new literature[11] focused on different approaches including new and sharper targets and timetables for countries, agreements on technologies and standards for technologies and agreements on research and development. Kuik et al. analyzed around 40 proposals. Most of these proposals focus on market-based instruments, while some focus on regulatory instruments. Most call for measures within the UN framework, while a few call for "John Wayne" type unilateral measures.[12] The increased urgency of the climate change problem as highlighted by the Fourth Assessment Report of IPCC, the growing political awareness and needs and the formal institutionalised processes of climate negotiations provided the setting for serious decisions to be taken at Bali.

3. OUTPUTS OF BALI

3.1 Introduction

After tough negotiations, on the 15th of December, a Bali Roadmap was adopted at the 13th Conference of the Parties (COP) to the Climate Convention. Until the last day, negotiators were linking different issues to each other leading to "an all-or-nothing" situation as the Executive Secretary put it. Ultimately, COP-13 adopted fourteen decisions including the Bali Roadmap. The Third Conference of the Parties serving as the Meeting of the Parties to the Kyoto Protocol (COP/MOP-3) adopted eleven decisions and the Ad Hoc Working Group (AWG) adopted one decision. This section elaborates first on the Bali Action Plan, then the other decisions taken by COP-13, discusses the decisions of COP/MOP-3 and finally the AWG decision.

---

[11] See e.g. Torvanger, A. et al. 2004. *Climate Policy Beyond 2012: A Survey of Long-term Targets and Future Frameworks*. CICERO Report 2004:02. Oslo: Center for International Climate and Environmental Research. Winkler, H., B. Brouns and S. Kartha 2006. 'Future Mitigation Commitments: Differentiating Among Non-Annex I Countries', *Climate Policy* 25(5), 469-486. Aldy, J.E., S. Barrett and R.N. Stavins 2003. 'Thirteen Plus One: A Comparison of Global Climate Policy Architectures' *Climate Policy* 3(4):373-397. Baer, B., T. Athanasiou and S. Kartha, 2007. *The Right to Development in a Climate Constrained World. The Greenhouse Development Rights Framework*. Available at: www.ecoequity.org/docs/TheGDRsFramework.pdf (accessed 30 September 2007).

[12] Kuik, O., J. Aerts, F. Berkhout, F. Biermann, J. Bruggink, J. Gupta and R. Tol, 'Post-2012 Climate Change Policy Dilemmas: How Do Current Proposals Deal With Them?', *Climate Policy*, forthcoming.

3.2 COP-13 Decisions: The Bali Action Plan

The key decision of COP-13 was the Bali Action Plan.[13] This Plan states that in order to achieve the ultimate objective of the Convention in accordance with its principles and recognizing that "deep cuts in global emissions will be required to achieve the ultimate objective of the Convention" and emphasizes "the urgency to address climate change as indicated in the Fourth Assessment Report" of the IPCC. The key decision is to "launch a comprehensive process to enable the full, effective and sustained implementation of the Convention through long-term cooperative action, now, up to and beyond 2012, in order to reach an agreed outcome and adopt a decision at its fifteenth session". Such a process should focus on (a) shared vision of cooperative action, (b) action on mitigation with measurable commitments and actions for the developed countries and nationally appropriate mitigation actions by developing country Parties, measures on deforestation, cooperative sectoral action, market based approaches, economic and social consequences of response measures and strengthening the catalytic role of the Convention; (c) enhanced action on adaptation; (d) enhanced action on technology development and transfer, and (f) enhanced action on the provision of financial aid and investment to support action.

The process will be undertaken within the Ad Hoc Working Group on Long-term, Cooperative Action, which will meet in April 2008 and another three times before the next COP. All countries are to submit their views on the Work Programme by 22 February 2008 and the Plan recommends that richer countries should support the Trust Funds that finance, among others, the participation of poorer countries in the process.

A brief assessment of the document shows that it clearly avoids any explicit reference to a quantitative elaboration of a long-term objective and does not go further than calling for "deep cuts" and referring via a footnote to specific pages in the most recent IPCC report (see also 3.5). The decision also does not refer in quantitative terms as to what such a long-term obligation implies in terms of short and medium-term goals. It is however to be welcomed that the document focuses on "measurable, reportable and verifiable" commitments for the developed countries, although the text is somewhat softened by the use of the term "nationally appropriate commitments or actions, including quantified emission limitation and reduction objectives". The use of "appropriate" opens up new doors for weakening goals; and the continued use of "limitation" implies that it is not yet certain that all developed countries will focus on

[13] Bali Action Plan, available at http://unfccc.int/files/meetings/cop_13/application/pdf/cp_bali_action.pdf.

reducing their emissions. Furthermore, it is unclear whether the negotiations will eventually lead to "commitments", comparable to the Kyoto Protocol, or merely "mitigation actions". This is a pity. Another key element of note here is that although there was considerable pressure on developing countries to also adopt "measurable, reportable and verifiable nationally appropriate mitigation actions", the developing countries were ultimately successfully able to renegotiate the text such that the term "measurable, reportable and verifiable" is now used more in connection with technology financing and capacity building to be provided to developing countries to enable them to implement their obligations. These were among the most contentious issues discussed at Bali.[14] Finally, the establishment of a process with clear recommendations for the content of the process and the need to complete the process within a set timetable operationalises the process; and this process includes the participation of the US, the single major producer of greenhouse gases apart from China.

3.3 Other COP Decisions

Twelve other COP decisions were taken and these are briefly elaborated here. A key decision was taken on reducing emissions from deforestation in developing countries. Although this decision is couched in very diplomatic language ("invites" and "encourages", requests, rather than "decides" and "adopts"), it puts pressure on countries to try and invest in reducing deforestation and forest degradation and encourages Parties to report on these using the Good Practice Guidelines for Land-Use, Land-use Change and Forestry. The decision requests countries to make recommendations on how to improve methodological issues, which will then be compiled and synthesised by the secretariat and may provide the basis for a follow-up decision.[15]

Two decisions were taken in relation to technology transfer. The first focuses on reconstituting an Expert Group on Technology Transfer under the Subsidiary Body for Scientific and Technological Advice for another five years. This Group should focus on adequate and timely financial support for technology transfer and the development of performance indicators for monitoring and evaluating effectiveness. The Group is expected to focus on enhancing the implementation of the framework for

[14] For details on the negotiation around this article, see Müller, B. 2007. Bali 2007: On the Road Again! Impressions from the Thirteenth UN Climate Change Conference, available at available at http://www.oxfordclimatepolicy.org/publications/mueller.html.

[15] COP 13 Decision on Reducing Emission from Deforestation in Developing Countries: Approaches to Stimulate Action, see http://unfccc.int/files/meetings/cop_13/application/pdf/cp_redd.pdf

meaningful and effective actions to implement technology transfer by promoting technology needs assessments and their inclusion in the second national communications of the developing countries; to maintain, update and improve TT:CLEAR (Technology transfer clearing house), to focus on creating enabling environments for technology transfer including "[t]o encourage Parties to avoid trade and intellectual property rights policies, or lack thereof, restricting transfer of technology", capacity building for technology transfer, and very importantly to take measures to convert such ideas into project proposals that can be financed. The Expert Group consists of 19 members - three each from Asia, Africa and Latin America, one from the small island sates, one from another non-Annex I party and eight from the Annex I countries. The group may also invite four resource persons from international bodies working in the field.[16]

A related decision was taken on technology development and transfer under the Subsidiary Body for Implementation. Although there was discussion about whether decisions should be taken in the context of two separate bodies, the developing countries were successfully able to get two related decisions. It was decided that the Expert Group on Technology Transfer shall make recommendations to help the COP to take decisions. It decided that funding should be made available for technology needs assessment, joint R&D programmes, demonstration projects, enabling environments for technology transfer; incentives for the private sector; North-South and South-South cooperation, endogenous capacity building; issues associated with meeting the agreed full incremental costs, licenses to support access to and transfer of low carbon technologies and a window for a venture capital fund. It recommends that the Expert Group should specify the gaps and barriers to these resources and whether new resources are needed and to develop a set of performance indicators to monitor and evaluate the effectiveness of these measures. It recommends that the Global Environment Facility (GEF) should especially look at addressing the financing needs of the developing countries.[17]

The Conference of the Parties was appreciative of the work by the IPCC and requested Parties to support its work and to build on its outputs.[18] A

---

[16] COP 13 Decision on Development and Transfer of Technologies under the Subsidiary Body for Scientific and Technological Advice, see http://unfccc.int/files/meetings/cop_13/application/pdf/cp_tt_sbsta.pdf

[17] COP 13 Decision on Development and Transfer of Technologies under the Subsidiary Body for Implementation, see http://unfccc.int/files/meetings/cop_13/application/pdf/cp_tt_sbi.pdf

[18] COP 13 Decision on Fourth Assessment Report of the Intergovernmental Panel on Climate Change, see http://unfccc.int/files/meetings/cop_13/application/pdf/cp_arfour.pdf

decision on the Fourth Review of the financial mechanism calls on all parties to submit their views on the technical paper on the experiences with international funds on climate change and on the report on the funds needed to deal with climate change, the report on the analysis of existing and potential investment and financial flows and the options for scaling up funding. Based on these comments, it requests the Subsidiary Body on Implementation to make recommendations to the Conference of the Parties on how to enhance complementarity and consistency between different financial flows in order to meet the needs of the developing countries.[19]

In an effort to be able to control the activities of the GEF, the COP made a decision requesting the GEF to take several measures such as continuing with country dialogues, using national experts where possible, simplifying the incremental cost principle, taking the lessons on Piloting an Operational Approach to Adaptation, to improve access to funds, to report to the Conference in time for it to be able to examine the report carefully before the meetings start, to ensure that the agreed full costs of developing countries are covered in relation to Article 12(1) and to report on these as part of the regular reports it makes to the Conference.[20]

Given the importance of public awareness and education, the Conference decided to amend the New Delhi Work Programme on Article 6.[21] The amended Programme focuses on developing policies on public awareness and education, identifying needs and gaps in countries, and recommends that a country driven approach that is cost effective and takes a phased approach to integrating such activities into climate policy should be adopted. It requests all Parties to report on the measures they take as part of their National Communications, NGOs to continue to share information through information network clearing houses and the GEF to fund, where appropriate, such activities.[22]

A decision was taken to request developed countries in Annex I to submit their fifth National Communication by 1 January 2010 and on the importance of the reviews synthesising this work.[23] Another decision

---

[19] COP 13 Decision on Fourth Review of the Financial Mechanism, see http://unfccc.int/files/meetings/cop_13/application/pdf/cp_arfour.pdf

[20] COP 13 Decision on Additional Guidance to the Global Environment Facility, see http://unfccc.int/files/meetings/cop_13/application/pdf/cp_guid_gef.pdf

[21] COP 13 Decision on Amended New Delhi Work Programme on Article 6 of the Convention, see http://unfccc.int/files/meetings/cop_13/application/pdf/cp_cs_ncfour.pdf

[22] COP 13 Decision on Extension of the Mandate of the Least Developed Countries Expert Group, see http://unfccc.int/files/meetings/cop_13/application/pdf/cp_leg.pdf

[23] COP 13 Decision on Compilation and Synthesis of Fourth National Communications, see http://unfccc.int/files/meetings/cop_13/application/pdf/cp_cs_ncfour.pdf

called on Parties to adopt the revised reporting guidelines on the global climate change observing system in preparing their National Communications.[24] Decisions were taken on the programme budget of the UNFCCC and each country needs to pay in accordance to the UN Scale of Assessments. It was also decided to invite the UN to finance the conference services from its own budget. Funding for the three Trust Funds is also discussed.[25] A decision on the dates and venues of the next two meetings was also taken.[26]

Briefly assessing these decisions, one can state that even though these tend to get less importance in the media and in the journals, these decisions demonstrate, first, the degree of persistence in following up on each of the COP related issues and ensuring not only that incrementally there is substantive improvement made to the procedures and rules of the Convention, but also that the processes and bodies established have the resources and mandate to move further. Second, it is also critical to note, I believe, that for the first time a serious effort has been made to give technology transfer provisions some teeth by inserting a clause on the need to convert technology needs assessment reports into project proposals that can be submitted for funding, and by calling on Parties to try and reduce the impediment that intellectual property rights often pose to technology transfer issues.

3.4 Decisions of COP/MOP-3

Eleven decisions were taken by the Parties to both agreements as COP/MOP-3 decisions. The most critical of these decisions is on adaptation. The Adaptation Fund will become operational in 2008 and will fund concrete adaptation projects in countries that are particularly vulnerable to climate change. It has an Adaptation Fund Board, a secretariat (GEF on an interim basis) and a trustee (The World Bank). The Board consists of 16 members with regional distribution and decisions must be taken by consensus at meetings held at least twice a year. Quorum calls for simple majority. The meetings are meant to be open and decisions transparent by making decisions available in all UN languages. The Adaptation Fund Board has

[24] COP 13 Decision on Reporting on Global Observing Systems for Climate Change, see http://unfccc.int/files/meetings/cop_13/application/pdf/cp_rso.pdf

[25] COP 13 Decision on Budget Performance and Functions and Operation of the Secretariat, see http://unfccc.int/files/meetings/cop_13/application/pdf/cp_budget_funct.pdf and COP 13 Decision on Programme Budget for the biennium 2008-2009, see http://unfccc.int/files/meetings/cop_13/application/pdf/cp_budget_funct.pdf

[26] COP 13 Decision on Date and Venue of the Fourteenth and Fifteenth Sessions of the Conference of the Parties and the calendar of Meetings of Convention Bodies, see http://unfccc.int/files/meetings/cop_13/application/pdf/cp_budget_funct.pdf

several functions, including developing strategic priorities for the COPs to adopt, operational policies and guidelines, criteria for project selection, and rules of procedure, monitoring and review of activities, establishing committees, panels and working groups as required and being responsible for the monetisation of certified emission reductions.[27] One of the most controversial issues at Bali was the role of the GEF and the World Bank in the Adaptation Fund. Consensus was ultimately found by asserting that meetings of the Board would take place in Bonn, even if the secretariat activities were undertaken by the GEF and that applications for funding would be directed at the Board and not via one of the implementing agencies of the GEF.

A decision was taken on the Clean Development Mechanism, noting that 128 Designated National Authorities exist, 825 CDM projects have been registered and 85,049,697 million certified emission reductions have been issued. It commends the Executive Board on its decision to streamline procedures, on setting up a CDM bazaar and encourages it to simplify and improve its procedures and ensure a fair and equitable regulatory system. Among a number of recommendations it suggests that the quality and consistency of verification work should be improved and minor issues should also be addressed in a transparent and timely manner in order to allow time to focus on major issues. It takes note of all the efforts made by the Board to make methodologies for assessing emission reductions and lists best practices. In an effort to increase projects in least developed countries, the Board abolished payment of registration fees and share of the proceeds at issuance of credits for projects hosted in these countries.[28]

On Joint Implementation, a decision was taken to request the secretariat to develop a web-based interface to encourage greater transparency and access to information on joint implementation projects. It also commends the Joint Implementation Supervisory Committee for its Joint Implementation Management Plan and endorses the revision to the fee structure on JI projects to cover the administrative costs of the Committee.[29]

In relation to the second review of the Kyoto Protocol pursuant to its Article 9, the decision invites all Parties to submit by 7 March 2008 their views on: (a) extending the share of proceeds to assist in meeting the costs

[27] CMP 3 Decision on Adaptation Fund, see http://unfccc.int/files/meetings/cop_13/application/pdf/cmp_af.pdf.

[28] CMP 3 Decision on Further Guidance Related to the Clean Development Mechanism, see http://unfccc.int/files/meetings/cop_13/application/pdf/cmp_af.pdf.

[29] CMP 3 Decision on Guidance on the Implementation of Article 6 of the Kyoto Protocol, see http://unfccc.int/files/meetings/cop_13/application/pdf/cmp_art_six_kp.pdf.

of adaptation to joint implementation and emissions trading, privileges and immunities for individuals who work in the bodies established by the Kyoto Protocol, the scope and functioning of the financial mechanisms, and the minimization of the adverse effects, including the adverse effects of climate change, effects on international trade, and social, environmental and economic impacts on other parties to the treaties. Such information should be compiled and synthesized leading to recommendations for decisions to be taken in 2009.[30] A decision to express appreciation of the work of the Compliance Committee was taken[31] and a draft decision on good practice guidelines for land use, land use change and forestry activities were adopted.[32] A draft decision expressed appreciation of the decrease in total aggregated greenhouse gas emissions of Annex I countries although it acknowledged that most reductions had taken place in the economies in transition (e.g. Lithuania - 66.2%; Latvia 58.5% reduction in 2003/2004 in relation to the base year). It recognized that some countries would have to intensify policies in order to be able to achieve their own targets since their emissions were much higher than in the base year (e.g. Austria: 16.6%, Finland: 14.4%; Spain: 40.6%).[33] A draft decision requests countries to provide supplementary information under Article 7(2) of the Kyoto Protocol in their National Communications.[34] A decision was also taken with respect to small-scale afforestation.[35] Draft decisions were taken on budget performance and the programme budget including a decision to establish a Trust Fund in which resources for CDM administration should be put.[36]

---

[30]CMP 3 Decision on Scope and Content of the Second Review of the Kyoto Protocol pursuant to its Article 9, see http://unfccc.int/files/meetings/cop_13/application/pdf/cmp_art_nine.pdf.

[31]CMP 3 Decision on Compliance Under the Kyoto Protocol, see http://unfccc.int/files/meetings/cop_13/application/pdf/cmp_art_nine.pdf.

[32]CMP 3 Decision on Good Practice Guidance for Land Use, land-Use Change and Forestry Activities under Article 3, paragraphs 3 and 4, of the Kyoto Protocol, see http://unfccc.int/files/meetings/cop_13/application/pdf/cmp_gpg_lulucf.pdf.

[33]CMP 3 Decision on Demonstration of Progress in Achieving Commitments Under the Kyoto Protocol by Parties included in Annex I to the Convention, see http://unfccc.int/files/meetings/cop_13/application/pdf/cmp_gpg_lulucf.pdf.

[34]CMP 3 Decision on Compilation and Synthesis of Supplementary Information Incorporated in Fourth National Communications Submitted in Accordance with Article 7, paragraph 2 of the Kyoto Protocol, see http://unfccc.int/files/meetings/cop_13/application/pdf/cmp_cs_supp.inf.pdf.

[35]CMP 3 Decision on Implications of Possible Changes to the Limit for Small-Scale Afforestation and reforestation clean development mechanism project activities, see http://unfccc.int/files/meetings/cop_13/application/pdf/cmp_ssc_ar_cdm.pdf.

[36]CMP 3 Decision on Budget Performance for the Biennium 2006-2007 and Programme Budget for the Biennium 2008-2009, see http://unfccc.int/files/meetings/cop_13/application/pdf/cmp_ssc_ar_cdm.pdf.

These decisions too illustrate the continuous monitoring of the process of implementation and science and show how incrementally the regime is moving forward. A critical decision here is the abolishment of the fees for CDM projects hosted in least developed countries to enable them to become more attractive as host countries. Another interesting recommendation is the beginning of a discussion on also setting fees on Joint Implementation and Emissions Trading for providing funds to the Adaptation Fund. This is particularly important to me since there was a tax on North-South cooperation but not on North-North cooperation and this discriminatory feature may be remedied in the future. The fact that the Adaptation Fund finally has been operationalised is of course the most important step, as many countries have been waiting for resources with respect to adaptation for more than 15 years.

3.5 AWG 4 Decisions

Finally, the Ad Hoc Working group on Further Commitments for Annex I Parties also finalised its conclusions.[37] Unlike the Bali Action Plan, which is vague about the quantitative objectives of the climate regime, this document states:

> "It noted the usefulness of the ranges referred to in the contribution of Working Group III to the Fourth Assessment Report (AR4) of the Intergovernmental Panel on Climate Change (IPCC) and that this report indicates that global emissions of greenhouse gases (GHGs) need to peak in the next 10–15 years and be reduced to very low levels, well below half of levels in 2000 by the middle of the twenty-first century in order to stabilize their concentrations in the atmosphere at the lowest levels assessed by the IPCC to date in its scenarios. Hence the urgency to address climate change. At the first part of its fourth session, the AWG recognized that the contribution of Working Group III to the AR4 indicates that achieving the lowest levels assessed by the IPCC to date and its corresponding potential damage limitation would require Annex I Parties as a group to reduce emissions in a range of 25–40% below 1990 levels by 2020, through means that may be available to these Parties to reach their emission reduction targets. The IPCC ranges do not take into account lifestyle changes which have the potential of increasing the reduction range. The ranges

[37] AWG 4 Decision on Review of Work Programme, Methods of Work and Schedule of Future Sessions, see http://unfccc.int/files/meetings/cop_13/application/pdf/awg_work_p.pdf.

would be significantly higher for Annex I Parties if they were the result of analysis assuming that emission reductions were to be undertaken exclusively by Annex I Parties. The AWG also recognized that achievement of these reduction objectives by Annex I Parties would make an important contribution to overall global efforts required to meet the ultimate objective of the Convention as set out in its Article 2."

As is well known, much of the literature on future emission trajectories tends to focus on stabilizing at 450 ppmv and above. However, there is also literature that argues that even stabilizing at 450 ppmv may not be adequate to protect the most vulnerable and marginalized people from the impacts of climate change. It is thus interesting to note that the AWG reflects on this concern as follows:

"4. The AWG noted the concerns raised by small island developing States and some developing country parties with regard to the lack of analysis of stabilization scenarios below 450 ppmv of carbon dioxide equivalent. In line with the iterative approach to the work programme, the information referred to in paragraph 3 above will be reviewed in the light of information received by the AWG, including from possible further scientific work on stabilization scenarios."

The AWG decision calls on all Parties to submit by 15 February 2008 their views on how to achieve the mitigation objectives of Annex I countries and that this should be compiled by March 2008. Specific tasks have been assigned for the fifth to eighth meetings of AWG, leading to a final set of decisions in 2009.

3.6 A Brief Integrated Assessment

The above outline gives you a brief but systematic account of all decisions taken at Bali, and these include decisions that appreciate ongoing work and incrementally push the process further through a number of decisions. As stated above, these decisions show the persistence and determination of all Parties to move the negotiations further on substantive, procedural and financial issues and incrementally the process moves ahead.

Further, the Bali Road Map and the AWG Conclusions provide two routes to moving the process of identifying new targets for the developed countries; one with the US and one without. It is unclear as yet as to whether the two processes will merge or whether one will put pressure on the other, or whether they will in fact compete. But it appears as if Parties are not willing to take chances and are keeping all options open at present.

Such competition can be seen in the unwillingness of the Bali Action Plan to explicitly mention quantitative targets, while the AWG conclusions do; although clearly the nature of the two processes is quite different.

The other key decisions are the decision on the Adaptation Fund and to operationalise it in 2008, the decision on Article 9 of the Kyoto Protocol, the decision on technology transfer and the need to turn technology needs into concrete proposals and that on deforestation.

It is curious to note that the president of the Conference thanked three super achievers at the Conference - the US "for their flexibility and participation in a spirit of cooperation", the secretariat for its continuous support, and the organizing committee!![38]

4. CONCLUSION

Bali has provided a two year road map for the preparation of targets for the developed countries and policies and measures on a range of other issues in order to ensure that there is no gap between the first and second commitment period. It has also moved the AWG process further. The question is will the political context change in the next two years to provide this process the substantive content it needs?

One could argue that with the presidential elections, things will change in the US. Most of the presidential candidates have far-reaching goals on climate change. Candidates like Hillary Clinton and Barack Obama support a 80% emission reduction by 2050 compared to 1990 levels in combination with market-based mechanisms. The US government supports the extension of Activities Implemented Jointly, the predecessor of CDM and JI since it has not ratified the Kyoto Protocol and cannot participate in the latter mechanisms. The US states consistently that all measures it takes, even though outside the framework of the UNFCCC are consistent with it. The active participation of the US delegation in the Bali negotiations and the recognition of the "constructive role" they played by the President of the Conference (see 3) there despite the limits imposed by the White House show that the US is not able to stop such a development and may even be inclined in the post-Bush era to be more proactive.

The EU will consistently move forward as it has done in the last 18 years, incrementally convincing its growing members to adopt the *acquis communitaire* and to develop its climate policy further. It will probably continue in its leadership role to push the climate change process further. The recent ratification of the Kyoto Protocol following the 2007 elections in

[38] http://unfccc.int/files/meetings/cop_13/application/pdf/close_stat_cop13_president.pdf

Australia shows that even the most recalcitrant developed country in the area of climate change is now trying to search for constructive ways of participating in the regime.

China and India remain important actors in the climate change process, especially with respect to the future.[39] The Indian Prime Minister discussed climate change with the Chinese Premier in January 2008 and a follow-up strategy. While both have extensive policies at home to promote energy efficiency and the further development of renewables they are also investing large scale in fossil fuel. However, both countries are actively exploring the choices before them and the future meetings between the Leaders may be critical for determining the attitude of the two countries in the coming two years.

Against this changing political context, and the increasing evidence of the impacts of climate change, and the vast number of decisions taken in Bali, the question is: Does all this activity get us any closer to the ultimate goal of keeping the climate change problem within manageable limits? Possibly not, but the wheels of the institutional process dealing with climate change continue to move forward. Whether the acceleration in policy will emerge in the next two years remains to be seen. But a message has been sent to all actors that climate change remains a serious political challenge and that all social actors have to prepare to search for solutions to this problem.

Acknowledgements

The author has worked on this essay as part of two projects - the Netherlands Organization of Scientific Research VIDI project on International and Private Environmental Governance: Sustainable Development, Good Governance and the Rule of Law (contract number: 452-02-031), and the European Commission financed Adaptation and Mitigation (ADAM) Project (contract number: 98476). The author acknowledges the comments of Harro van Asselt on a previous draft of this paper.

[39] See, for more details, J. Gupta, 'De Rol van China en India in het Mondiale Klimaat Beleid', *Nederlands Juristenblad,* 45-46: 2888-2892.

46

CHAPTER

Bali and Beyond

Donald M. Goldberg
Executive Director of the Climate Law & Policy Project and an Adjunct Professor at the American University Washington College of Law, Washington, D.C., USA

First there is a mountain, then there is no mountain, then there is.[1]

Was Bali a success?[2] At the end of the grueling two-week negotiating round, delegates, observers, and press generally gave the agreement that emerged - the Bali "Roadmap" - high marks. We really won't know whether those marks are deserved, however, until negotiation of the post-2012 period is completed, in 2009. Given the dire warnings contained in the IPCC 4th Assessment Report (AR4), released earlier in the year,[3] no one can doubt the seriousness of purpose with which negotiators labored. Nevertheless, the decisions taken at COP 13 and COP/MOP 3 have a disconcerting Rorschach quality about them. They are susceptible to a wide range of interpretations, permitting each Party to see what it wants. On the one hand, they reference AR4's most ambitious short and medium-

[1] This phrase comes from a Zen expression describing states of perception on the path to enlightenment.

[2] Bali, Indonesia hosted the 13th Meeting of the Conference of the Parties to the UN Framework Convention on Climate Change (COP 13), the 3rd Meeting of the Parties to the Kyoto Protocol (COP/MOP 3), and several subsidiary meetings, 3-15 December 2007. Report of the Conference of the Parties on its thirteenth session, FCCC/CP/2007/6, 14 March 2008 [*hereinafter* COP 13 Report]; Report of the Conference of the Parties serving as the meeting of the Parties to the Kyoto Protocol on its third session FCCC/KP/CMP/2007/9 [*hereinafter* CMP 3 Report].

[3] Intergovernmental Panel on Climate Change, IPCC Fourth Assessment Report: Climate Change 2007 (Cambridge Press).

term scenarios: 25-40% reductions by 2020 and 80-95% by 2050, with substantial deviation from baseline in most developing country regions.[4] On the other hand, the reference to these scenarios in the Bali Action Plan – the centerpiece of the Roadmap – is so oblique that if one did not know where to look, one probably would not find it.

Ambiguity on certain key points, such as the level of commitment to AR4's safest scenario, is to be expected.[5] After all, the Roadmap is merely an agreement to negotiate the post-2012 international regime, and it would not be appropriate for it to specify or prejudice the results of those negotiations. On the other hand, if the Roadmap does not point Parties in the right direction, they may not reach their destination.

Reading the Roadmap optimistically, the outline of a grand bargain may be discerned: developed countries will adopt legally binding mitigation commitments consistent with AR4's safest scenario if developing countries adopt "[n]ationally appropriate mitigation actions" that are "measurable, reportable and verifiable." For their part, developing countries expect developed countries to provide significant new technology and financial resources to help them mitigate and adapt. For many developing countries, this must include incentives to reduce emissions from deforestation and forest degradation. While the text is fraught with loopholes and back doors that permit wildly different interpretations of many points, both the Roadmap and the context in which it was adopted seem to reflect a genuine desire to do what must be done to minimize and protect against the worst impacts of global warming.

THE ROAD TO BALI

To understand the process that emerged from Bali, it is important to understand the process that brought negotiators to Bali. Work on the post-2012 period began in earnest in 2005 at COP 11 and COP/MOP 1, in Montreal, Canada. Technically, the terms of the Kyoto Protocol require negotiations to begin no later than seven years before the end of the first commitment period, which runs from 2008-2012.[6] This meant something had to be done in Montreal to get the ball rolling.

[4] Working Group III to the Fourth Assessment Report of the IPCC, Technical Summary, at 39 and 90, and Ch. 13, at 776.

[5] Ambiguity in treaty text is rarely accidental. It usually reflects the failure of negotiators to reach a meeting of the minds. For example, the author of Article 4.2(a) and (b) of the UNFCCC candidly admits he intended to write text so ambiguous that it could neither be interpreted nor enforced. (Personal communication with author).

[6] Kyoto Protocol to the United Nations Framework Convention on Climate Change, Art, 3, 9p. Available at http://unfccc.int/essential_background/convention/background/items/1349.php.

The thorniest question to be resolved in Montreal was whether negotiations for the post-2012 period should take place under the UNFCCC, to which the United States is a Party, or the Kyoto Protocol, to which it is not.[7] Some Parties feared that, because the United States could vote on UNFCCC decisions, but not Kyoto Protocol ones, it could more easily block progress in the former forum, if it chose to. By promoting the Asia Pacific Partnership and other alternative venues for negotiating climate and energy issues, the United States appeared to be laying the groundwork for a soft alternative to the Kyoto Protocol, despite U.S. protestations to the contrary. If previous U.S. intentions were to lead the world in a different direction, Bali suggests it may be abandoning this strategy or simply running out of time.[8]

The Montreal meeting created two negotiating tracks, one that allowed all Parties to negotiate under the authority of the UNFCCC (the "Dialogue"), and one under the Protocol that addressed future Annex I commitments only.[9] The Dialogue, the broader of the two tracks in both substance and participation, would meet only four times before its mandate expired at COP 13. The mandate for a second track was more difficult to achieve, though it was essential, in retrospect. This was an ad hoc working group with a mandate to explore the targets and architecture of the Kyoto Protocol after the expiration of the first commitment period in 2012.[10] A third element, a review of the Protocol under Article 9, while not itself a negotiating track, can influence the tracks, for example, by issuing a finding that more ambitious targets will be needed.[11]

Other significant events on the road to Bali included a G8 meeting with global warming high on the agenda, a meeting of the 17 Major Economy countries convened by the United States, and a high-level global warming

---

[7] An even more fundamental question prior to Montreal was whether the Marrakesh Accords, also known as the Kyoto rule book, could be adopted by decision so that the Protocol could start to operate. This decision was adopted quickly.

[8] On the final (extra) day of negotiations, during which the United States was hectored for its opposition to some provisions, the head of delegation, Paula Dobriansky, announced that the United States wanted to go forward as part of the new framework and be part of the Bali roadmap.

[9] Report of the Conference of the Parties on its eleventh session, FCCC/CP/2005/5/Add.1, 30 March 2006. Decision 1/CP.11; Report of the Conference of the Parties serving as the meeting of the Parties to the Kyoto Protocol on its first session, FCCC/KP/CMP/2005/8/Add.1, 30 March 2006, Decision 1/CMP.1.

[10] Most of the architecture was already codified in the Marrakesh Accords, but new issues and several unresolved old ones were in need of attention. For example many experts regard contiguity between commitment periods as essential.

[11] *See* CMP 3 Report, Decision 4/CMP.3.

meeting convened by the UN Secretary General.[12] COP 12 and COP/MOP 2 in Nairobi, Kenya were also very important in setting the stage for Bali and beyond. This negotiation, dubbed the "Africa COP," ratcheted up attention to adaptation and funding mechanisms.[13] Adaptation was also the subject of a special workshop in September 2007.[14]

THE BALI ROADMAP

COP 13 and COP/MOP 3 created a set of agreements that together form the Bali Roadmap. The term Roadmap refers both to the negotiating process and to the substance to be negotiated, the essence of which is contained in the four "building blocks"—mitigation, adaptation, technology and finance. The Roadmap also includes at least one important substantive issue not contained in the building blocks—reducing deforestation and forest degradation in developing countries.

The Roadmap creates a process that is in many ways an elaboration of the process set up in Montreal. Presented with several options, negotiators elected to stay with the two-track process, one under the UNFCCC that covers all Parties and one under the Kyoto Protocol, affecting Protocol Parties only. The COP adopted the Bali Action Plan, which creates a new UNFCCC negotiating body known as the *Ad Hoc Working Group on Long-Term Cooperative Action under the Convention.*[15] The Protocol negotiating body created at CMP 1 in Montreal—the *Ad Hoc Working Group on Further Commitments for Annex I Parties*—reviewed and further elaborated its own work program. To the extent possible, the two working groups will meet concurrently to facilitate consistency and a free flow of ideas between them. Both are to complete their work in time for consideration at COP 15 and COP/MOP 5 in 2009. Some anticipate possible tension between these tracks, with Parties moving from one to the other depending on which is more favorable to their national circumstances.

The Bali negotiations were fierce at times, particularly on the last day of negotiations, when it sometimes seemed as if the entire world was venting

---

[12] *See* Joint Statement by the German G-8 Presidency and the Heads of State and/or Government of Brazil, China, India, Mexico and South Africa on the Occasion of the G-8 Summit in Heiligendamm, Germany, 8 June 2007; Council on Environmental Quality, Final Chairman's Summary: First Major Economies Meeting On Energy Security and Climate Change, Sept. 27-28, 2007; United Nation, The Future in our Addressing the Leadership Challenge of Climate Change, 24 September 2007.

[13] *See* Report of the Conference of the Parties on its twelfth session, held at Nairobi from 6 to 17 November 2006, FCCC/CP/2006/5, 26 January 2007.

[14] Report on the workshop on adaptation and planning practices: Note by the secretariat, FCCC/SBSTA/2007/15, 25 October 2007.

[15] COP 13 Report Decision 1/CP.13, 2.

its frustration with U.S. intransigence.[16] The U.S. go-slow approach shows itself clearly in the contrast between the two mandates. The Action Plan is cautious to a fault, reflecting the unresolved differences in approach to global warming of key players, whereas the report by the *Ad Hoc Working Group on Further Commitments* (the Kyoto track) appears almost to be straining to take more aggressive action against global warming.

If the documents display a divergence of views on key matters, such as the depth and timing of emissions cuts, they contain equal evidence of the potential for consensus. Despite their differences in emphasis and enthusiasm, both documents contain the hope, if not the promise, that all the elements needed to combat global warming will be included in the final package: emissions cuts by Annex I Parties in line with the safest of the IPCC Working Group III scenarios; quantifiable and verifiable mitigation measures by developing countries; and access by developing countries to the technology and financing they will need to mitigate the impacts of global warming and adapt to those impacts that cannot be avoided.

The UNFCCC Track: The Ad Hoc Working Group on Long-Term Cooperative Action

The Bali Action Plan provides the mandate of the UNFCCC working group, the *Ad Hoc Working Group on Long-Term Cooperative Action*. As already noted, it reflects the tensions of its major players, the United States, the EU, Russia, China, and other large emitters. While the specifics remain hazy, the general shape of the report that will emerge in 2009 can be discerned. It likely will contain the following key elements:

- long-term commitments by developed countries reasonably aimed at reducing emissions in those countries by 25-40% by 2020;
- mitigation actions by developing countries that are "measurable, reportable and verifiable." These actions will be "supported and enabled" by technology, financing and capacity-building provided by developed countries;
- "[p]olicy approaches and positive incentives" to reduce emissions from deforestation and forest degradation in developing countries with possible additional actions aimed at conservation and sustainable management of forests and "enhancement of forests carbon stocks";

[16]Several other countries, notably Canada, Saudi Arabia, and Japan, joined the United States in many of its objections and raised some of their own. While their interventions were perceived by many as obstacles to progress, it is difficult to gauge their effect on the outcome.

- new and stronger action on adaptation;
- "enhanced action" on technology development and transfer and provision of financial resources and investment for mitigation and adaptation in developing countries.

The Kyoto Track: The Ad Hoc Working Group on Further Commitments for Annex I Parties under the Kyoto Protocol

With the EU and vulnerable developing countries leading the charge (urged on by environmental NGOs), this document fully embraces the most stringent of the Working Group III scenarios and suggests that even more rigorous scenarios that contemplate stabilizing GHG concentrations below 450 ppm of CO_2 equivalent may be required.[17] It notes that the IPCC scenarios do not take into account, lifestyle changes that could lower emissions further and implies that the task would be far more difficult, if not impossible, if reductions were to be undertaken exclusively by Annex I Parties.[18]

The remainder of the document sets out the group's two-year program of workshops and roundtables, submissions from Parties and observers, technical papers from the secretariat, and conclusions and decisions to be forwarded to the COP/MOP for adoption.

Reducing Emissions from Deforestation in Developing Countries (REDD)

Deforestation, mainly in the tropics, accounts for approximately 20% of total CO_2 emissions, making it the second most important source of GHGs.[19] Despite its large contribution to global warming, deforestation proved to be so controversial in the Kyoto negotiations that its avoidance was not included as one of the forest-based activities that could generate credits through the CDM.[20] Several objections were put forward to the use of credits from avoided deforestation: the sheer number of credits

[17]Report of the Ad Hoc Working Group on Further Commitments for Annex I Parties under the Kyoto Protocol on its resumed fourth session, FCCC/KP/AWG/2007/5, 5 February 2008, 17.

[18]*Id.*, at 16.

[19]See, e.g., IPCC, Land Use Change and Forestry, Cambridge University Press, 2000.

[20]Currently Parties can get credit for afforestation or reforestation only. Annex I Parties may use forest-based CDM activities to offset no more than 1% of their total allowed emissions during any commitment period. Report of the Conference of the Parties on its Seventh Session, Held at Marrakesh from 29 October to 10 November 2001, FCCC/CP/2001/13/Add.1, 21 January 2002, at 61, 2.[*Hereinafter* Marrakesh Accords].

available and their presumed low cost could give emitters a cheap way to offset their emissions; activities that caused deforestation might "leak" from protected to unprotected areas; uncertainty about the quantity of carbon stored in forests would make it difficult to quantify benefits; the risk of forest-fires and other threats, some of them caused or exacerbated by global warming itself, meant that credits could not be regarded as permanent.

Much work has been done to answer these questions, and most of these disputes are resolved or appear to be approaching resolution. For the first commitment period, negotiators decided to allow a limited number of credits from afforestation and reforestation, while excluding deforestation, but agreed that the question of emissions from deforestation, forest degradation, agriculture, and other biotic sources would be reconsidered for future commitment periods. Given the enormity of the problem—in many developing countries forests are the main source of emissions—it was inevitable that negotiators would return to the issue of avoided deforestation.

They did so in 2005, when Papua New Guinea and Costa Rica proposed reconsideration of deforestation and degradation, with the stipulation that national baselines would be used to minimize additionality and leakage, and participation would be voluntary.[21] This time, environmental groups were more receptive to the idea and contributed to developing it further.[22]

As noted above, the issue has been included in the Bali Roadmap, and the COP passed a decision on "approaches to stimulate action," which affirms "the urgent need to take further meaningful action to reduce emissions from deforestation and forest degradation." It requests that the SBSTA establish a work program on policy approaches and positive incentives, invites Parties to submit their views on methodological issues, and request the secretariat to organize a workshop.

Despite objections from environmental groups, the Roadmap also considers the role of conservation, sustainable management of forests, and enhancement of carbon stocks. Environmentalists prefer to keep the focus on deforestation and forest degradation. They fear that activities such as sustainable management of forests and enhancement of carbon stocks could actually lead to the loss of natural forests, for example, by creating additional incentives for plantation forestry. Another issue that could

---

[21]Reducing Emissions from Deforestation in Developing Countries: Approaches to Stimulate Action, Submission by the Governments of Papua New Guinea & Costa Rica Eleventh to the 11th Conference of the Parties, May 20.

[22]*See, e.g.*, Climate Action Network, Reducing Emissions from Deforestation and Forest Degradation (2007).

prove contentious is the source of incentives. Some countries and environmental groups believe the only adequate source of incentives is the emissions trading system, whereas others prefer, for many of the reasons stated above, that incentives to protect forests not be linked to the trading system at all.

Technology Transfer

After fifteen years of inaction on the issue of technology transfer, significant gains were made in Bali. The traditional view of wealthy countries is that technology is proprietary and cannot simply be transferred by governments. The two COP decisions on technology transfer, one laying out an agenda for SBSTA, the other instructing SBI, contain a number of solutions to this problem. The mandate of the Expert Group on Technology Transfer (EGTT) to oversee technology transfer was expanded and extended for another five years.[23] There is a large dollop of "bootstrap," i.e., actions to help developing countries create environments more conducive to the transfer of environmentally sound technology. These include improved needs assessments, learning centers for capacity building, technical studies on good practice, information sharing, and—a favorite of industrialized countries—removal of policies concerning intellectual property rights that inhibit technology transfer and creation of policies that enhance it.

More concretely, the EGTT's mandate includes ensuring "adequate and timely financial support" and development of performance indicators to monitor and evaluate the effectiveness of the technology transfer program. An interim funding mechanism managed by the Global Environment Facility (GEF) was informally agreed early in the negotiations. The subsidiary bodies will oversee the process, and the SBI, in particular, will conduct monitoring and evaluation based on the performance indicators developed by the EGTT. The presumption is that, as developing countries strengthen their mitigation commitments, developed countries will increase their levels of financial support.

Adaptation

As noted above, COP 12 and COP/MOP 2 in Nairobi, Kenya put the spotlight on adaptation and funding mechanisms. The Nairobi work programme continued work begun at COP 9 to help countries improve

[23]The Expert Group on Technology Transfer was created to enhance the implementation of technology transfer activities under the Convention. Terms of Reference for the Expert Group are available at
http://unfccc.int/essential_background/convention/convention_bodies/constituted_bodies/items/2581.php.

their understanding of climate change impacts and vulnerability and to increase their ability to make informed decisions on how to adapt successfully. It contains nine areas of work to be implemented by Parties, intergovernmental and non-governmental organizations, the private sector, communities, and other stakeholders.[24]

Work on adaptation began well before Nairobi, however. COP 7 established national adaptation programmes of action (NAPAs) to help least developed countries identify priority adaptation activities – those for which further delay could increase vulnerability or lead to increased costs – and set up an LDC Expert Group to provide guidance on preparation and implementation of NAPAs.[25]

An Adaptation Fund, funded by proceeds from the clean development mechanism (CDM), also was established at COP 7.[26] An initial list of guidelines for the operation of the Fund was developed at COP/MOP 1 and was further elaborated at COP/MOP 2.[27] As mentioned above, adaptation was also the subject of a special workshop in September 2007.[28]

In Bali, COP/MOP 3 made the Adaptation Fund operational and decided that developing country Parties that are particularly vulnerable to the adverse effects of climate change are eligible for funding. Because the Fund is financed primarily by the proceeds of CDM projects in developing countries, the question of Fund management was a particularly sensitive one. The GEF sought the job of managing the Fund, creating a controversy between developed and developing countries. Developing countries have expressed some ambivalence about the GEF, arguing, for example, that it is inefficient.[29]

As a compromise, the task was given to an Adaptation Fund Board comprised of representatives of 16 Parties. The GEF will serve as

---

[24]Report of the Subsidiary Body for Scientific and Technological Advice on its twenty-fifth session, held at Nairobi from 6 to 14 November 2006, FCCC/SBSTA/2006/11, 32-71.

[25]Marrakesh Accords Decision 5/CP.7. NAPAs should include short profiles of projects and/ or activities intended to address urgent adaptation needs. They should require no new research, but rely primarily on existing information obtained from community-level input, be action-oriented, country-driven, straightforward and easily understood. *Id.*, Decisions 28/CP.7 and 29/CP.7

[26]Marrakesh Accords, Decision 10/CP.7

[27]CMP 1 Report, Decision 28/CMP.1; CMP 2

Report, Decision 5/CMP.2.

[28]Note by the secretariat, *supra* note 12.

[29]Report of the Conference of the Parties serving as the meeting of the Parties, from concept identification to implementation. Craig Hart, "The Bali Action Plan: Key Issues" (CIEL 2008), at http;//www.ciel.org/Climate/Bali_KeyIssues_21 Dec07.html.

secretariat to the Board, and the World Bank was made trustee of the Fund.[30]

ASCENDING THE MOUNTAIN

The announcement on the last day of negotiations that the United States would drop its remaining objections to the Bali Action Plan suggests that it wants to be back in the international game, even if that game is played largely by Kyoto rules, in which case a consensus agreement could emerge in 2009. If this is the view of President Bush, who made opposition to Kyoto a cornerstone of his administration's policy, it would be difficult for future administrations to backtrack.[31] Rumors have been extant for some time, however, that President Bush will attempt to wrap up an agreement with the Major Economies before leaving office, so the outcome may not be certain until the next President takes office in January 2009. Another key player is the U.S. Congress, which could be a wild card, much as it was during the Kyoto negotiations.

In no event should the Bali Action Plan be taken as an assurance that the United States, Japan, Russia, or several other Annex I countries have agreed to join or remain in the Kyoto Protocol as currently structured. The Plan is peppered with phrases like "long-term cooperative action," "now, up to and beyond 2012," "long-term global goal for emissions reduction," "nationally appropriate," and "comparability of efforts" that do not suggest an approach consistent with Kyoto in its present form. Some of them suggest elements that, arguably, should have been included in the original Protocol. Some Kyoto critics claim that its targets failed to reflect national differences and demanded more effort from some developed countries than others.[32] Kyoto also has been criticized for lacking a long-term concrete objective. It is possible that the United States and others will seek to correct these "flaws" in a new universal agreement for 2012 and beyond.

The *Ad Hoc Working Group on Further Commitments* is also taking a longer view. Neither Working Group mandate discusses the length of future commitment periods, but, despite U.S. efforts to have them deleted from the Action Plan, the two milestone periods discussed by the IPCC, the next 10-15 years and 2050, will surely guide negotiations. Unless the science changes, the cuts suggested by AR4 will be the yardstick by which the world judges the success of this negotiation.

[30] CMP 3 Report, Decision 1/CMP.3, 23.COP 13 mandated a broad review of the GEF in its capacity as the "operational entity" of the UNFCCC financial mechanism. Its outcome could influence whether the GEF will continue to operate the Adaptation Fund and could also affect GEF's management of other climate funds. *Id.*

[31] All the U.S. Presidential candidates are already on record supporting some form of cap-and-trade.

[32] In fact, Europe was thought to be about 10% points closer to its target than was the United States when Kyoto was adopted.

An unexpected but very positive development is the inclusion of developing country commitments to take concrete actions that should lower emissions from business-as-usual.[33] Although there is no way to predict what levels of developing country reduction might be feasible, it seems an important procedural matter has been tacitly agreed upon: mitigation actions by developing countries are to be "measurable, reportable and verifiable."[34]

Another Bali milestone is the inclusion in the Roadmap of incentives to reduce emissions from deforestation and forest degradation in developing countries.[35] Until now, deforestation has been an intractable problem for many countries, made all the more so by the inability, for lack of resources, of many governments to enforce their own forest protection laws. The loss of these forests can begin a cycle that leads to erosion and desertification, making restoration difficult and reducing the ability of the planet to remove carbon from the atmosphere. Many experts believe that giving financial value to the vast amounts of carbon stored in tropical and other natural forests holds the only hope for saving them.

Protecting their forests is the most important contribution many developing countries can make to mitigating global warming and adapting to its impacts. In addition to containing valuable commodities, which countries are learning to utilize sustainably, natural forests provide vital ecosystem services, such as preventing erosion and desertification, cleaning and protecting freshwater, providing habitat for biodiversity, and even stabilizing microclimates. The importance of protecting and restoring natural forests cannot be overstated.

More controversial is the inclusion of conservation, sustainable forest management, and enhancement of carbon stocks. The CDM rules appear to permit projects to generate CERs through such undesirable activities as removal of natural forest to provide land for plantations of fast growing tree species. While there is no doubt that many countries already engage in such activities have occurred in the past, there is no evidence that the CDM has approved any projects of this type.[36] Nevertheless, this issue cannot be

[33]COP 13 Report, Decision 1/CP.13, 1(b)(ii).

[34]Two proposals that appear to fit this language are sustainable development policies and measures (SD PAMs) and action targets (ATs). Under SD PAMs developing countries would take measures to promote sustainable development. While the aim would be to achieve GHG reductions in the process, they might not be adequately verified to allow incorporation into the international emissions trading system. Action targets, would have countries specify a quantity of reductions, rather than measures, and they would be verified to make them amenable to full or partial incorporation into the trading system.

[35]COP 13 Report, Decision 2/CP.13, 1(b)(iii).

[36]As of May 2008, CDM forest projects represented only 0.07% of registered projects, making generalizations hightly speculative. Rejected forest projects outnumbered accepted ones nearly three to one. Two projects recently registered in India appear to utilize mainly or solely waste (waste wood, sawdust, rice husk, agricultural waste, etc.) UNFCCC Clean Development Mechanism Home. Available at http://cdm.unfccc.int/index.html.

ignored. Rules must be developed to ensure that such perverse outcomes don't occur.[37]

Making the Adaptation Fund operational was a significant step, because it can now begin to provide financing to eligible developing countries. It remains to be seen whether the stipulation that funding will go to "particularly vulnerable" countries will prove problematical. The Fund is unique in that, unlike other UNFCCC or Kyoto Protocol funds, it is not financed by donor countries, but, as noted above, by the proceeds of CDM projects. This gives developing countries considerably more say in the management of the Fund.

A very important issue that has not been discussed by either negotiating body is how the agreement will deal with possible surprises. What if the science, or the climate itself, reveals a threat that requires a very rapid response? The problem of reducing emissions has often been compared to turning a battleship, but the possibility of such surprises demands an agreement that can turn on a dime. Neither the negotiating history nor the Protocol itself provides any assurance that the process is capable of a rapid response. Altering a Protocol target currently requires that an amendment be adopted at an ordinary session of the COP/MOP, which occurs only once a year. Furthermore, it must be submitted six months before it can be voted on. Thus, under current rules if an emergency were to arise it could take almost two-and-a-half years before an amendment could be adopted. Several more years might be required for ratification and entry into force. Clearly, an expedited process is needed for responding to emergencies.

The decision of the United States to allow a consensus agreement to emerge from the Bali negotiations virtually ensures that global warming will not be "Balkanized," but will be addressed as a global issue. This is particularly important from an economic standpoint, as a unified global market with an internationally agreed set of rules should be more cost-effective than a patchwork of segmented markets. On the other hand, a global market managed by, and answerable to, an international body that gives every country an equal vote could prove unwieldy and, ultimately, inefficient. The fact that the United States may have given its blessing to a universal agreement does not necessarily imply a "one size fits all" approach. It is possible that sub-agreements, possibly along the lines of a Major Economies agreement or the EU "bubble," may yet emerge. While Bali was undoubtedly an important step in the right direction, much hard work remains before the optimal format for addressing global warming can be identified and, hopefully, adopted.

---

[37]Some NGOs claim that carbon-driven forest protection has already deprived people of their land and livestock. See, e.g., Fern and Sinkswatch, *Human rights abuses, land conficts, broken promises – the realiry od carbon 'offset' projects in Uganda,* Feb. 11, 2007.

47

CHAPTER

Beyond Bali and Bush: The Future of Climate Policy

Joshua W. Busby
Lyndon B. Johnson School of Public Affairs,
The University of Texas at Austin
P.O. Box Y, Austin, Tx 78713-8925, USA
E-mail: busbyj@mail.utexas.edu

In December 2007, delegates from more than 180 countries met in Bali, Indonesia to map out the future of the climate regime. Supporters of multilateral climate negotiations need a new agreement to take effect after the first commitment period of the Kyoto Protocol expires in 2012.[1] Unless the more technical aspects of the negotiations are concluded by 2009, the rules for how to credit action on climate change may lapse in 2012 without sufficient guidance or clarity. For those committed to this process, Bali was thus a critically important summit.

This short concluding chapter assesses the outcome of Bali and discusses the possibilities for action looking ahead.

[1] It was the 13th Conference of Parties (COP) for signatories to the original 1992 United Nations Framework Convention on Climate Change (UNFCCC) and the third meeting of the parties (MOP) for those countries that elected to ratify the 1997 Kyoto Protocol. Whereas the former encompassed most of the world's countries by virtue of its non-binding and general character, the latter had binding commitments only for so-called Annex I advanced industrialized countries. After Australia ratified in December 2007, 176 countries and the European Economic Community had ratified the Kyoto Protocol. The UNFCCC has been ratified by 192 countries.

Bali provided familiar storylines of U.S.-European rivalry over timelines and binding emissions reductions. Despite the media hype over U.S. recalcitrance and the final compromise agreement, this outcome was neither surprising nor the most important one of the summit. With the Bush Administration still in power but winding down, the Europeans went through the motions of trying to get the Americans to commit to deep emissions reductions and the obligatory outrage when the United States failed to move. As David Sandalow of the Brookings Institution described it, "This dispute was as predictable as it was meaningless."[2] More significant were breakthroughs on avoided deforestation, adaptation, and recognition by developing countries that they would accept some sort of action on climate change in the next agreement.[3]

Notwithstanding these developments, advocates may have vaunted expectations for what is possible in a meeting of nearly two hundred countries and more than ten thousand attendees. As the *New York Times* columnist Tom Friedman wrote of the Bali deliberations, "I'm not opposed to forging a regime with 190 countries for reducing carbon emissions, but my gut tells me that both the North and South Poles will melt before we get it to work."[4] Other smaller, multilateral fora and national and sub-national level decision-making arenas will be increasingly important in the coming years. Moreover, as governments send markets signals to price carbon, the terrain for action will, and indeed must, shift to research labs and firms where the emissions savings technologies of the future will be generated.

As long as international negotiations like Bali are seen as the most important arenas for progress, this is not good news. The real work will have to take place at a more localized level as firms and consumers respond to incentives for action.[5] To that point, producers and consumers of energy will have those incentives (1) when the U.S. adopts a mandatory national-level carbon constraint and (2) when the major emerging emitters, including China and India, get serious about (and serious help) investing in clean energy technology.

---

[2] Sandalow 2007.

[3] My pre-summit podcast anticipated much of what ultimately occurred. Busby 2007c. For a post-summit summary, see Levi 2007. These developments are also described in Pew Center on Global Climate Change 2007b.

[4] Tom Friedman expressed a similar tone in his Bali wrap-up. Friedman 2007.

[5] David Victor and his co-authors describe this bottoms-up process as "Madisonian" climate policy in Victor, House and Joy 2005.

BALI'S PARADOX: PARALYSIS AND PROGRESS

Bali provided some modest progress while at the same time displaying the same sort of political gridlock that plagued earlier negotiations such as The Hague in 2000.

Paralysis

The Americans came implacably opposed to making specific commitments on medium-run, binding emissions reductions while Europe and a number of developing countries demanded an ambitious set of targets and timetables for reducing greenhouse gas emissions.[6] Proponents wanted rich countries to pledge to reduce greenhouse gas emissions below 1990 levels between 25% and 40% by 2020. The Americans for their part were having none of it; the Bush Administration was never going to agree to this. Long-time observers of the Bush Administration and U.S. climate policy were unsurprised.

Despite this replay of U.S. intransigence and, depending on your perspective, European leadership/grandstanding, the summit delivered some hopeful advances. A change in government in Australia directly preceded the summit and brought to power a new pro-Kyoto Administration led by Kevin Rudd. This left the United States isolated as the sole remaining advanced industrialized country not to have ratified Kyoto. That said, while Australia's domestic politics had become more Kyoto friendly, both Canada and Japan, faced with soaring emissions and less environmentally-oriented governments, have increasingly become more skeptical of the existing climate architecture despite having ratified Kyoto.

With the Bush Administration balking at making specific, medium-run legal commitments at Bali, climate negotiators in Bali were thinking ahead to President Bush's successor.[7] Negotiators sought language that was strong enough to satisfy domestic constituencies and weak enough to keep the Americans engaged in the next round of negotiations. While squabbling over language nearly led to a complete breakdown, European and American negotiators found a compromise they each could half-heartedly support. The main text has no target or timetable but a footnote

[6] The Americans were more willing to countenance a long-run target, provided comparable action by other countries, but they judged it premature to commit to specific, medium-run binding targets at the beginning of the negotiations.

[7] The Bush Administration judged it premature to start the negotiations over the post-Kyoto period with binding medium-run targets and timetables. The U.S. administration is more supportive of a longer-run target, as long as other countries make comparable commitments.

references pages in the Third Working Group report to IPCC's Fourth Assessment Report that suggest cuts of 10-40% below 1990 levels are likely needed by 2020.[8]

Progress

Despite the deadlock over targets and timetables, the Bali conference made modest progress on a number of important areas: developing country commitments, avoided deforestation, and adaptation.

Developing Country Commitments

Bali provided the first important signal of the willingness by China, India, and other emerging economies to consider "measurable, reportable and verifiable" mitigation actions looking ahead. This new flexibility was qualified, provisional upon them receiving ample financial incentives to adopt clean energy technology.[9] As discussed below, the appropriate way forward on technology transfer is not straightforward.

Avoided Deforestation

In addition to the new willingness by developing countries to countenance mitigation commitments, Bali also moved the agenda forward on avoided deforestation.

The Bali roadmap makes it possible for the successor agreement to the Kyoto Protocol to allow heavily forested countries to be compensated for preventing and reducing deforestation.[10] Under the Kyoto Protocol, countries can be compensated for replanting after a forest has been cut down but not for preventing deforestation in the first place.

This agreement on avoided deforestation, also known as Reduced Emissions from Deforestation and Degradation (REDD), potentially marks an important breakthrough. Deforestation is responsible for about a 1/5 of the world's total greenhouse gas emissions. Deforestation and forest fires in COP-13 host country Indonesia helped make it the third largest contributor of greenhouse gases behind the United States and

[8] See page 90 in IPCC 2007.

[9] Among the enhanced mitigation strategies that will be part of the post-Kyoto agreement, the Bali roadmap text includes: "Nationally appropriate mitigation actions by developing country Parties in the context of sustainable development, supported and enabled by technology, financing and capacity-building, in a measurable, reportable and verifiable manner." UNFCCC 2007a.

[10] The Bali meeting decided the successor agreement would include: "Policy approaches and positive incentives on issues relating to reducing emissions from deforestation and forest degradation in developing countries; and the role of conservation, sustainable management of forests and enhancement of forest carbon stocks in developing countries." UNFCCC 2007a.

China.[11] Paying countries to keep their forests would likely be a much cheaper way for rich countries to avoid emitting greenhouse gases than retrofitting existing industrial infrastructure or seeking a rapid change in transportation alternatives.[12] The World Bank estimates that an expanded forest protection plan could result in 1 billion tonnes in avoided emissions by 2015.[13]

At Bali, the World Bank announced that it had $160 million in commitments from donors for an ambitious $250 million pilot project on avoided deforestation.[14] Interestingly, the Bush Administration, despite professing support for the initiative, did not contribute to the Bank's pilot project. Though he received commitments from other governments, the new president of the Bank, Bob Zoellick, failed to get any money from his own, the richest country in the world.[15]

Some big issues on avoided deforestation remain for the Bank to iron out as it implements the pilot project: (1) the accounting procedure for how to track emissions reductions from avoided deforestation; (2) the compensation mechanism – will proceeds go to governments, local communities, or get captured by large commercial interests;[16] and, (3) should there be a fund to support this or market mechanisms.[17]

Adaptation

In Bali, the parties also reached agreement on an Adaptation Fund. 2% of the proceeds from transactions through the Clean Development

[11] Wright 2007. Other countries with large forest reserves include Brazil and the Democratic Republic of Congo.

[12] Pacala and Socolow 2004 outline fifteen potential options, including avoided deforestation, for reducing emissions over the next half-century. Each option reduces emissions by about 25 billion tonnes. Pacala and Socolow 2004.

[13] The Bank has proposed has an ambitious Global Forest Alliance (GFA), partnering with large environmental NGOs like the Nature Conservancy to implement the program, the so-called Forest Carbon Partnership Facility. World Bank 2007b.

[14] The pilot program had been blessed earlier in the year at the G8 summit hosted by Germany. Nine countries made commitments to the new initiative at Bali, including Germany (US$59 million), the United Kingdom ($30 million), the Netherlands ($22 million), Australia and Japan ($10 million each), France and Switzerland ($7 million each), and Denmark and Finland ($5 million each). The Nature Conservancy also pledged $5 million. World Bank 2007a.

[15] Mongabay.com 2007.

[16] Some NGOs have warned of large-scale inequities over the distribution of these resources and that forest dwellers and small-scale actors may find themselves cut off, both from the funds and access to the forests. Griffiths 2007.

[17] Many heavily forested developing countries favor market approaches, but Brazil, worried about sovereignty, notably has supported a fund. Other developing countries think large transfers are unlikely to be forthcoming through a fund. Bali leaves this unresolved for now.

Mechanism have been set aside to support efforts in developing countries to adapt to climate change.[18] Adaptation projects include coastal defenses, water conservation, investments in drought resistant crops, and other risk reduction measures. The value of those credits was estimated to rise to between $80 million and $300 million a year from 2008 to 2012.[19]

The major source of contention at Bali on these funds was over how these funds would be administered. Developed countries wanted the fund to be run by the Global Environmental Facility (GEF) at the World Bank, which already administers two other adaptation funds—the Special Climate Change Fund (SCCF) and the Least Developed Country Fund (LDCF)—that have about $200 million in commitments.[20] Developing countries do not much care for the GEF because they think the funds are too difficult to access and/or bound up with larger issues of World Bank conditionality. They preferred a new stand-alone institution like the Global Fund for AIDS, TB, and Malaria that would be seen less a tool of the great powers.

After much debate, the Bali conference resolved the debate by naming the GEF the administrator of the Fund but providing greater developing country representation on the Fund's sixteen-member board.[21]

Despite this progress, the scale of resources available for risk reduction is wholly inadequate. Developing countries will likely need tens of billions of dollars to protect themselves from extreme weather events and other effects of climate change.[22] However, until the scale of resources for adaptation is dramatically increased (and actually starts to get spent), developing countries will remain incredibly vulnerable.

LOOKING AHEAD

Despite European leadership on climate change through their emissions trading scheme, the United States remains critical to the future of climate policy. The United States is the second largest emitter of greenhouse gases. Other countries and major emerging emitters in the developing world currently await U.S. action before making significant commitments of

[18] These are the projects where companies in rich countries get emissions credits by financing emissions reductions projects in the developing world.

[19] UNFCCC 2007b.

[20] In April 2007, for example, the LDCF had total pledges of $115.8 million and the SCCF had pledges of $62 million. Another $50 million was available for the Strategic Priority on Adaptation under the GEF Trust Fund. Global Environmental Facility 2007.

[21] My pre-Bali podcast foreshadowed this debate and result.

[22] I discuss the adaptation and risk reduction policy agenda in Busby 2007b; Busby 2007a; Busby 2008, forthcoming-a.

their own.[23] Once the Bush Administration leaves office at the beginning of 2009, a new U.S. president will have an opportunity to change his or her country's policies on climate change and, in turn, those of major emitters in the developing world.

At home, the new president will have an opportunity to steer through Congress a national carbon constraint, most likely a cap-and-trade system. In December 2007, the Senate Environment and Public Works Committee on a 11-8 vote sent the Lieberman-Warner cap-and-trade bill to the full Senate for consideration. As of this writing in April 2008, the bill's fortunes look bleak. The Democrats likely lack the sixty votes needed to withstand a filibuster, and even if the Senate passes the measure, President Bush will likely veto the bill.

That said, the politics of climate change have changed in the United States, giving a new president some leeway to be able to get a modest carbon constraint passed in Congress. There is much broader bipartisan, cross-regional, multi-sectoral, faith-based, and business-backed support for a more vigorous and robust U.S. climate policy. No longer is the issue *if* the United States adopt controls on carbon but *when*.[24]

Even if domestic action is increasingly likely in the United States, international supporters of climate mitigation may have outsized expectations for what the next U.S. president will do. In 2007, most of the cap-and-trade bills pending before the U.S. Senate mandated returning U.S. emissions to 1990 levels by 2020, significantly less than what Europeans suggested would be acceptable.[25] The next U.S. president, even if he or she wants to, will find it hard to commit to a 25% or 40% reduction by 2020. The U.S. Senate may balk at deeper short-run commitments. Since getting started has been the hardest part for the Americans, getting too hung up over the magnitude of the commitment could lead to another lost decade of no federal policy by the U.S. government.

Moreover, as discussed in my earlier chapter in this volume, it may not be especially productive to focus so much diplomatic energy on achieving breakthroughs in a 190-country conference. The collective action problems of having so many negotiating players undermine the incentives for significant action.

Looking at the Bali deliberations, so much work went into drafting elaborate rules for technology transfer but the summit reached no significant agreement on funding sources. Rich countries are reluctant to

[23] I discuss the possibilities for European leadership on climate policy in Busby 2008, forthcoming-b.

[24] I discuss the politics of this bill and broader energy policies in Busby 2008.

[25] Pew Center on Global Climate Change 2007a.

make elaborate promises to transfer technology to poor countries for a number of reasons. First, many of the technologies are in private hands so "transfer" must involve some sort of compensation mechanism by which a country's own firms have an incentive to do so. At the same time, the real possibilities for intellectual property theft—of reverse engineered versions of clean energy technology showing up soon after transfer—makes firms reluctant to sell the latest and most efficient versions of their technology. Finally, clean energy is potentially such a lucrative arena for a country's own firms that countries likely prefer not to multilateralize the process. As Michael Levi of the Council on Foreign Relations has argued:

> I would give [an] incredibly small chance of a country committing to subsidize these sorts of things with any quantitative commitments as part of something that's binding, [something] that they can't alter depending on the relationship with particular countries, depending on how the world evolves, depending on how cheap, for example, these technologies become. But they may be more willing to make particular steps through unilateral measures.[26]

Once the United States has a carbon constraint, this will make it politically possible for developing countries to make some commitments, provided advanced industrialized countries make it worth their while through technological incentives. As suggested above, the process for technological transfer may take place either through a large multilateral forum like the UNFCCC, smaller decision-making arenas like meetings of major emitters and the G8 Summit, or through unilateral approaches.

The Bush Administration's parallel efforts to convene major economies outside the UNFCCC process may have given the idea a bad name. In the first of these meetings in September 2007, the U.S. offered major emerging emitters little in the way of new technology or incentives. In the waning days of Bali, the Europeans, upset over U.S. intransigence on targets and timetables, threatened to boycott planned meetings of the major emitters in early 2008 in the lead up to the G8 Summit in Japan. This is unfortunate. A smaller meeting of major economies offers great potential to simultaneously reduce collective action problems and cover most of the world's emissions in a single gathering. By making these into sidepiece distractions rather than real opportunities for substantive breakthroughs, President Bush may have made it difficult for his successor to resuscitate the process. Nonetheless, the Japanese want to make the 2008 G8 summit a success so will likely press the Americans to join with them in some symbolic gesture.

[26] Levi 2007.

There will be considerable pressure on the next U.S. president to be ready for the fifteenth COP in December 2009 in Copenhagen, Denmark, where proponents would like to finalize negotiations on the successor agreement to the Kyoto Protocol. However, with less than a year to prepare and make key appointments, we should not be surprised if the new American administration is not quite up to speed by COP-15. The rest of the world will likely have to go ahead and muddle through whether or not the Americans have their act together.

More important indicators of U.S. seriousness will be when the country enacts its own carbon constraint and reaches out to China, India and other major economies to facilitate clean energy exports from American companies. Once that happens, the landscape of climate policy will be decentered. No longer will multilateral or national level decision-making be the most important arenas for progress. The action will move finally move to the firm, factory, and local levels where it should, as thousands more businesses, innovators, and consumers begin to alter their behavior. That day can come none too soon.

Bibliography

Busby, Joshua. 2007a. *Climate Change and National Security: An Agenda for Action*. New York: Council on foreign Relations. Available at: <www.cfr.org/publication/14862>

Busby, Joshua. 2007b. Climate Change and Security: A Credible Connection? *Disarmament Times* (Fall): 2-3, 8.

Busby, Joshua. 2008, forthcoming-a. Who Cares About the Weather? Climate Change and U.S. National Security. *Security Studies*.

Busby, Joshua. 2007c. *Busby: On the Bali Climate Change Conference*, Council on Foreign Relations. Available at: <http://www.cfr.org/publication/14918/busby.html >

Busby, Joshua. 2008, forthcoming-b. "The Hardest Problem in the World: Leadership in the Climate Regime." In *The Dispensable Hegemon: Explaining contemporary international leadership and cooperation*. Stefan Brem and Kendall Stiles, ed. London: Routledge.

Busby, Joshua. 2008. *Overcoming Political Barriers to Reform in Energy Policy*: Centre for a New American Security. Available at: <http://www.cnas.org/en/cms/?1608/>

Friedman, Thomas. 2007. What Was That All About? *New York Times*. December 19.

Global Environmental Facility. 2007. *Status Report on the Climate Change Funds*, GEF. Available at: <http://thegef.org/Documents/Council_Documents/GEF_C28/documents/C.28.4.Rev.1ClimateChange.pdf >

Griffiths, Tom. 2007. *Seeing 'RED': 'Avoided deforestation' and the rights of Indigenous Peoples and local communities*, Forest Peoples Programme. Available at:

<http://www.forestpeoples.org/documents/ifi_igo/avoided_deforestation_red_jun07_eng.pdf >

IPCC. 2007. *Contribution of Working Group III to the Fourth Assessment Report of the Intergovernmental Panel on Climate Change: Technical Summary*. Available at: <http://www.ipcc.ch/pdf/assessment-report/ar4/wg3/ar4-wg3-ts.pdf>

Levi, Michael. 2007. *Levi: Modest Results From Bali's Heated Climate Conference*, Council on Foreign Relations. Available at: <http://www.cfr.org/publication/15063/levi.html>

Mongabay.com. 2007. *U.S. contributes $0 to World Bank's new $300m forest carbon fund*. Available at: <http://news.mongabay.com/2007/1211-world_bank.html>

Pacala, S. and R. Socolow. 2004. Stabilization Wedges: Solving the Climate Problem for the Next 50 Years with Current Technologies. *Science*. **305**: 968-972.

Pew Center on Global Climate Change. 2007a. *Economy-wide Cap-and-Trade Proposals in the 110th Congress*. Available at: <http://www.pewclimate.org/docUploads/110th%20Congress%20Economy-wide%20Cap&Trade%20Proposals%2012-18-2007%20-%20No%20Chart.pdf>

Pew Center on Global Climate Change. 2007b. *Summary of COP 13 and COP/MOP 3 prepared by the Pew Center on Global Climate Change*. Available at: <http://www.pewclimate.org/docUploads/Pew%20Center_COP%2013%20Summary.pdf >

Sandalow, David. 2007. *Climate Change: Beyond Bali*, Brookings Institution. Available at: <http://www.brookings.edu/opinions/2007/1217_climate_change_sandalow.aspx >

UNFCCC. 2007a. *Bali Action Plan*. Available at: <http://unfccc.int/files/meetings/cop_13/application/pdf/cp_bali_action.pdf >

UNFCCC. 2007b. *UN Breakthrough on climate change reached in Bali*. Available at: <http://unfccc.int/files/press/news_room/press_releases_and_advisories/application/pdf/20071215_bali_final_press_release.pdf >

Victor, David G., Joshua C. House and Sarah Joy. 2005. A Madisonian Approach to Climate Policy. *Science*. **309**(5472): 1820-1821.

World Bank. 2007a. Forest Carbon Partnership Facility Launched at Bali Climate Meeting.

World Bank. 2007b. *Global Forest Alliance (GFA) and the Forest Carbon Partnership Facility (FCPF) Presentation to ProFish Board*. Available at: <http://siteresources.worldbank.org/EXTARD/Resources/profishppt4.pdf >

Wright, Tom. 2007. World Bank Targets Forest Preservation – Climate Link. *Wall Street Journal*. June 14.

48

CHAPTER

From Kyoto to Copenhagen by Way of Bali

Jean Crête
Département de science politique
Université Laval, Québec GIK 7P4
Québec, Canada
E-mail: Jean.Crete@pol.ulaval.ca

At the end of 2007, countries of the world were invited to participate in a conference at Bali, Indonesia, to draft a roadmap for negotiating cuts in heat-trapping carbon emissions from 2012, when current pledges under the Kyoto Protocol run out. The conference was held under the United Nations Framework Convention on Climate Change (UNFCCC). The roadmap was designed to bring all countries to Copenhagen in 2009 where a new protocol would address global warming after 2012.

What are the odds that the world political leaders will deliver a masterplan to effectively combat the threat of climate change? Relying on Kingdon's framework in his book *Agendas, Alternatives, and Public Policies,* we suggest the following guidelines to answer the question.

First, what is the problem? Environmentalists know, or claim to know, that action must be taken now to reduce greenhouse gas. How do they know? Basically through indicators. The United Nation Intergovernmental Panel on Climate Change has produced a huge amount of data showing both global and regional changes as well as the sources of these changes. Yet, not all decision-makers have been convinced yet of either what the situation is or of the causes of this situation. It appears to have been the case of the coalition of the unwilling, namely the USA, Canada and Australia, three countries which could have afforded to diminish the growth of their emissions of GHG between 1997 and 2007 but chose not to do it. To attract their attention, focusing events, crises and symbols are required. The attribution of the 2007 Nobel Peace Prize to the Intergovernmental Panel on Climate Change and to Al Gore is such a symbol. Catastrophe and crises are often what is needed to get decision-makers acting. Local catastrophes like the inundation of the city of New Orleans in the USA did attract the attention of the decision-makers in the USA to the state of the climate. In some other countries, like former units of the Soviet Union, economic disorganization diminished production and, as a consequence, relieved the stress on the environment. As a consequence, the urgency of the problem of gas emission diminished for a while. In some other countries GHG have never been a hot issue. The indicators of the UNFCCC describe a situation. A situation becomes a problem only if one sees a mismatch between his or her value and the situation. Even if the leaders of the world were convinced that the situation is a problem, solutions would not necessarily flow from it.

SOLUTIONS

The most obvious solutions to solve the problem of the greenhouse gas are not morally acceptable. Indeed, if the problem has its sources in the industrialization of the world and the tremendous rise of the population since the XVII century, then one solution could be to curb one or the other phenomenon or both. Obviously the «termination» of hundreds of millions of human beings is not an option. The denegation of the benefit of industrialization to mankind is not an acceptable proposition either. Furthermore, the migration of population from poorer and less polluting areas to richer but more polluting areas is highly valued. Which solutions are left to political leaders to choose from?

Decision-makers have to look at technical solutions, alternative sources of energy, less polluting techniques in extracting fossil fuel and transforming it in energy, less energy-consuming machines, tools and artefacts used for manufacturing, transportation, housing, and so on. But how to get there?

POLITICS OR BRINGING SOLUTIONS TO THE PROBLEMS

The charter of the United Nations (article 55) urges everybody to promote «higher standards of living, full employment, and conditions of economic and social progress and development». The control of the human environment should target these objectives.

Humans do already control much of what is on earth. The control of what is around the earth however is still in its infancy. The first great success is probably the control, if the word is not too strong, of the ozone layer. Few years after reaching an agreement on the problem and the solution, countries of the world have substantially reduce the production and dissemination of chlorofluorocarbon (CFC) which was seen as a major contributor to the depletion of the ozone layer. This has been possible because a less harmful product, hydrochlorofluorocarbon (HCFC), was available. In the case of the greenhouse gas the solution is trickier because the problem is even more complicated and the sources of the problem much more numerous.

One way to achieve the goal of reduction of greenhouse gas would be to fix a low target of gas emission, to regulate directly the polluters, inspect their behaviours with constancy and, when necessary punish the non compliant with very heavy fines. This would correspond to the «command and control» way to manage public policies. No jurisdiction has yet implemented a rigorous plan in this fashion.

Europe has followed another route. The European Union has started to implement a major tool for cutting carbon emission. The mechanism, called Emissions-Trading-Scheme or ETS, works roughly like this. Firms in the dirtiest industries are issued with permits to emit a certain amount of carbon dioxide; if they want to emit more pollution, they need to buy more permits. If they succeed in diminishing their emissions, they can sell permits accordingly. The over all effect should be to internalize the cost of producing whatever they are producing. When countries met at Bali, the European scheme had yet to produce a reduction of released pollutants. The price of aluminium, steel and cement, three big polluters, has not rise more in Europe than elsewhere, plants have not been closing more than elsewhere and, within a so short period between Kyoto and Bali, the plants producing these commodities have not been replaced either. Why did it not work? Because the allowances to pollute were so high that they didn't really trouble the industries. Furthermore allowances cannot be reduced in one area of the world, such as the European Union, if other jurisdictions do not implement a similar scheme. Otherwise plants in Europe will close to the benefit of plants in areas where standards are lower or where pollution is lower. For example, if emissions of CO_2 are now roughly 5 tonnes per person in China, 10 tonnes in Europe and 25 tonnes in the USA and Canada, moving big polluters from North America or from Europe to China may smooth distribution over areas and ease international tensions, but would not diminish the total amount of CO_2 released. The diminution of greenhouse gas is presumed to benefit all humans on earth which makes it a pure public good since nobody can be excluded from the benefit.

Hence the importance of the Copenhagen meeting where nobody is excluded. Politics, it is said, is the allocation of values. Who shall get what? Who should pay? The dominant ideology of the United Nations is democracy which should leave the power to decide to the majority of humans. If it were the case, this majority would, if properly informed, probably vote to have the rich pay. Two values would support this wish: the polluter pays principle and the democratic principle that redistribution of wealth would command the rich to pay a higher amount than the poor. Fortunately, as it has been observed, rich people value a clean environment and are ready to pay for it. Richer people want to *add* clean environment to what they already have, not to subtract from what they have. Hence the key solution in the relatively short term rests with technological advancement aimed at controlling the harm done to nature by industrialization.

One other part of the solution is the promotion of women rights and welfare which is also an objective of the United Nations charter article 55. When women get richer and gain control over their bodies, the growth of population tends to stabilize itself. Hence, humanity has a vested interest in making sure that girls get proper schooling and that women become equal to men in society. Such a social program would probably bring population growth at a more sustainable level than what is now projected. While the promotion of women's equality is in itself a highly praised value, it meets also, as a by-product, environmental objectives.

Industrialization can be controlled. Many technologies exist either to reduce the amount of CO_2 emitted with current automotive and industrial processes or to replace altogether the processes causing the gas emissions. In so far as gas emissions are not only a problem for the entire planet but also a local problem, rich people should easily be convinced to clean their own local environment. Furthermore, highly technically advanced societies do favour the research, development and implementation of new technologies which make them richer yet. The internalization of costs, polluter pays principle, appears to be one of the few possible ways to go.

Deforestation, which has also been highlighted as an important cause of GHG, will be more difficult to control because it depends on a long causal chain of events in economic development in the developing countries.

From Kyoto to Bali the situation of the atmosphere has been more and more perceived as a problem by more and more people. From Bali to Copenhagen solutions have to be found and agreed upon. Not an easy task. From Copenhagen to a cleaner atmosphere decades will need to go by.

49

CHAPTER

Looking Ahead from 2007

G.A. McBean
Institute for Catastrophic Loss Reduction
Departments of Geography and Political Science
The University of Western Ontario
1491 Richmond Street
London, ON, N6G 2M1, Canada
E-mail: gmcbean@uwo.ca

1. INTRODUCTION

The year 2007 was the tenth anniversary of the Kyoto Protocol's signing and the fifteenth anniversary of the signing of the United Nations Framework Convention on Climate Change (UNFCCC). If for no other reasons it would be an important year in the progress towards addressing global climate change. However, it was a very important year regardless of anniversaries because of a sequence of major events and milestones that occurred in 2007. It was the year of the Fourth Assessment Report of the Intergovernmental Panel on Climate Change that placed important, definitive and alarming scientific assessments before the global community – climate science conclusions could no longer be denied. 2007 was the year that the IPCC was awarded, jointly with former US Vice-President Al Gore, the Nobel Peace Prize, which gave climate change not only further recognition but also placed it clearly in the context of global peace and security. And 2007 was the year of the 13th Conference of the Parties under the UNFCCC and the 3rd Conference of Members of the Kyoto Protocol leading the important Bali Declaration and associated decisions. This short paper will put these in the context of where we are and where the global community may be going as we look ahead to the next decade.

2. THE CLIMATE SCIENTISTS REPORT

In February 2007, the IPCC Working Group I[1] on the physical science basis reported that global atmospheric concentration of carbon dioxide had increased to 379 ppm in 2005, higher than any value over the past many thousands of years. The planet has been warming with the linear rate of warming being 0.07°C per decade over the past 100 years and now increased to 0.18°C per decade over the past 25 years. Their analysis of global observations of the climate system led to the conclusion that:

> *"Warming of the climate system is unequivocal, as is now evident from observations of increases in global average air and ocean temperatures, widespread melting of snow and ice, and rising global average sea level."*

In attributing these changes, the IPCC reported that *"most of the observed increase in global average temperatures since the mid-20th century is very likely due to the observed increase in anthropogenic greenhouse gas concentrations."*

In April 2007, IPCC Working II on impacts, adaptation and vulnerability[2] added to the concern with their report that observational evidence from all continents and most oceans showed many natural systems being affected by regional climate change. From their assessment of data since 1970: *"it is likely that anthropogenic warming has had a discernible influence on many physical and biological systems."*

In essence, climate change is happening, it is affecting natural systems and human activities are the main cause.

Both Working Group reports then looked to the future. WGI noted that: *"For the next two decades, a warming of about 0.2°C per decade is projected for a range of SRES emission scenarios"* a small acceleration in the warming compared to the last 25 years resulting in about 1°C warming compared to the year 2000 or about 1.6°C warming compared to pre-industrial values. After mid-century, the climatic warming associated with different

[1] IPCC. 2007. Summary for Policymakers. In: Climate Change 2007: The Physical Science Basis. Contribution of Working Group I to the Fourth Assessment Report of the Intergovernmental Panel on Climate Change [Solomon, S., D. Qin, M. Manning, Z. Chen, M. Marquis, K.B. Averyt, M.Tignor and H.L. Miller (eds.)]. Cambridge University Press, Cambridge, UK and New York, NY, USA.

[2] IPCC. 2007. Summary for Policymakers. In: Climate Change 2007: Impacts, Adaptation and Vulnerability. Contribution of Working Group II to the Fourth Assessment Report of the Intergovernmental Panel on Climate Change [M.L. Parry, O.F. Canziani, J.P. Palutikof, P.J. van der Linden and C.E. Hanson (eds.)]. Cambridge University Press, Cambridge, UK, 7-22.

scenarios diverge, with warming by the end of century of ranging from 1.5°C to 4°C compared to 2000, or about 2-5°C warmer than pre-industrial global temperature. For any of the scenarios, warming continues for centuries to follow.

For each of these scenarios there is the scientific uncertainty which is typically +/- 1°C or more[3]. Including this uncertainty, the difference in possible temperatures by the end of the century ranges from about 1.5-7°C relative to the pre-industrial global temperature. This range is what might be called the "human choice" or "our" impact.

With this century's climatic changes will likely come more intense tropical cyclones with larger peak wind speeds and more heavy precipitation. Drought-affected areas will likely increase in extent and more frequent heavy precipitation events will augment flood risk. Approximately 20-30% of plant and animal species assessed so far are likely to be at increased risk of extinction if increases in global average temperature exceed 1.5-2.5°C. They also noted that poor communities can be especially vulnerable since they tend to have more limited adaptive capacities, and are more dependent on climate-sensitive resources such as local water and food supplies.

Based on this kind of information, European Union and some others have adopted a 2°C target, 2°C warmer than pre-industrial global temperature as the target for avoiding dangerous climatic change. However, since the publication of the IPCC Reports, the IPCC Chair, R. Pachauri, has stated: "*People are actually questioning if the 2°C benchmark that has been set is safe enough.*" He went on to note that "*Rising temperatures this century could bring risks for the extinction of up to 30% of the world's species. A creeping rise in sea levels could threaten Pacific islands and many coastlines.*"

The Working Group III on mitigation[4] of climate change through emission reductions was the last to report, in May 2007. Their analysis showed that there was "*substantial economic potential for the mitigation of global GHG emissions over the coming decades that could offset the projected*

---

[3] Meehl, G.A., T.F. Stocker, W.D. Collins, P. Friedlingstein, A.T. Gaye, J.M. Gregory, A. Kitoh, R. Knutti, J.M. Murphy, A. Noda, S.C.B. Raper, I.G. Watterson, A.J. Weaver and Z.-C. Zhao. 2007. Global Climate Projections. In: Climate Change 2007: The Physical Science Basis. Contribution of Working Group I to the Fourth Assessment Report of the Intergovernmental Panel on Climate Change [Solomon, S., D. Qin, M. Manning, Z. Chen, M. Marquis, K.B. Averyt, M. Tignor and H.L. Miller (eds.)]. Cambridge University Press, Cambridge, UK and New York, NY, USA.

[4] IPCC. 2007. Summary for Policymakers. In: Climate Change 2007: Mitigation. Contribution of Working Group III to the Fourth Assessment Report of the Intergovernmental Panel on Climate Change [B. Metz, O.R. Davidson, P.R. Bosch, R. Dave and L.A. Meyer (eds.)], Cambridge University Press, Cambridge, UK and New York, NY, USA.

growth of global emissions or reduce emissions below current levels". Looking to 2030, the economic models projected economic costs between a 3% decrease and a small increase in global GDP with significant regional differences in costs. Looking towards longer term stabilization targets, the IPCC noted that emissions would need to peak and then decline, with the lower the stabilization level, the more quickly this peak and decline would need to occur. Emissions increases over the next two to three decades will have a large impact on opportunities to achieve lower stabilization levels.

In October, 2007, the Global Carbon Project[5] reported on global fossil fuel emissions up to 2006. The report was not encouraging. During the 1990's, global emissions had been increasing at 1.3% $year^{-1}$; for the period 2000-2006, the rate of increase was 3.3% $year^{-1}$. In 2006, global emissions were 8.4 GtC (billions of tonnes carbon or carbon equivalents), which was higher than even the most pessimistic of the IPCC emission scenarios. Clearly emission reduction strategies have been ineffective on a global scale.

3. THE NOBEL PEACE PRIZE AND CLIMATE CHANGE

On October 12, 2007, the Norwegian Nobel Committee awarded the Nobel Peace Prize for 2007 to the Intergovernmental Panel on Climate Change (IPCC) and Albert Arnold (Al) Gore Jr. *"for their efforts to build up and disseminate greater knowledge about man-made climate change, and to lay the foundations for the measures that are needed to counteract such change."* The Norwegian Nobel Committee noted that it *"is seeking to contribute to a sharper focus on the processes and decisions that appear to be necessary to protect the world's future climate, and thereby to reduce the threat to the security of mankind. Action is necessary now, before climate change moves beyond man's control."*

4. THE ROAD TO BALI

Within Canada, the occurrence of many weather-related events, the IPCC reports and other factors had resulted in an upsurge in concern for the environment with polls showing the environment as the highest single issue of concern doing the last half of 2007[6]. On September 7, 2007, Prime

---

[5] Raupach et al., 2007, PNAS; Canadell et al., 2007, PNAS.

[6] Harris Decima poll of August 2007 had environment being selected by 30% of Canadians as the most important issue facing Canada. Similar results were seen in Ipsos-Reid polls in February and July 2007.

Minister Stephen Harper speaking to the APEC Business Summit, in Sydney, Australia, stated: "*...one of the most important international public policy challenges of our time: the growing menace of climate change. The weight of scientific evidence holds that our atmosphere is getting hotter, that human activity is a significant contributor, and that there will be serious consequences for all life on earth.*"

He then went on to say that: "*When I say that we must balance environmental protection and economic prosperity, I do so quite deliberately*" and on Sept. 24, 2007, he announced that Canada would be formally joining the Asia-Pacific Partnership on Clean Development and Climate, the U.S.-led group with China, India, Australia, South Korea, Japan and now Canada. The Asia-Pacific Partnership has been criticized by many for its lack of specific targets and as being parallel to and probably weakening the Kyoto Protocol[7].

Prime Minister Howard was the host of the APEC meetings and a long time opponent of the Kyoto Protocol; Australia and the United States were the only two Annex I countries that signed but did not ratify the Kyoto Protocol. Soon after the Australian general election was held and Howard was defeated and the new government announced soon after that it would ratify the Kyoto Protocol. This was the first national election where climate change was a major issue.

The road to Bali included other intergovernmental meetings. UN Secretary-General Ban Ki-moon had made climate change a major issue for the UN and a high-level meeting with the participation 80 heads of state or government was held on 24 September 2007. The need to halve emissions by 2050 in order to limit temperature rise to 2°C was supported by many countries.[8] A few days later, representatives from 16 major economies were hosted in Washington for discussions on a new post-2012. A month later, almost 40 environment ministers gathered in Bogor, Indonesia and agreed on general building blocks of mitigation, adaptation, technology, and investment and finance as the core of a post-2012 framework. Notably they agreed that equal weight must be given to adaptation and mitigation.

[7] CBC News. Kyoto alternative. What is this new Asia-Pacific Partnership all about? Last Updated September 27, 2007.

[8] Earth Negotiations Bulletin. A Reporting Service for Environment and Development Negotiations, Online at http://www.iisd.ca/climate/cop13/Published by the International Institute for Sustainable Development (IISD).

5. THE 13TH CONFERENCE OF THE PARTIES AND THE BALI ACTION PLAN

Coming into the meetings, the UNFCCC Secretariat presented basically a report card[9] on how the Parties of the Kyoto Protocol were doing with respect to emissions reductions compared the base year of 1990. National reports for 2003 or 2004 showed a wide range. Within the former Soviet Union and related states, the emissions have greatly reduced due to economic changes, with Lithuania having the largest decrease (–66.2%) and the Russian Federation being –33%. The European Community, as a whole, reporting a decrease, –1.7%, based on a range from Spain and Portugal both being near 40% increase while the two largest emitters in an absolute sense both reported significant reduction Germany (–16.7%) and the United Kingdom (–14.6%). Two other major emitters did not do well: Japan at 8.3% increase and Canada at 26.5% increase.

The 13th Conference of the Parties (COP) to the UNFCCC and the 3rd Meeting of Parties to the Kyoto Protocol was held in Bali, 3-14 December 2007, with a major focus being post-2012 regime. At the beginning there were marked differences in views. Australia, speaking for the Umbrella Group, called for a comprehensive global agreement including a long-term aspirational goal to which all can contribute. Pakistan, for the G-77/China, emphasized an approach based on key principles stated in the Convention and Protocol. The Association of Small Island States representative noted the impacts of sea level rise on small island developing states and stressed the need for comprehensive global response leading to stabilization well below 445 ppm. Nigeria, for the African Group, urged developed countries to fulfill existing commitments. Switzerland, for the Environmental Integrity Group, highlighted IPCC AR4 and called for urgent action while the Maldives, representing the Less Developed Countries, focussed on the Adaptation Fund and its needs for funds; Bangladesh supported this. Portugal, on behalf of the European Union, said growth in global emissions must be halted in the next 10-15 years and urged a comprehensive global agreement by 2009.

In the end, after much, sometimes acrimonious debate, there was agreement on the Bali Action Plan[10]. The key outcomes were the launch of a process to lead to decisions at the 15th COP in 2009 based on *"(a) a shared vision for long-term cooperative action, including a long-term global goal for*

---

[9] Bali Documents – CMP.3 – Demonstration on progress in achieving commitments under the Kyoto Protocol by Parties included in Annex I to the Convention. See www.unfccc.int

[10] See the UNFCCC website for the documents and reports – www.unfccc.int. See the Earth Negotiations Bulletin for reports on sessions and analysis.

emission reductions, to achieve the ultimate objective of the Convention, in accordance with the provisions and principles of the Convention, in particular the principle of common but differentiated responsibilities and respective capabilities, and taking into account social and economic conditions and other relevant factors; (b) Enhanced national/international action on mitigation of climate change, including, inter alia, consideration of: (i) Measurable, reportable and verifiable nationally appropriate mitigation commitments or actions, including quantified emission limitation and reduction objectives, by all developed country Parties, while ensuring the comparability of efforts among them, taking into account differences in their national circumstances; (ii) Nationally appropriate mitigation actions by developing country Parties in the context of sustainable development, supported and enabled by technology, financing and capacity-building, in a measurable, reportable and verifiable manner."

This wording for developed and developing countries was a point of considerable negotiations. In order to move forward, an Ad Hoc Working Group on Long-term Cooperative Action was established to complete its work in 2009 and present the outcome of its work to 15th COP. It was also agreed to move effectively ahead with the first session of the Working Group to be held no later than April 2008.

So the Bali roadmap has been laid out, without the specific targets that many countries wanted but with at least agreement in principle on where the global community is going – the question is how far and how fast.

The IPCC had stated that emission cuts in the range of 25-40% by Annex I parties by 2020 are needed to meet stabilization targets. The EU and developing countries favoured including reference to this in the Declaration but the US, Canada, Japan and the Russian Federation were opposed based on it being to prescriptive. In the end, it was not included.

6. LOOKING AHEAD

However, there are encouraging signs. The Bali Action Plan included both *"quantified emission limitation and reduction objectives, by all developed country Parties"* and *"appropriate mitigation actions by developing country Parties"*. Thus, developing countries agreed, with provisions, to undertake mitigation actions. In the United States, many of the candidates in the Presidential race have made strong statements on the need to address climate change and perhaps more directly important are the actions being taken at the state level. More than one-half of the states have adopted targets[11] and the US cap-in-trade legislation may pass after the 2008 presidential election. What emission reductions will be achieved by 2020

[11] USA Today 21 January 2008 – front page story – www.usatoday.com

and 2050 is not yet clear. The Prime Minister of Japan[12], speaking at the Science and Technology in Society Forum in Kyoto in October, 2008, proposed a 50% reduction by 2050 target. The Canadian government's plan, called the Turning the Corner Action Plan is to reduce greenhouse gas emissions by 20% by 2020 and 60-70% by 2050, but against a 2006 base level, when Canadian emissions were more than 27% higher than the internationally agreed base year of 1990. Hence, a 60% target for 2050 is a 33% target for a comparable baseline as used by Japan and most other countries.

Another step forward, in a sense, is the recognition of the unfortunate reality that the climate is changing now and will continue to do so for decades to come, independently of emission reductions. The Bali Action Plan included *"Enhanced action on adaptation"*, with *(i) international cooperation to support urgent implementation of adaptation actions (ii) Risk management and risk reduction strategies... (iii) Disaster reduction strategies ...(iv) Economic diversification to build resilience"* as steps towards addressing the very significant impacts that climate-related hazards are having and will continue to have in the developing world.

It was also agreed technology development and transfer of information for both mitigation and adaptation was needed. This will be coupled with *"(e) Enhanced action on the provision of financial resources and investment to support action on mitigation and adaptation and technology cooperation..."*.

7. SOME LAST THOUGHTS

The next two years, until COP15 in Copenhagen in late 2009, will be very busy and critical times for climate change negotiators. Hopefully they will listen to and rely on science (natural, social, economic, health,...) and technological expertise in deciding what is needed and what is possible; and listen to people and see the already occurring impacts around the world. The world awaits with concern what they will put forward and what governments will agree to in 2009. It is not very far away. At least we now see climate change positioned rightfully as an issue of international security and one of international and intergenerational equity and ethics. We need action now for our grandchildren and children and grandchildren around the world.

[12] STS-Forum – see www.stsforum.org

50

CHAPTER

Post-2012 Institutional Architecture to Address Climate Change: A Proposal for Effective Governance

Norichika Kanie
Associate Professor, Department of Value and Decision Science
Graduate School of Decision Science and Technology
Tokyo Institute of Technology
Tokyo, Japan
E-mail: kanie@valdes.titech.ac.jp

INTRODUCTION

In the past few years, there have been dramatic changes in the international political situation surrounding climate change, which is in the process of transformation from being a "low politics" to a "high politics" issue—placing it at the center of international politics. Since the Group of Eight (G8) Gleneagles Summit hosted by the United Kingdom in 2005, climate change has become one of the most important topics addressed at this annual meeting of heads of state of the eight leading economies, and indeed, host country Japan announced that climate change will be the most prominent topic at the G8 Hokkaido Toyako Summit in 2008. On April 17, 2007, the United Nations Security Council discussed the issue of climate change for the first time. In early September that year, the Asia-Pacific Economic Cooperation forum (APEC) also addressed climate change as a key topic, and it was also the theme of the Special Session of the United Nations General Assembly at the end of that month. The selection of Nobel Peace Prize laureates in 2007 is a clear symbol of these developments, as the award went jointly to Al Gore, who had awakened public concern about the issue with his documentary film "An Inconvenient Truth", and the Intergovernmental Panel on Climate Change (IPCC). It would not be exaggerating to say that climate change is now recognized as one of the most important international political issues today.

An overview of overall international political trends of recent years makes it clear that the political importance of climate change has made it a key topic in a great number and diversity of political fora. No one should forget that the United Nations Framework Convention on Climate Change (UNFCCC) and the Kyoto Protocol are the international regimes at the center of these various consensus-building processes. As reaffirmed by both the G8 Summit and the United Nations General Assembly in 2007, it is clear that universal global fora are most suitable as the point of convergence for solving problems, because the issue of climate change is a universal global issue that knows no national borders.

At first glance, from the perspective of economic efficiency, one may get the impression that greater impacts can be achieved from economic fora such as the G8, APEC, or MEM (Major Economies Meetings on Energy Security and Climate Change). Or perhaps from bilateralism and "mini-lateralism" (relations between small numbers of partners) approaches that might consider the activities of the major economic powers (currently the major emitters) and seek to achieve balance in the international economic competitiveness of the major emitters. Because their primary focus is on coordinating economic interests between major economic powers, however, one cannot deny that these fora tend emphasize the *economic* dimensions of climate change rather than on the real essence of the problem—*preventing dangerous climate change*. Meanwhile, if we look at the essence of the problem, the logical conclusion is that multilateralism provides the most important and most effective fora for discussion, as can bring together not only the present and past emitters of greenhouse gases (GHGs), as one would expect, but also the countries that are vulnerable to damage, as well as developing countries whose emissions will grow in the future.

Multilateralism takes time to build consensus, and developed countries may often become frustrated by debates about global equity concerns, but problems will not be solved if we avoid looking at the current realities of international society.[1] While fully acknowledging that the UNFCCC and

[1]See for example Ruggie (1993), Zartman (1994).

Kyoto Protocol framework are the right fora to determine the international framework to respond to climate change, we should make use of a myriad of initiatives under that umbrella, functioning as supplementary mechanisms to promote further responses to climate change. As explained below, this institutional framework makes good sense if we also consider recent scholarship on institutions. It is also necessary to consider this in the context of the coordination of economic interests. The commonly-used phrase of "balance between economy and the environment" is important, but it would be misguided to use this phrase simply as an excuse to justify an overemphasis on economic activities when dealing with the environment. We must admit that the problems of energy and climate change are two sides of the same coin, and when discussing climate change responses, we must give the greatest priority to the *prevention of dangerous climate change*.

Based on the above points, I offer the following proposal. The main four points are listed below.

1) Set long-term aspirational targets: global (e.g., 50% reduction by 2050), developed countries (e.g., 80% reduction by 2050, etc.). [Non-binding long-term targets, essentially to positively influence technology development]

- Establish scientific dialogue processes to help meet those targets (international dialogue processes on emissions reduction models, emissions allocation models, possibly as G8 or UNFCCC processes).

2) Set 2020 targets and allocations for developed countries [Binding, short-term targets, essentially to positively influence policy]

- Next commitment period is the eight years to 2020. Developed countries set targets for all developed countries, and also set country-specific short-term reduction targets. These will ultimately be decided through international negotiations.
- In principle, the commitment period is the next eight years. Targets for subsequent commitment period are to be agreed three years before that period begins.
- Establish a dialogue process between science and policy on reduction targets, to indicate (as basis for negotiations) the range of necessary ("top down" approach) and potential emission reductions ("bottom up"). Process also produces indicative 2030 targets, as reference for future reductions/allocations/negotiations.
- Clearly state criteria for Annex I countries as being OECD members (Korea and Mexico to be included).

3) Set sector-specific emission reduction targets and performance targets (policies and measures) for specific sectors in major emitting countries (both developed and developing). [Outside the UNFCCC, binding among members to the agreement]

- Implement sector-specific initiatives in frameworks outside the UNFCCC and Kyoto Protocol (for example, based on the Major Economies Meeting on Energy Security and Climate Change [MEM] or Asia-Pacific Partnership on Clean Development and Climate [APP]), but require that the status of implementation and results be reported to the UNFCCC.
- In developing countries, count sectoral portions the country's emission reduction target as a sectoral target of the country (to the extent that a given sector meets its sector-specific commitments, it can be excused from being bound by the national target).
- Similarly, sectors in developing countries that are major emitters set new sector-specific targets alongside schemes that provide incentives for technology development, technology transfer, etc.

4) Developing countries set emissions reductions targets for developing countries as a whole (excluding Least Developed Countries). As for performance, they make reduction efforts based on selection of no-lose targets or SD-PAM depending on capacity. [Non-binding]

- Developed countries provide assistance for policy implementation in both cases.
- Continue a variety of processes (partnerships) outside the UNFCCC. Establish a body to quantitatively assess country-specific reduction efforts in a variety of forums. The Japanese government should propose that such an institution be established under the UNFCCC (or SBSTA), UNEP or G8.

1.1 COMMITMENTS (TARGETS)

1.1.1 Long-Term Targets

The current institutional architecture has the shortcoming of focusing on a short-term time scale, as it lacks measures to promote models that will encourage long-term technology development and the creation of a low-carbon society as a long-term challenge. If we consider the "chain of innovation," we see from previous studies that the research and development phase takes from one to several decades and diffusion takes at least a decade, while service life of equipment and products is from a few years to over three decades (Figure 1, Table 1).

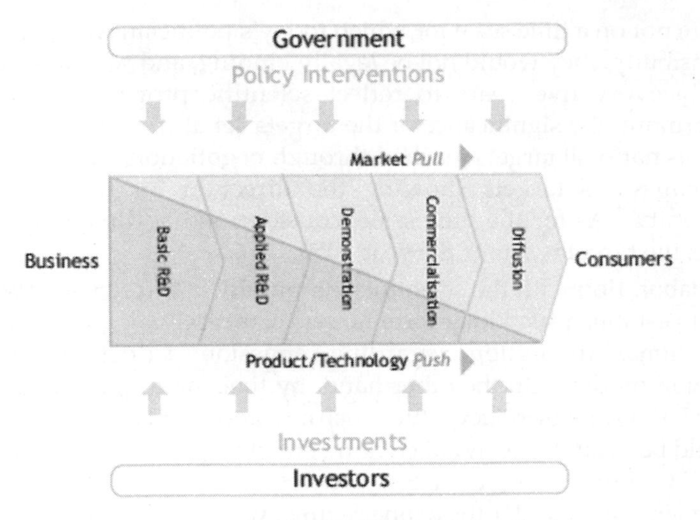

Figure 1 Chain of technology innovation.
Source: J.T. Foxon (2003), p. 18.

Table 1 Service life and replacement period of key items

Service Life(Equipment & Machinery)		*Replacement Period(Consumer Durables)*	
Asset	*Service life (years)*	*Item*	*Average use(years)*
Office and computing	7	Electric refrigerator	10.4
Steam engines and turbines	32	Room air conditioner	10.4
Internal combustion engines	8	Color television	9.4
Metalworking machines	16	Electric clothes washer	8.6
Electrical transmission, distribution, and industrial apparatus	33	Passenger car (new)	7.0
Aircraft	20	Video camera	6.0
Ships and boats	27	Personal computer	4.6
Railroad equipment	28	DVD player/recorder	4.4
Industrial building	31	Digital camera	3.5
Office building	36	Mobile phone	2.7

Sources: Fraumeni (1997), Matsumoto et al. (2007), Cabinet Office Statistics of Consumption Trend Study (2007), other materials.

In other words, if we consider technological factors, a time scale of 20 to 30 years is appropriate for dealing with climate change. This contrasts with the period for which today's politicians can take political responsibility — typically about eight years at most.

To compensate for this gap, long-term targets should be established at the global level, and these should include long-term targets for developed countries as a whole. These would be aspirational targets, and because they are not on a time scale for which today's politicians can take political responsibility, they would not be legally binding, and would be reviewed roughly every five years to reflect scientific progress (Kanie 2005). Furthermore, the significance of the targets set at the global level would be not as national targets reached through negotiation, but rather as long term emissions targets showing the direction for the international community.[2] As for the time scale, considering the direction of current debate, the target is set for the year 2050.

Collaboration with the scientific community is extremely important in order to establish such long-term targets. Currently, no process exists for international discussion regarding emission reduction models or allocation models. On the other hand, by their nature, these models are highly influenced by policy. Thus, such an international discussion process it would be desirable to avoid creating an inter "governmental" panel like the IPCC, but rather, to use a process similar to the model of the assessment panel under the ozone regime, which only considers scientific contributions (and geographical balance).

For example, it might be possible for G8 Summits or other fora to propose a process that brings together scientific knowledge (e.g., models for proposed emission targets and burden-sharing schemes for climate stabilization), present a range of targets, and present options for consideration in negotiation processes. Such an effort could be a valuable initiative, in the sense of moving one step further with processes like one to define "dangerous climate change" (at the start of its G8 Presidency in 2005, the United Kingdom hosted an international conference titled "Avoiding Dangerous Climate Change," the year of the G8 Gleneagles Summit)[3], and in the sense of Japan leading the way to new progress in international discussions. Another approach might be for the UNFCCC Secretariat and UNEP to jointly promote such a process.

1.1.2 Commitments of Developed Countries: Short-Term Targets

For the reasons explained above, the establishment of short-term targets for developed countries is a commitment that cannot be avoided if momentum is to be maintained in climate change policies today. The eight years until 2020 would be an appropriate commitment period, for a number of reasons: political acceptability (politicians cannot commit to an excessively long commitment period); economic acceptability (long-term

[2]For individual countries, it is recommended that each country voluntarily establish its own targets.

[3]http://www.direct.gov.uk/en/Nl1/Newsroom/DG_10031725

commitment periods are preferable for predictability of investments and business as well as signals to the market); and consideration of various criticisms about the Kyoto Protocol. Developed countries would establish emission reduction targets for the year 2020 for all developed countries overall, and allocate the reductions within that frame. These quantified targets would be legally binding, and implemented by amending Annex B of the Kyoto Protocol. The criteria for "developed countries" referred to here would be clarified as the current Annex B countries and all other member countries of the OECD. Accordingly, Korea and Mexico would be added to this group. These countries would establish binding national targets similar to those in the Kyoto Protocol. Short-term emission reduction targets would be established for developed countries overall, and legally binding short-term national reduction targets would also be established.

For these emission reduction targets, it is important to consider the amount of reduction necessary in terms of environmental effectiveness (so-called top-down approach), but it is also important to consider the potential for emission reductions in terms of economic acceptability and fairness among developed countries (so-called bottom up approach). Therefore, both approaches should be considered when establishing the targets. Other existing approaches to calculations would be worth considering, such as the Triptych approach used by the European Union in discussions about the distribution of intra-regional emission reductions for Kyoto Protocol negotiations. Basically, for consideration of emission reduction amounts for all developed countries overall and for each individual country, this approach considers the relationship of *possible* emission reductions to *required* emission reductions (Philipsen et al. 1998). For reference, we consider the results of global calculations of the Triptych approach by Groenenberg et al. (2001).

Ultimately, targets should be set through international negotiations, because a large number of factors are considered in the targets and political considerations are also involved. Nevertheless, the need for indicative target numbers for discussion purposes, based on scientific considerations, is the same as described earlier in the context of long-term targets. It is important that the scientific parallel process produce a target range: both the *required* emission reductions ("top down" approach) and the *potential* emission reductions ("bottom up"). After considering these figures, countries could, of course, proceed with negotiations for setting national targets, and countries that are capable could make pledges for additional reductions. Because international opinion and other factors will also be reflected in this process, the ultimate quantified targets will end up at the compromise point where science and politics interact. For quantified targets, it is also important to ensure institutional continuity and a sense of the future prospects for targets. Therefore, by providing indicative emission reduction targets for the year 2030, the scientific process should offer reference values for future emission allocations, and these should be presented in advance for consideration in negotiations for the subsequent commitment period. In principle, the commitment period should be the eight years from 2021 to 2028, and targets for the subsequent commitment period should be agreed three years before that period begins. Considering the fact that the COP meetings have customarily been held near the end of each year, this means that the time frame of 10 years is secured considering lead time toward commitments for the next period, so this is probably a suitable compromise considering political, economic and technological acceptability.

1.1.3 Commitments of Both Developed and Developing Countries

Binding sector-specific emission reduction targets and performance targets (i.e., policies and measures) should be established for a number of sectors in countries that are major emitters, regardless of whether they are in developed or developing countries (e.g., aluminum, cement, coal mining, power generation and transmission, steel, etc.—the sectors covered by the APP).[4] The countries considered to be "major emitters" would differ depending on the sector, it would be reasonable to include the G15 or G20 countries. Admittedly, the selection of those countries was subjective, without precise selection criteria. Thus, sector-specific initiatives should be implemented outside UNFCCC and Kyoto Protocol framework, with the MEM and APP, or if they are terminated, any subsequent framework to these being important candidates as host. The main reasons for proposing these initiatives being outside the UNFCCC and Kyoto Protocol framework include the following: (1) the focus on major emitting countries is not suitable for multilateralism; (2) the participating countries themselves are not universal, so if this is implemented as a multilateral solution, it would be used for a source of bargaining; (3) it would be more effective to make the most of the strengths of the decentralized network systems that are currently being formed; and (4) what is being proposed here is basically a *partnership* approach rather than an *inter-national* approach involving national governments. That

[4]A useful approach to examine sector-specific initiatives is to use a matrix of policies and measures, as shown in the policy matrix under Category I in Sawa (2007).

being said, in order to effectively promote networking, there should be a reporting requirement to the UNFCCC Secretariat on the status of implementation and quantitative outcomes.

In developed countries, under the national emission reduction targets, the allocations for the relevant sectors should be counted as this part of sectoral commitment (in other words, to the extent that a given sector meets the core commitments, it could be excused from obligations to meet national targets). In developing countries that are major emitters, this would mean that sectors set new sector-specific targets and performance targets that take into account technology development, technology transfer schemes, and so on.

Sectors (and the relevant authorities) in developed countries can anticipate that by working with sectors in developing countries they can maintain international competitiveness, while sectors in developing countries can expect to benefit from technology transfers through the international framework. Environmental effectiveness is ensured by a cap on emissions from sectors in developed countries under the UNFCCC. Considering the current institutional architecture, this framework can be expected to function satisfactorily outside the UNFCCC, and to be effective in terms of institutional flexibility (new sector-specific targets can be established separately as new technologies are developed).

1.1.4 Commitments of Developing Countries

On the assumption that developed countries will achieve their Kyoto Protocol targets, developing countries overall should establish their own aspirational targets for emissions in the year 2020. These targets would not be binding. The following options would be available from which to choose.

- No-lose targets

  As in Argentina in the years around CoP4, for countries that can voluntarily establish targets and have the capacity to monitor national emissions (i.e., countries about to enter the ranks of developed countries), no-lose targets would only be beneficial and would have no downside. This approach would also lead to further capacity development, and for countries experiencing significant economic development, it would constitute a form of preparation to "graduate" by accepting the same commitments as those of developed countries. Because countries without the capacity to make these types of commitments would not choose this option, however, it is important to note that disparities in capacity could grow larger between countries.

- Sustainable Development Policies and Measures (SD-PAMs)
  Some countries lack the capacity to establish no-lose targets and also strongly resist the idea of considering targets. Many countries are seeking official support from developed countries in the form of official development assistance (ODA) and so on. For countries with domestic circumstances that have not progressed to the point of establishing no-lose targets, a suitable option would be to work to reduce emissions by implementing SD-PAMs.

Generally, the first option (no-lose targets) is preferable, but because this would require the capacity to develop emission targets, the option of SD-PAMs is offered as an alternative choice. To facilitate implementation, developed countries would offer capacity development and technical assistance. In terms of environmental effectiveness there is no question that it would be desirable to have developing countries set binding targets. It is difficult to expect that developing countries will establish binding targets for the next commitment period, if we consider (1) that the UNFCCC principle of common but differentiated responsibilities is reflected in Annex I, and the G77 and China coalition have strong solidarity, and (2) that in the process from the UNFCCC to the Kyoto Protocol, developed countries initially proposed non-binding targets, but it was decided to turn to binding targets after it became clear that the non-binding targets would not be met.

In addition, it is essential to note that "developing country" is a catch-all term. The "G77 plus China," for example, consists of over 130 countries that obviously have a wide range of circumstances in terms of their stages of economic development, level of education, culture, geopolitical context, international political position, and so on. When considering measures, therefore, it is important to keep the above points in mind and leave some room for flexibility.

That said, however, it will be important to evaluate to what extent developing countries make an effort to mitigate emissions when it comes to consideration about limiting future emissions, and also when we think about environmental effectiveness, especially for major emitters. We live in a world in which international relations are ultimately based on the sovereign state, and the fundamental role of international regimes and international institutions is raising the capabilities of participating countries through information exchange and the creation of norms. It is exactly this kind of role that is the essence of what international institutions have to offer.[5] Thus, I propose that a new system be established

[5]For example, refer to P.M Haas, R.O. Keohane and M.A. Levy (1993)

within the UNFCCC Secretariat or UNEP (or possibly within the OECD) to provide a clearinghouse function for information relating to GHG emission reductions under a variety of frameworks relating to climate change, and that it organize and publish data on a country-by-country basis. Organizing the data on a country-by-country basis will facilitate comparisons of GHG emission reductions in a variety of programs and frameworks, this may help to create pressure on countries, sectors, and programs that have not made adequate efforts to reduce emissions. Furthermore and frameworks, this kind of assessment will facilitate flexible thinking about the architecture of the post-2012 climate institutions.

Conclusion

This paper focused on the issue of emission reduction and limitation commitments as a part of the post-2012 institutional architecture on climate change. They play a major role in the design of a regime and have been a key aspect of negotiations since the negotiations for the Kyoto Protocol, although there are admittedly many other factors besides commitments to consider in the post-2012 institutional architecture on climate change. Indeed, in the actual international negotiations, it seems likely that some type of compromise will be sought on issue linkages and trade-offs with factors other than commitments. Nevertheless, whatever the practicalities of actual negotiations may be, when considering what would create the most effective future institutions on climate change it is entirely worthwile to examine each factor separately. In fact, each factor in negotiations could be subject to the type of analysis presented in this paper or even more rigorous examination. These topics could include, for example, the design and role of carbon markets (emissions trading, CDM, etc.), the design and details of technical cooperation and transfers, the design of financial assistance (multilateral or bilateral funding flows, including those for adaptation measures), the treatment of forests, institutional frameworks for adaptation, and so on. Discussion of these factors will be left for future consideration.

References

Aggarwal, V.K. 1998. *Institutional Designs for a Complex World*. Cornell University Press, USA.

Aldy, J.E., S. Barrett and R.N. Stavins. 2003. "Thirteen Plus One: A Comparison of Global Climate Policy Architectures." *Climate Policy* 3 (4): 373-397.

Aldy, J.E. and R.N. Stavins. 2007. "Architectures for an international global climate change agreement: lessons for the international policy community." In: J.E.

Aldy and R.N. Stavins (eds.) *Architectures for Agreement: Addressing Global Climate Change in the Post-Kyoto World.* Cambridge: Cambridge University Press, UK.

Ansell, C.K. and S. Weber. 1999. "Organizing International Politics." *International Political Science Review*, January 1999.

Bodansky, D. 2004. *International Climate Efforts Beyond 2012: A Survey of Approaches.* Pew Center on Global Climate Change.

Foxon, J.T. 2003. "Inducing Innovation for a low-carbon future drivers, barriers and policies."*The Carbon Trust.*

Fraumeni, B.M. 1997. The Measurement of Depreciation in the U.S. National Income and Product Accounts. Survey of Current Business.

Groenenberg, H., D. Philipsen and K. Block. 2001. "Differentiating commitments world wide: global differentiation of GHG emissions reductions based on the Triptych approach - preliminary assessment." *Energy Policy* 29, 1007-1030.

Haas, P.M., R.O. Keohane and M.A. Levy. 1993. *Institutions for the Earth: Sources of Effective International Environmental Protection.* The MIT Press, Cambridge, Massachussetts, USA.

Haas, P.M., N. Kanie and C.N. Murphy. 2004. "Conclusion: Institutional design and institutional reform for sustainable development." In: N. Kanie and P.M. Haas (2004).

Hashimoto K., A. Fujimoto, A. Umeda, K. Masui, N. Kondo, K. Matsumoto and H. Tsuchiya. 2007. "Study of technology selection models for global warming countermeasures" (in Japanese). Research Center for Advanced Science and Technology. The University of Tokyo.

Kanie, N. 2005. "Establishment of medium and long-term targets and issues for international acceptance: global greenhouse gas emission reductions and Japan's targets," (in Japanese). *Environmental Research*, No. 138, pp. 84-92.

Kanie, N. 2006. "International regime beyond the Kyoto Protocol: Toward decentralized climate change governance" (in Japanese). *Kokusai Mondai* (International Issues). No. 552, pp. 47-59.

Kanie, N. 2007. Creating an international order for climate security: How the environment became 'high politics' (in Japanese). *Gendai Shiko* (Contemporary Thought), October 2007.

Kanie, N. and P.M. Haas. 2004. *Emerging Forces in Environmental Governance.* New York and Tokyo: UNU Press.

Kanie, N. and K. Morita. 2005. "Triptych approach and multisectoral convergence approach" (in Japanese). Y. Takamura and Y. Kameyama (eds.), in *Direction of Global Warming Negotiations.* Daigaku Zusho, pp. 200-206.

Ostrom, E. 2001. "Decentralization and Development: The New Panacea." In: K. Dowding, H., Hughes and H. Margetts, *Challenges to Democracy: Ideas, Involvement and Institution.* Palgrave Publishers, pp. 237-256.

Philipsen, G.J.M., J.W. Bode, K. Blok, H. Merkus and B. Metz. 1998. "A Triptych sectoral approach to burden differentiation; GHG emissions in the European bubble." *Energy Policy.* Vol. 26, No. 12, pp. 929-943.

Rabe, B.G. 2004. *Statehouse and Greenhouse: The Emerging Politics of American Climate Change Policy,* Brookings Institution.

Ruggie, J.G. 1993. *Multilateralism Matters: The Theory and Praxis of an Institutional Form.* NY: Columbia University Press.

Sawa, A. 2007. "Proposal for a new Post-Kyoto Protocol framework." In: *Japanese strategies and international cooperation policy for Post-Kyoto Protocol,* an interim report of a research project of the Twenty-First Century Policy Research Institute (in Japanese).

Usui, M. 2003. "Sustainable Development Diplomacy in the Private Business Sector: An Integrative Perspective on Game Change Strategies at Multiple Levels." *International Negotiation* 8: 267-310.

Usui, M. 2004. "The private business sector in global environmental diplomacy." In: Kanie and Haas (2004): 216-259.

Zartman, I.W. 1994. *International Multilateral Negotiations: Approaches to the Management of Complexity,* Jossey-Bass Publishers, San Francisco, USA.

51

CHAPTER

A Gender-Sensitive Climate Regime?

Ulrike Röhr[1] and Minu Hemmati[2]
[1]Genanet – focal point gender justice and sustainability
LIFE e.V., Dircksenstr. 47, D-10178 Berlin, Germany
E-mail: roehr@life-online.de
[2]Ansbacher Str. 45, 10777 Berlin, Germany
E-mail: minu@minuhemmati.net

COP13 in Bali: from a gender perspective, a significant breakthrough was achieved.

For the first time in UNFCCC history, a worldwide network of women, **gendercc - women for climate justice,** was established. The group published several position papers articulating the women's and gender perspectives on the most pressing issues under negotiation. And for the first time a range of activities on women's and gender issues was organized by various organizations and institutions. And they met with interest, increasing awareness, and increased expression of commitment to gender justice from a number of stakeholders.

It was already at the SBSTA/SBI meeting in Bonn in May 2007, half a year before the COP, when the Indonesian Ministry for the Environment, preparing to host the COP, expressed their commitment to support women's involvement in the conference as well as the desire for integrating gender equality in the deliberations. Furthermore, the president of the conference, Indonesians Minister for the Environment Rachmat Witoelar, expressed his commitment to mainstream gender into the Bali Outcomes during a meeting with Indonesian NGOs. Although he did not succeed doing this, his statement constitutes a strong message.

Some days after the conference, the Bangkok Post published an article referring to Thailand's Minister of the Environment expressing his disappointment with the Bali outcome and calling on governments to support women's roles in combating global warming.

Beyond such statements, there were also a great variety of activities addressing women's and gender concerns going on around the Bali COP, for example:

- *Side events* of development organizations, women's and women ministers' networks, and governmental departments were the most visible manifestations of the new situation. Altogether, six side events had women's/gender issues as their main focus or integrated them in a broader framework. This is the highest number of such events ever held at a climate COP. The events addressed an impressive array of issues, including forestation/deforestation, adaptation, financing, energy, biodiversity, and future climate regime, among others.
- The *Women's Caucus* was cooperating closely with the *Climate Justice Caucus*, which was newly established at the COP. Issues of climate justice are proving to be excellent entry points for highlighting gender issues. Indeed, it seems that climate justice is one of the upcoming and may play an important role the debates over the next years.
- *Trade Unions*, traditionally a partner for campaigning on gender equality, were approved as a constituency to the UNFCCC for the first year. They expressed their interest in cooperation and included a paragraph on gender equality and gender mainstreaming, suggested by the Women's Caucus, in their lobbying document.
- A press briefing of the newly established global network *gendercc – women for climate justice** during the conference and media coverage in various countries generated additional attention to the issues beyond the closed conference area.
- And finally, daily Women's Caucus meetings helped to draft positions and statements and to coordinate lobbying efforts among the participating women. The Women's Caucus and the *gendercc network* was also the main node of contact between the women and gender advocates and the UNFCCC Secretariat.

The described activities and developments mark a step change in terms of gender and climate change issues in the international policy process

* The ***gendercc** network* is the global alliance of women and gender scholars and activists from Asia, Africa, America, The Pacific, and Europe working for gender and climate justice. www.gendercc.net

and arena: new connections between different issues have been made, and new alliances have been established.

Climate Justice: Entry Points for Gender Justice

It was not only NGOs, but also United Nations Organizations and International Organisations like IUCN who expressed their commitment to gender mainstreaming in climate change policies. Thus, it seems that 'gender equality' is finally beginning to be accepted as one of the core principles of mitigating climate change and adapting to its impacts. This may be due to the importance of climate justice in the future climate regime and the increased understanding among at least some of organizations forming the UNFCCC constituencies that the discourse on climate change needs to be widened beyond its current main focus on technologies and economic instruments. Root causes of climate change, like consumption patterns and lifestyles in industrialized countries and quickly developing societies must be brought onto the agenda immediately. Women and gender activists have been pointing out for some years that we need to question the dominant perspective focusing mainly on technologies and markets, and put caring and justice in the centre of measures and mechanisms. The lack of gender perspectives in the current climate process not only violates women's human rights, but it also leads to shortcomings in the efficiency and effectiveness of climate related measures and instruments. The notably increased attention paid to climate justice and gender mainstreaming is certainly the outcome of many, many conversations with individual delegates, the increasing presence, and other aspects of the multi-track advocacy strategy that a small group of women and gender experts has engaged in at the COPs over the years. It seems that these patiently continued activities, including through tough times, are finally paying off.

Some countries, and not least the UNFCCC Secretariat, are also appearing more open-minded towards gender equality. During the side event "Integrating gender into climate change policy: challenges, constraints, perspectives" and in various smaller debates they expressed their concerns about the lacking gender dimension and assured their support for future activities. And they asked for very concrete suggestions, in particular regarding language, to be used in upcoming negotiations. This will be one of the tasks, and challenges, for further collaboration in the *gendercc network*: to pay very close attention to the negotiations and work closely with like-minded parties towards appropriate agreements. The network is committed to doing engaging in this way without compromising the independent, and sometimes radical, stance that the *gendercc network* has developed. Taking gender aspects into account implies a radical move away from dominant, market-based to people-centred mechanisms. This is a message that is not warmly welcomed in most of the climate change community. Hence, while there has been a step change at Bali in terms of awareness of and public commitment to gender sensitivity, really integrating gender into climate protection will remain a big challenge.

Future Strategies

Activities during COP13 in Bali were supported by funding from UNDP, aiming to bring seven women from developing countries and countries in transition to the conference, and to organize and coordinate the activities. This included preparing women's position papers, which generated a lot of interest, providing arguments and recommendations for the debates. Developing further positions and suggestions and to provide necessary background information will therefore be essential for further developing a gender sensitive climate regime.

In the future, efforts similar to those before and at COP13 must be undertaken related to the process and institutional arrangements. Continuous representation of women and the *gendercc network* will be crucial. Furthermore, it will not be sufficient to participate in the annual COPs and SBSTA/SBI meetings. In order to succeed in integrating gender in climate change policy, it will be even more important and more promising to actively participate in the growing number of workshops organized in the context of pursuing the Bali Roadmap. Yvo de Boer, Executive Secretary of the UNFCCC, announced 4 to 5 additional annual meetings in order to discuss and negotiate the future climate regime. In addition, there is a significant number of related workshops, for example addressing, the review of articles and conventions, or methodological issues.

On the other hand, in order to prepare substantial input into the workshops, it is necessary to link discussions at local levels to those at the international level – feeding local realities and experiences into the general and abstract discussions at the global level, as well as 'translating' global changes and international policies so as to communicate what these will, or may, mean for local communities. Providing capacity development opportunities for women and gender activities who are prepared to raise their voices in the international policy arena will be key, so that they can become effective advocates on policy and effective communicators to communities and networks around the world.

In order to be able to meet said requirements, the *gendercc network* agreed to work towards institutionalizing its structure and activities by:

- Establishing *regional focal points,* aiming to communicate between international, regional and local levels (in both directions!) to improve women's capacity on climate policy as well as climate change experts' capacity on gender equality issues. Positions and text modules for workshops, meetings and conferences should be drafted by the focal points, using materials shared in the global network;
- Establishing a *gendercc network secretariat,* coordinating the activities of the focal points, discussing the most important issues and meetings, providing support and conducting outreach and advocacy globally; and
- *Raising funds to cover the costs of participating* in the UNFCCC process for a number of women from around world.

Such a structure will also serve the women's goal of being recognized as a constituency in the UNFCCC process. This recognition, in turn, will facilitate invitations to participate in workshops, give statements and submit positions.

Network members stand ready to do this work, and there are ideas and concepts to move it forward. However, the *gendercc network* will need more, and more sustained financial support. The growing interest and expressed commitment from government parties, IGOs and others give hope that such support will be available: further progress will indeed depend on potential funders putting their money where their mouth is.

Anecdotal Conclusion

In a meeting with NGOs in Bali, UNFCCC Executive Secretary was asked how he would ensure that women's perspectives and issues of social justice and human rights were being advanced in the Post-Kyoto Regime. His answer: "I have no idea, tell me how". Gender and climate change experts from all over the world are prepared to tell him what to do and how to do it, and are awaiting his request. Knowledge and skills are available. Time is ripe for open minds to learn and change.

Index

Q

R

S

T

W

Z

Colour Plate Section

Chapter 18

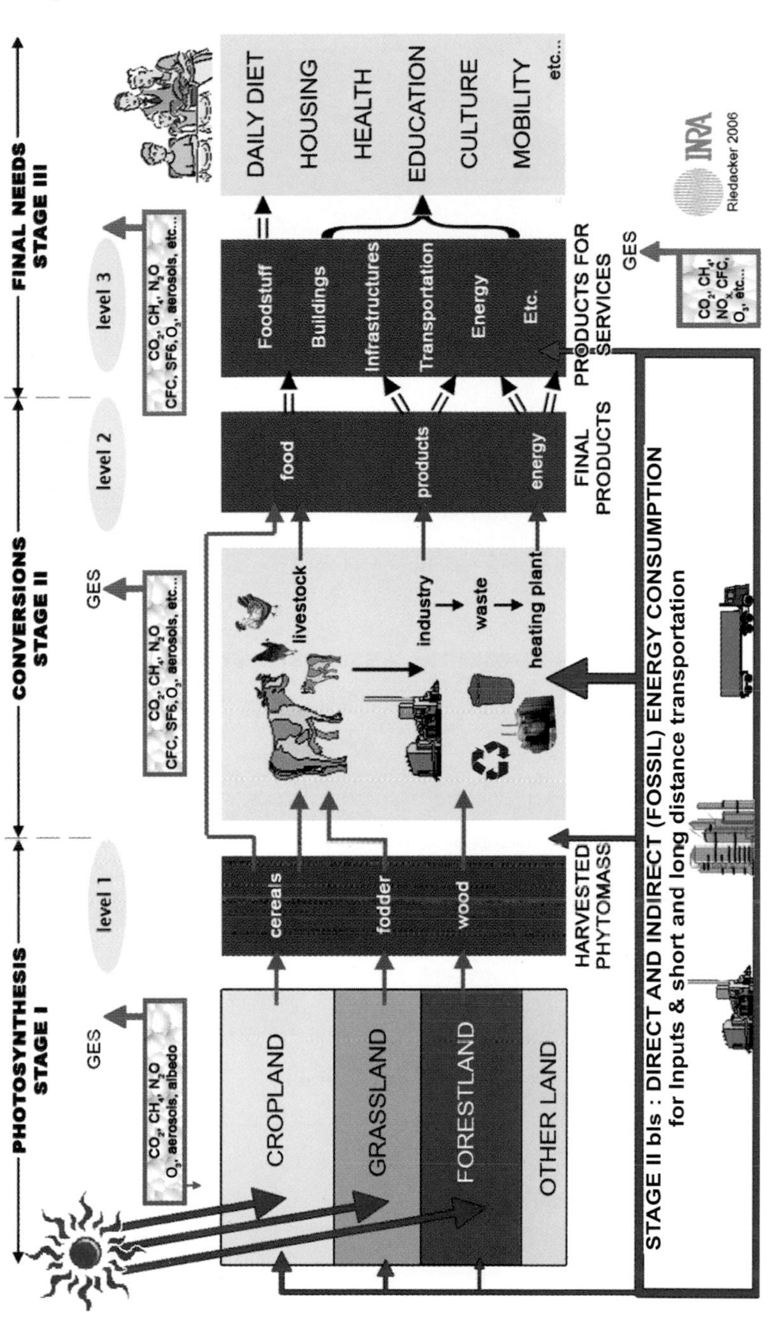

Fig. 3 Global Integrated Environmental Assessment (GIEA), from the sun to final use at the global level: Stage I: Solar energy bioconversion; Stage II: Conversion of phytomass and non-renewable products into final products and energy carriers; Stage III: Arrangements of final products to meet final needs.

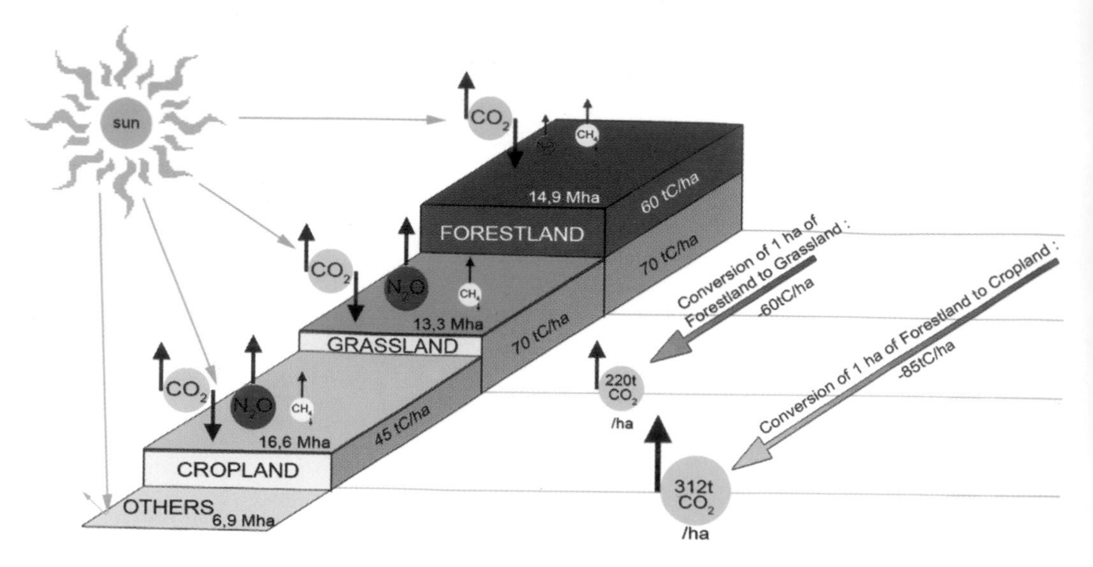

Fig. 6 French land-use and average carbon stocks, in cropland, grassland and forestland. Main GHG emissions from the different land are indicated, as well as carbon stocks. This is used to calculate CO_2 net emissions resulting from land-use changes in a country.

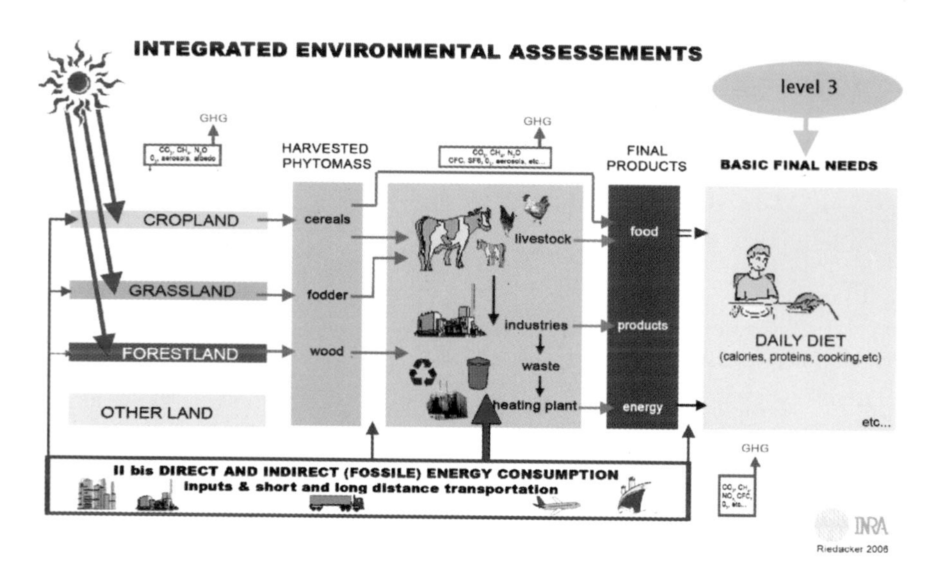

Fig. 8 Detail of a Global Integrated Environmental Assessment for the supply of human diet. (cf. Fig. 9).

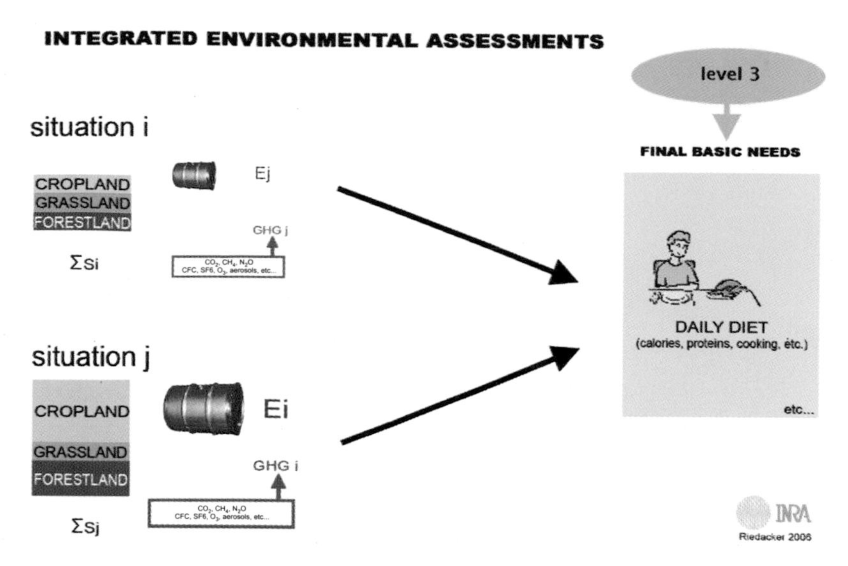

Fig. 9 Results of an assessment at Stage 3 for two typical human diets: Comparison of land and fossil energy requirements, and of GHG emissions for the daily diet (for instance 3000 kcal/day) under two situations i and j, with different land use (cropland, grassland and forest land for cooking energy), and different fossil fuel consumptions, for instance with more or less intensive production systems, with various amounts of fossil fuel inputs in fields, various proportions of animal proteins, high or low transportation distances, more or less energy consumption for conversion, storage and cooking and various waste disposal systems etc. More detailed preliminary data for such analysis have been made recently (Riedacker and Migliore 2008).

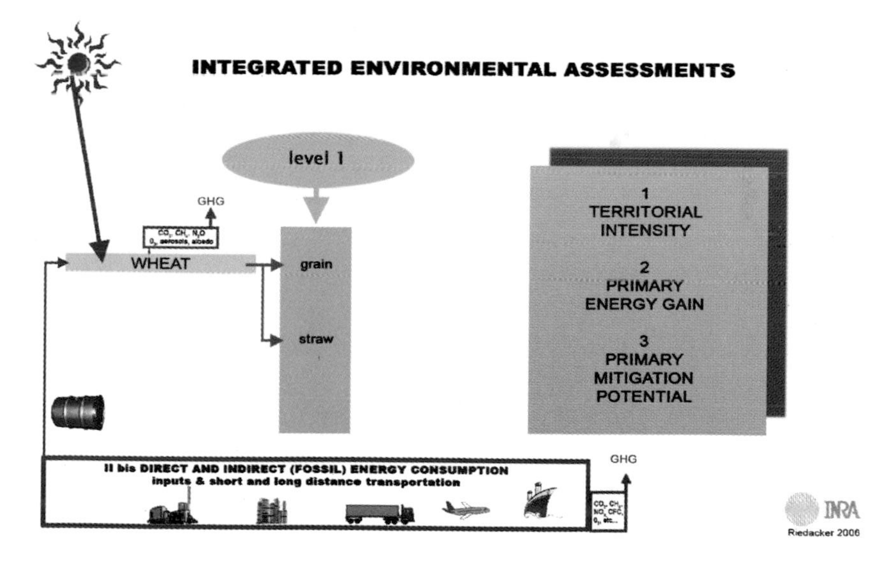

Fig. 10 A partial GIEA for wheat production at Level 1.

Fig. 14 Evolution, between 1950 and 2000, of world cereal production (in milled rice equivalent) land used (660 million hectares) and of land spared due to yield increase per hectare (1.1 billions ha) (Data FAO and Agrostat From Borlaug in IFDC 2006)

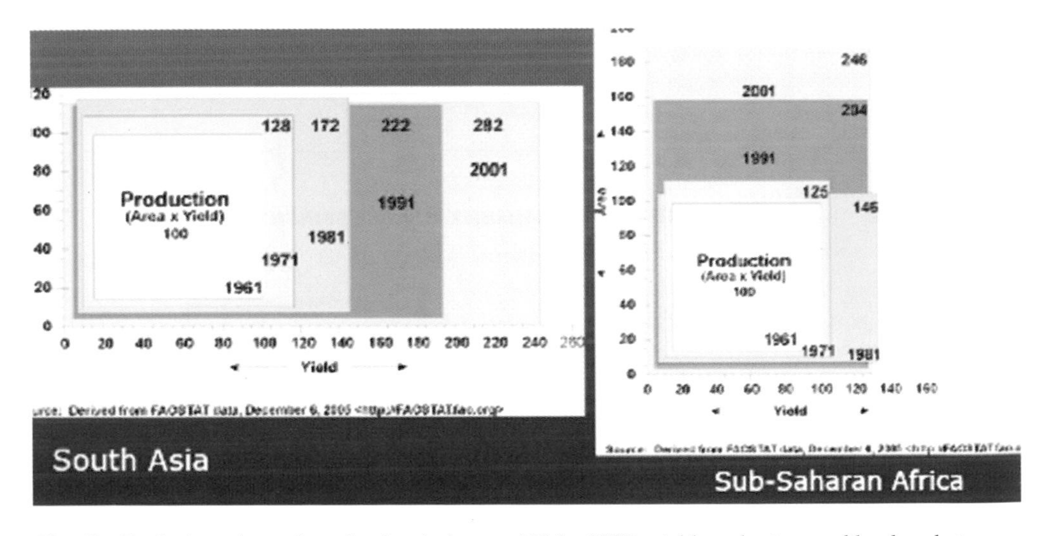

Fig. 15 Evolution of cereal production (reference 100 for 1960), yield per hectare and landuse between 1961 and 2001, in South Asia and Sub-Saharan Africa (Source IFDC 2006).

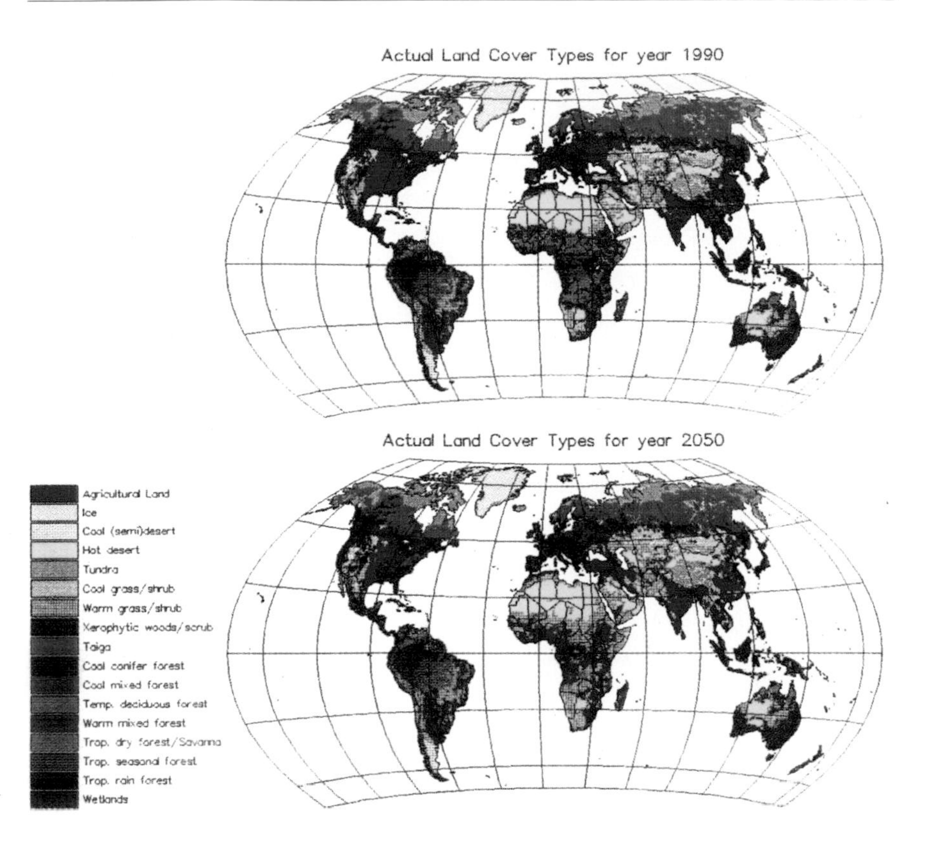

Fig. 16 Land cover change in the world between 1970 and 2050 if agricultural efficiencies of each continent were to remain the same as in 1990, and if each continent was to feed its population. In 2050 all African forests would have disappeared except some small forestland in the centre of the Congo Basin (Modelling exercise, personal communication from RIVM)

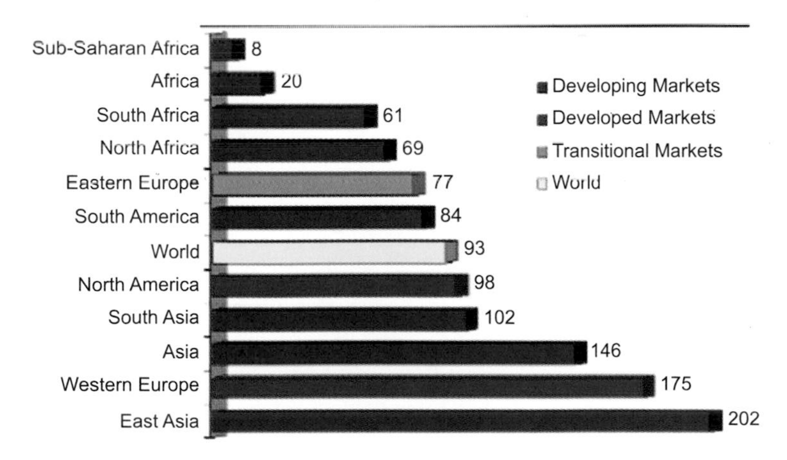

Fig. 18 Average fertilizer input per ha in 2002/2003 in various parts of the world IFDC 2006 www.ifdc.org

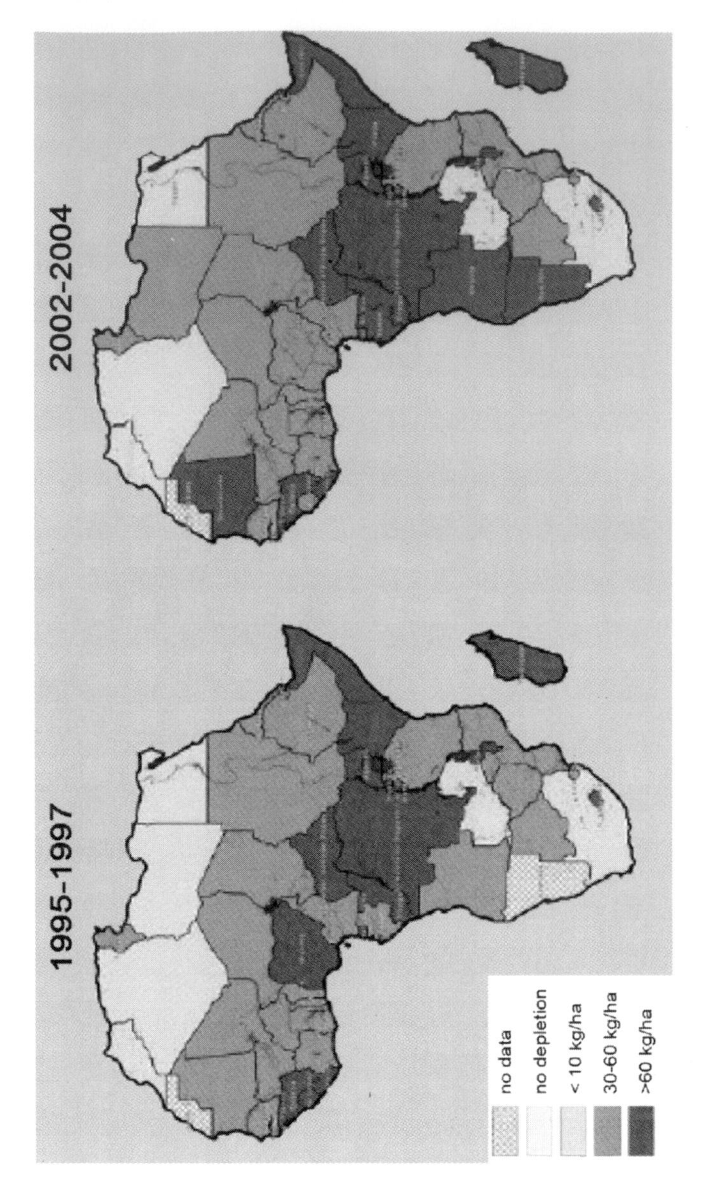

Fig. 19 Annual mineral nutrient depletion of African soils (in kg/ha). Source IFDC (2006).

Chapter 23

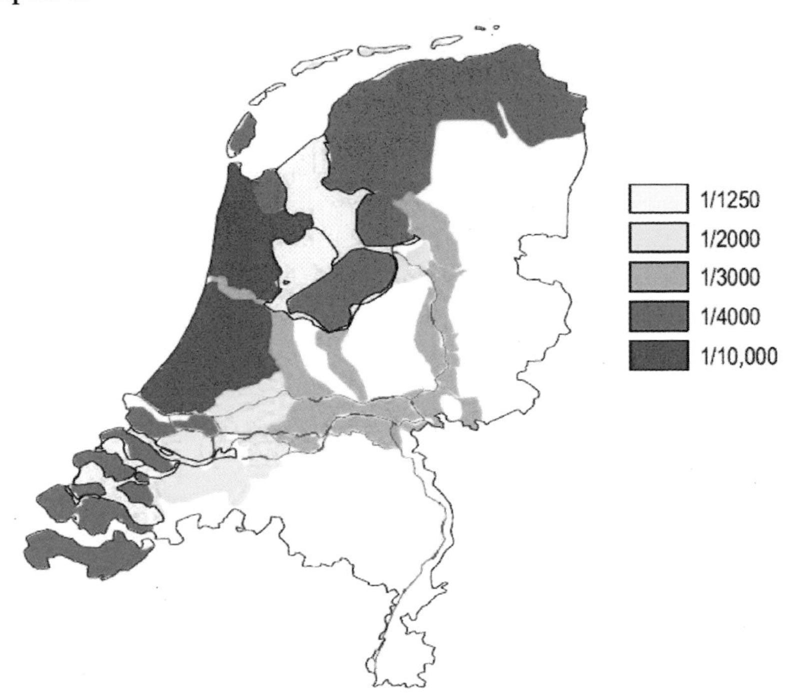

Fig. 1 Map of The Netherlands showing the different safety norms.

Fig. 2 Floating house in Middelburg, province of Zeeland (Huitema et al. 2003).

Chapter 29

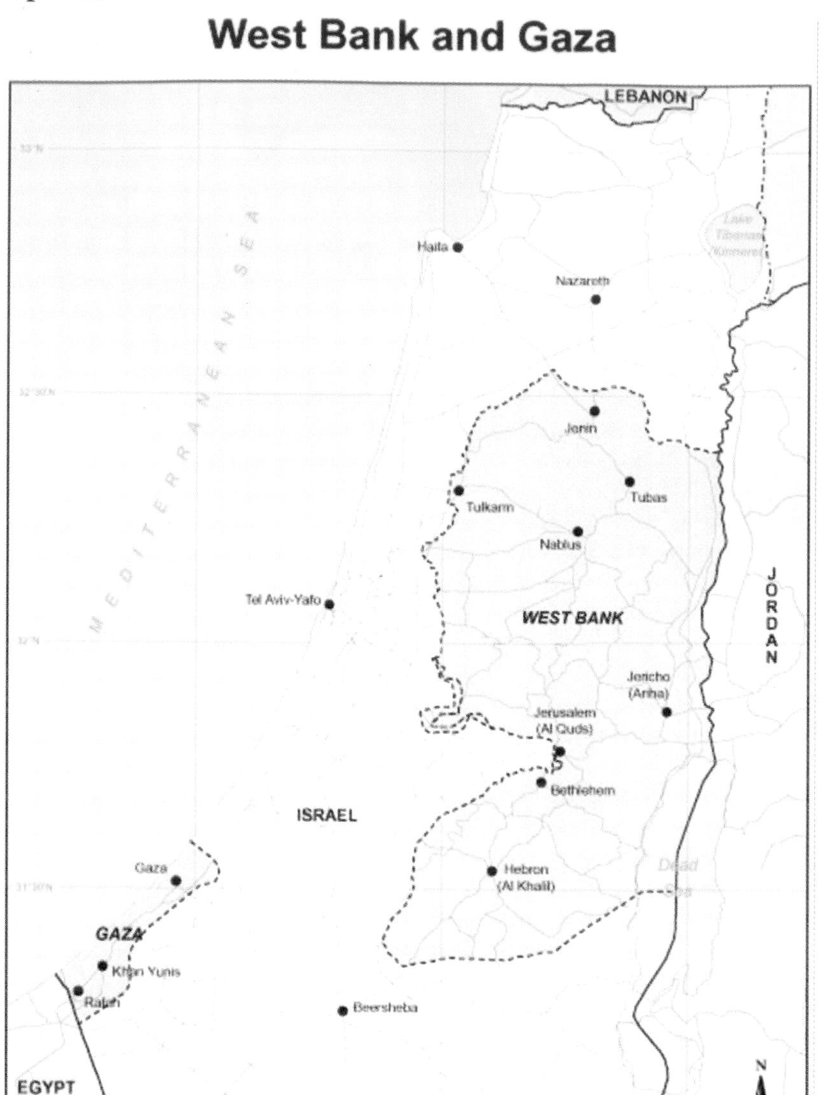

Fig. 1 General Location Map of the West Bank and Gaza Strip.

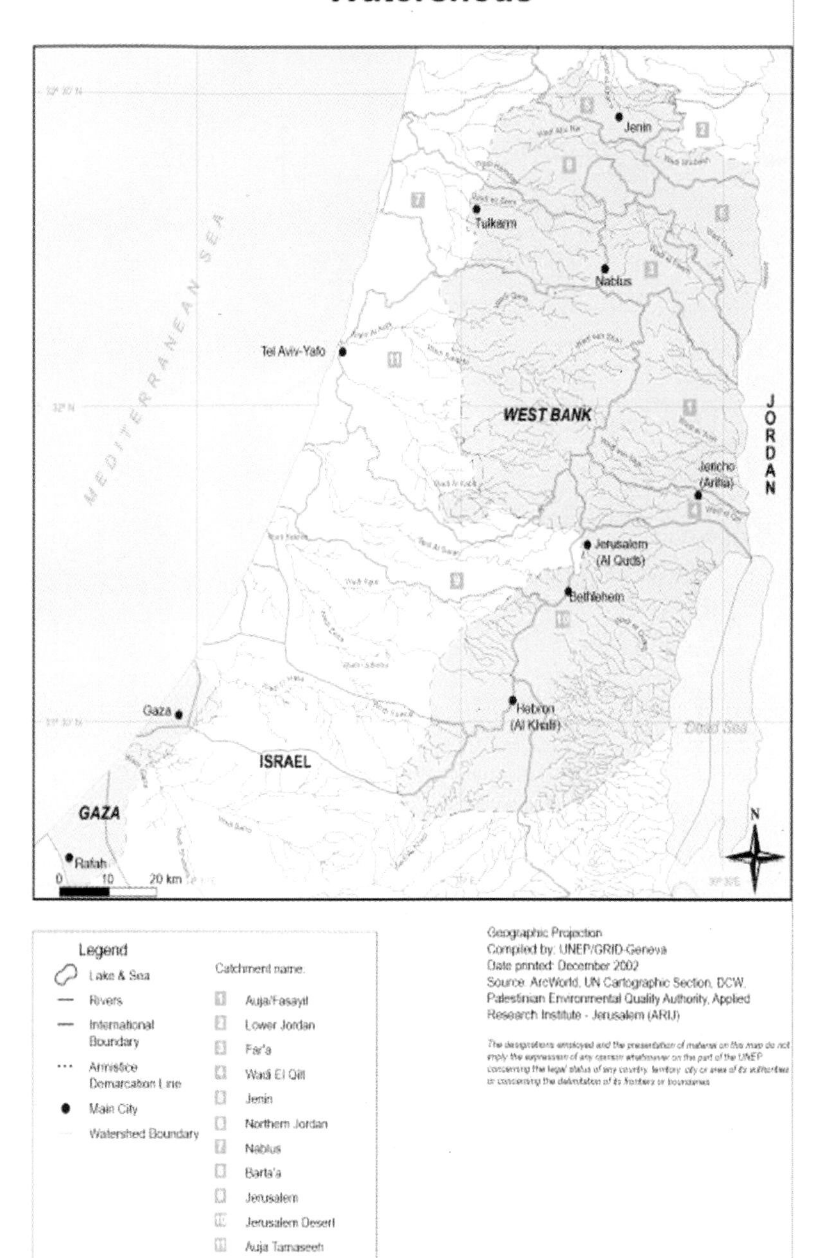
Watersheds
Jenin
Tulkarm
Nablus
Tel Aviv-Yafo
WEST BANK
JORDAN
MEDITERRANEAN SEA
Jericho
(Ariha)
Jerusalem
(Al Quds)
Bethlehem
Hebron
(Al Khalil)
Dead Sea
Gaza
ISRAEL
GAZA
Rafah
0 10 20 km
N
Legend
Lake & Sea
Rivers
International Boundary
Armistice Demarcation Line
Main City
Watershed Boundary
Catchment name:
Auja/Fasayil
Lower Jordan
Far'a
Wadi El Qilt
Jenin
Northern Jordan
Nablus
Barta'a
Jerusalem
Jerusalem Desert
Auja Tamaseeh
Geographic Projection
Compiled by: UNEP/GRID-Geneva
Date printed: December 2002
Source: ArcWorld, UN Cartographic Section, DCW, Palestinian Environmental Quality Authority, Applied Research Institute - Jerusalem (ARIJ)

Fig. 2 Watersheds in Israel and Palestine

Chapter 30

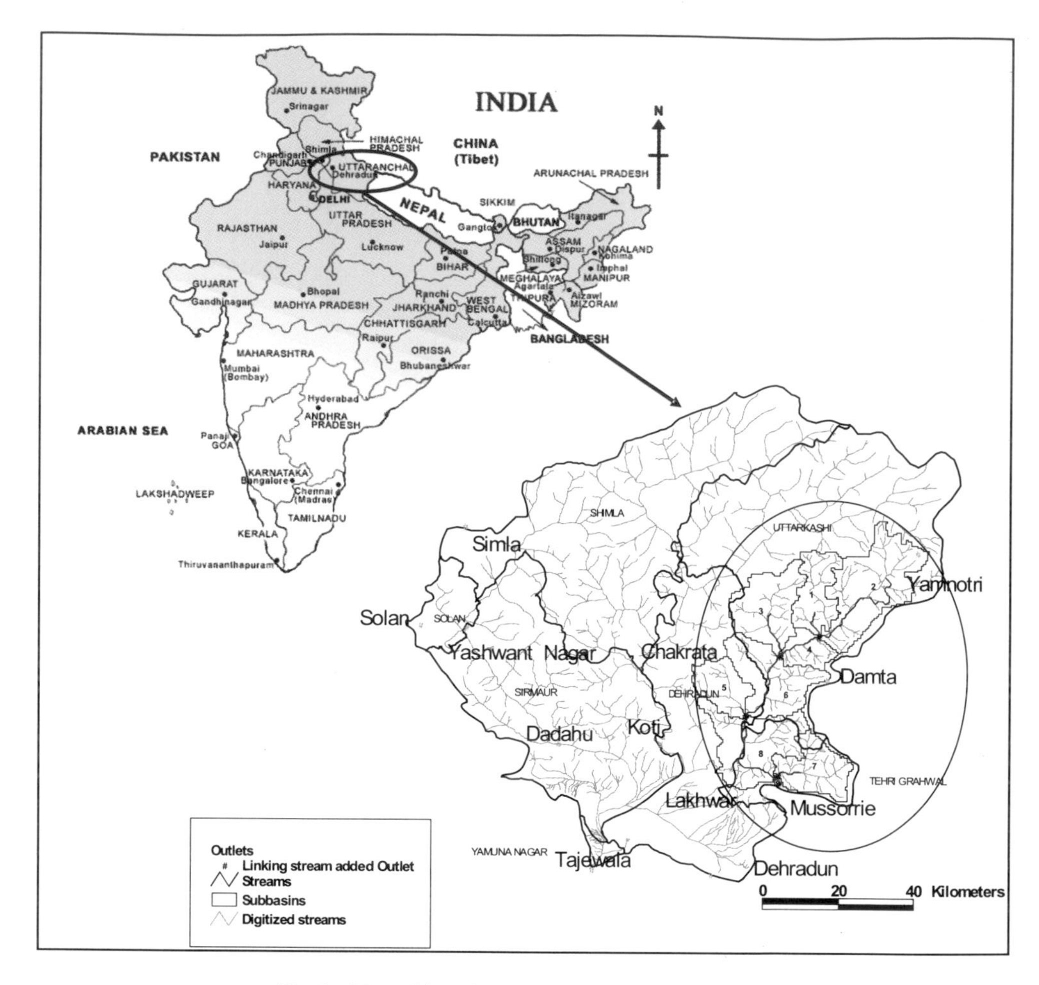

Fig. 2 Uttarakhand State: Case study area in India

The above is a rough sketch map which does not attempt to depict political boundaries.

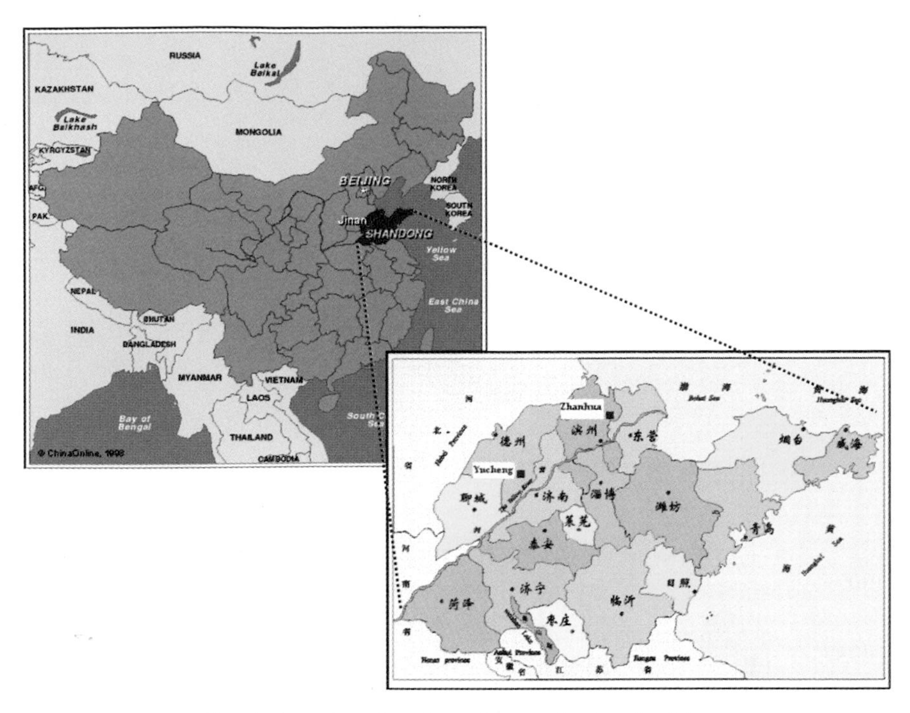

Fig. 4 Shandong Province: Case study Area in China

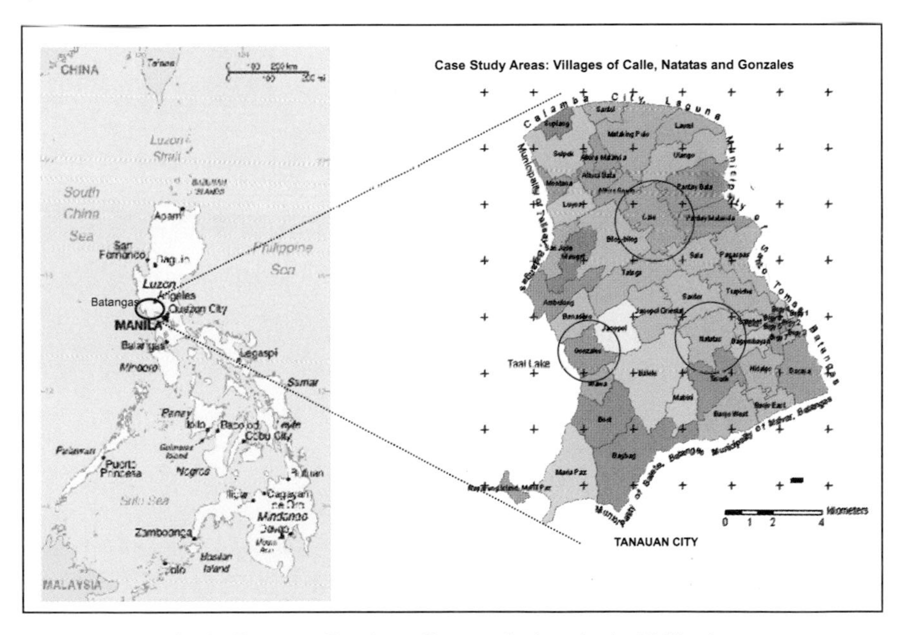

Fig. 6 Batangas Province: Case study Area in the Philippines

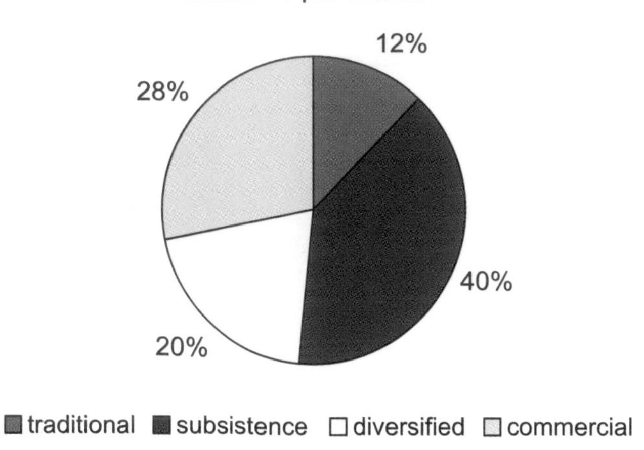

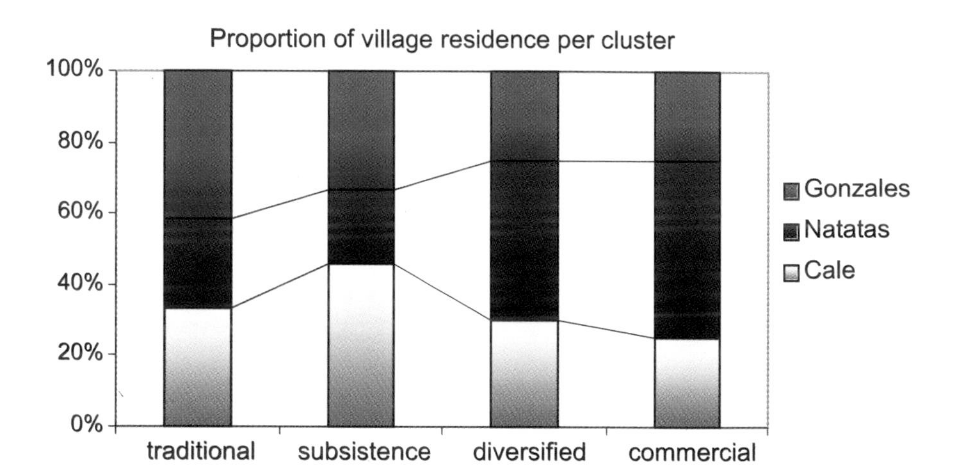

Fig. 7 Distribution of the farmer typologies in the case study areas

Chapter 31

(a)

(b)

Fig. 1 Over the last 20 years, natural forests have been logged and replaced with crops. In sites which are fragile, such as on steep slopes, the poor soil fixation by unsuitable species can lead to a) soil erosion in the Sichuan province (Photo: Y. Chen) and b) landslides in the Guizhou province; landslides such as this can be seen every few kilometers along major road axes (Photo: T. Fourcaud, CIRAD, France).

Fig. 2 Some species are not suitable for planting on steep slopes. Landslides are more frequent in plantations of big node bamboo (*Phyllostachys nidularia* Munro) compared to many other vegetation types (Stokes et al. 2007). Nevertheless, new environmental incentives can encourage stakeholders to plant unsuitable species at fragile sites (Photo: L. Jouneau, INRA, France).

Fig. 3 Sand arresters help prevent the movement of shifting sand onto the railway lines of the new Tibet-Qinghai plateau railway. These simple fences and quadrants of rocks can also be planted with native sand-adapted species to help fix sand and improve substrate quality (Photo: A. Stokes).

Fig. 4 New farming methods may help reduce soil loss on steep slopes. In trials in the Sichuan province of China, hedgerows were planted with a combination of pear trees (*Pyrus* spp.) and Chinese day lily (*Hemerocallis citrine* Baroni) flowers. Pears were planted with a within-row spacing of 1 m and day lilies were planted with a spacing of 13.3 × 13.3 cm (two rows in one hedgerow). On sloping land, soil loss was reduced by 80% using this technique (Photo: Y. Chen).

Fig. 5 After the original forest plantation was destroyed, terraces of Sabaigrass (*Eulaliopsis binata* Retz.) help to stabilize the slope and prevent landslides (Photo: Y. Chen).

Fig. 6 Mulberry (*Morus alba* L.) is planted at the edges of paddy fields to stabilize the risers.

Chapter 37

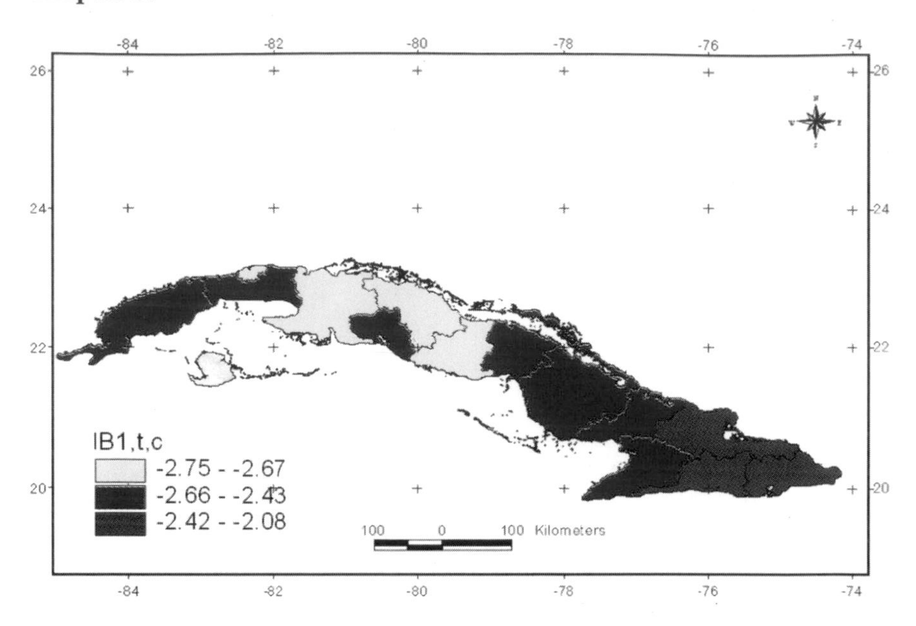

Fig. 1 Winter trend anomalies in the 1980s using the $IB_{1,t,c}$ index

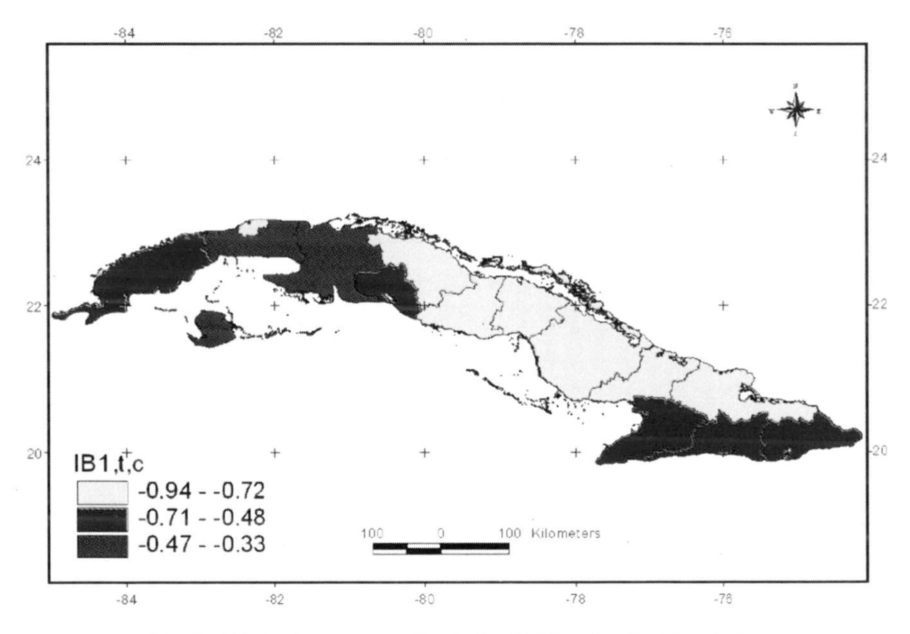

Fig. 2 Winter trend anomalies in the 1990s using the $IB_{1,t,c}$ index

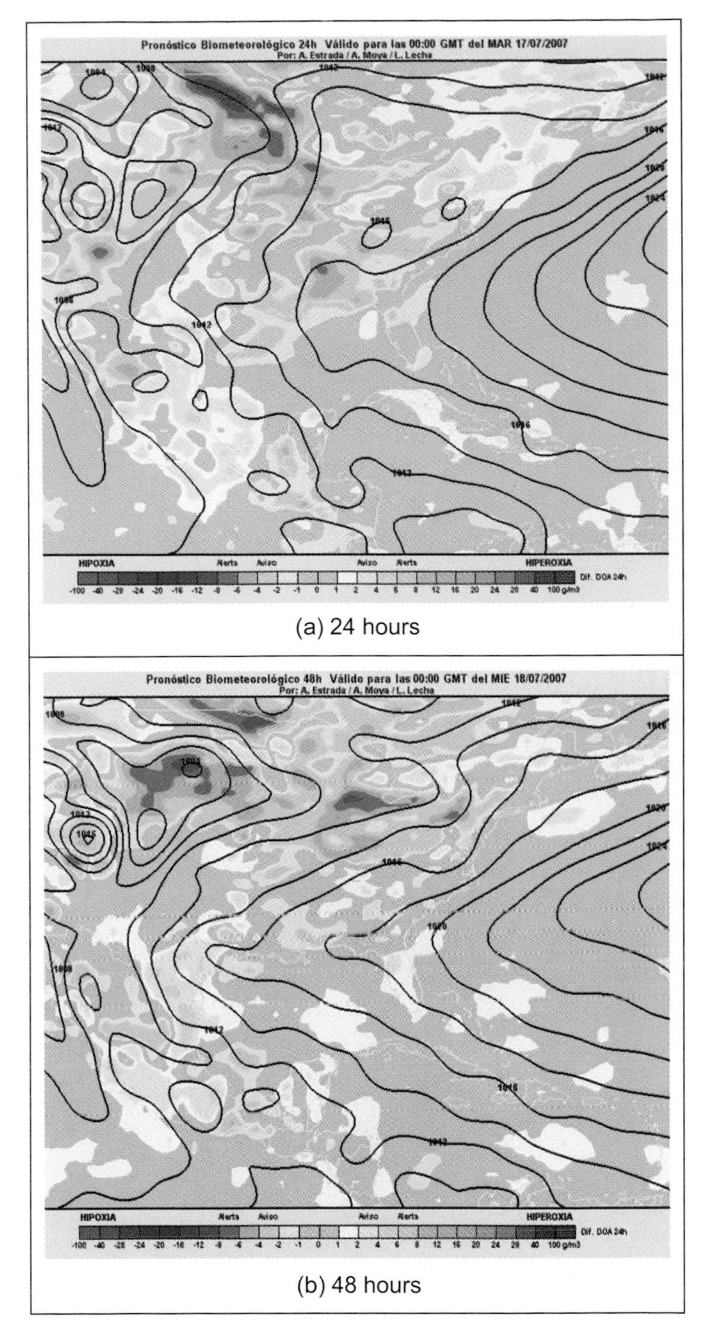

(a) 24 hours

(b) 48 hours

Fig. 4 Demonstrative output maps from the model SAAS version 2.0 initialized the day March 16, 2007 at 0000 GMT (19:00 hours, local time in Cuba).

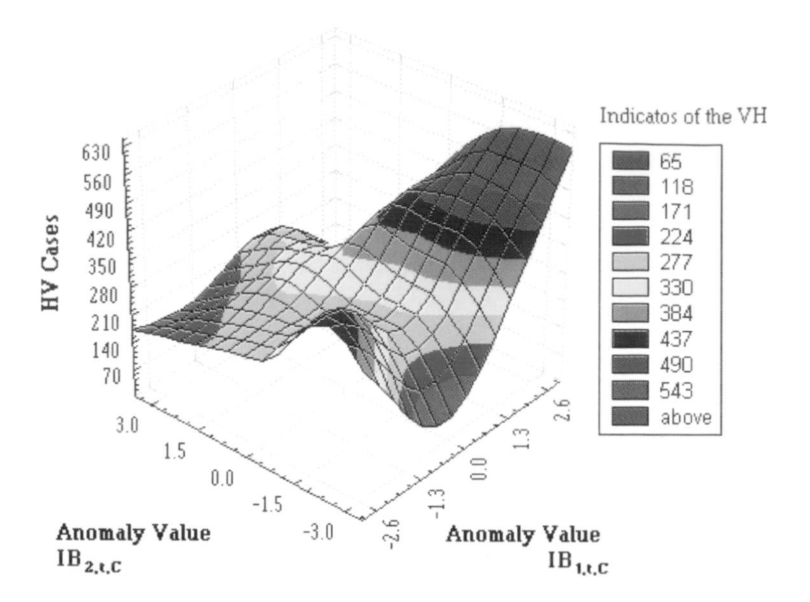

Fig. 5 Association between climate variability and viral hepatitis according to the indexes.

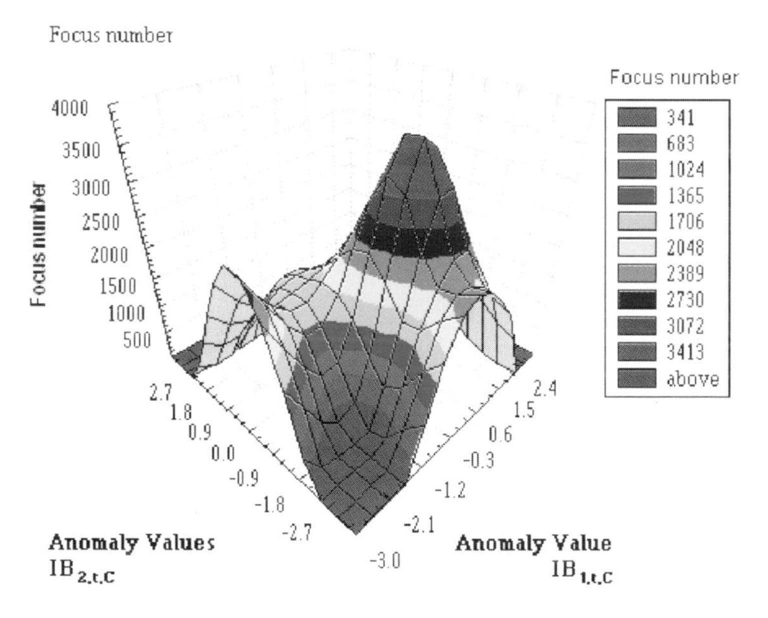

Fig. 7 Association between climate variability and the number of positive houses (hotspot) of the *Aedes aegypti* by climate variability according to indexes.

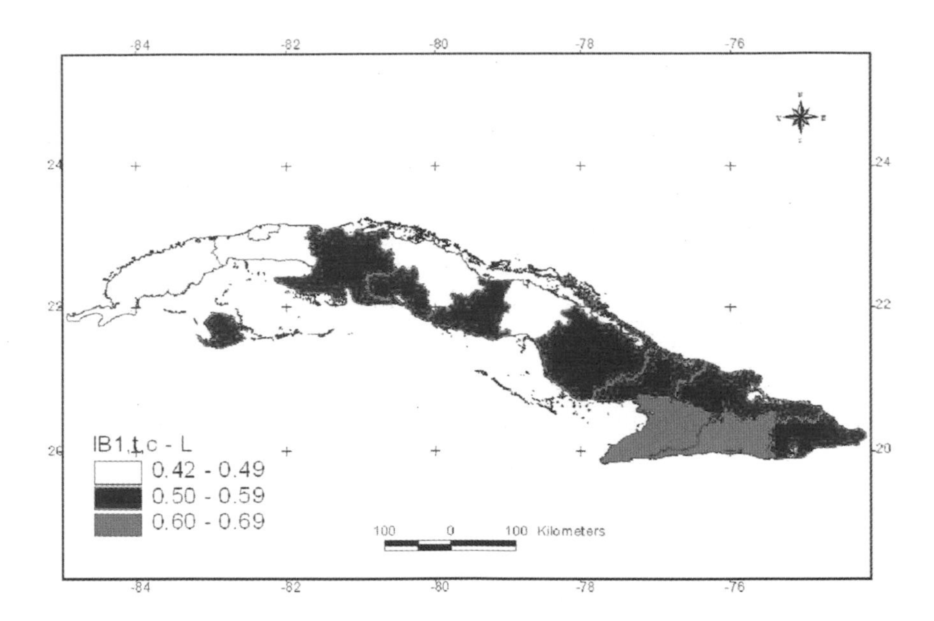

Fig. 8 Scenario of climate variability. Low sensibility range < 0.70 (change per decade).

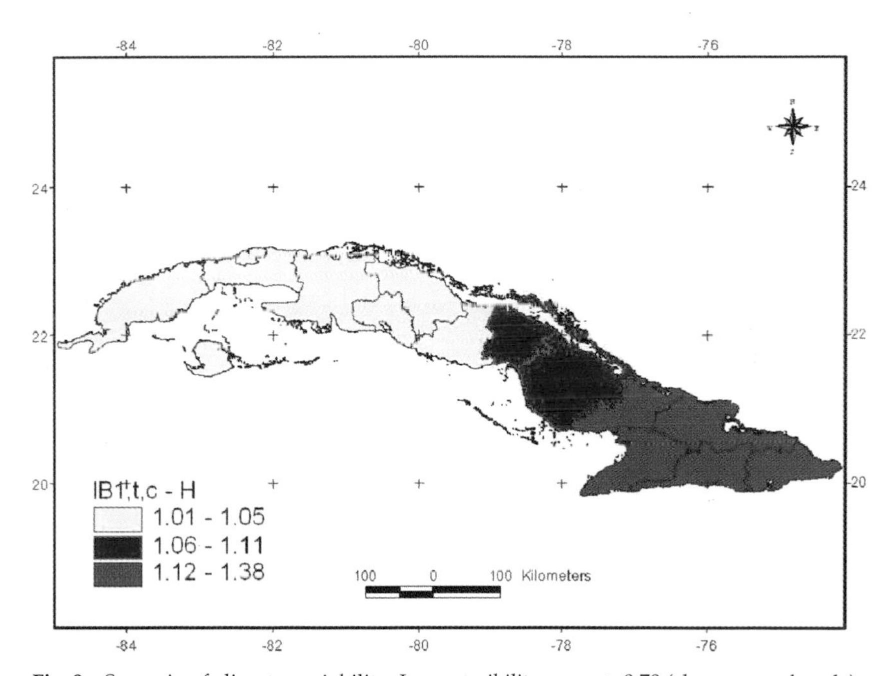

Fig. 9 Scenario of climate variability. Low sensibility range > 0.70 (change per decade).

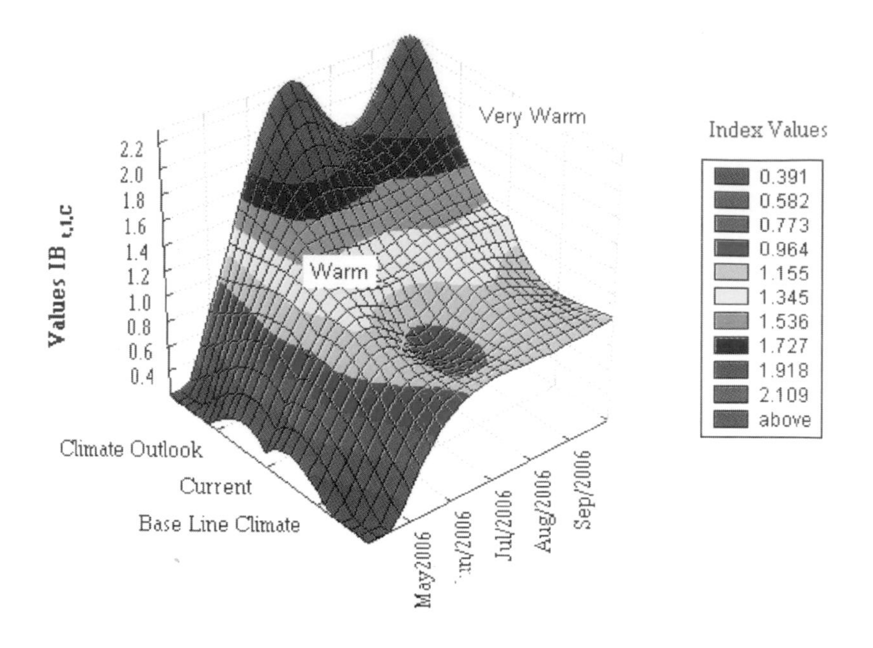

Fig. 13 Seasonal Climate Outlook. May–August/2006 Period of base line used 1961-1990 and current condition 1991-2005

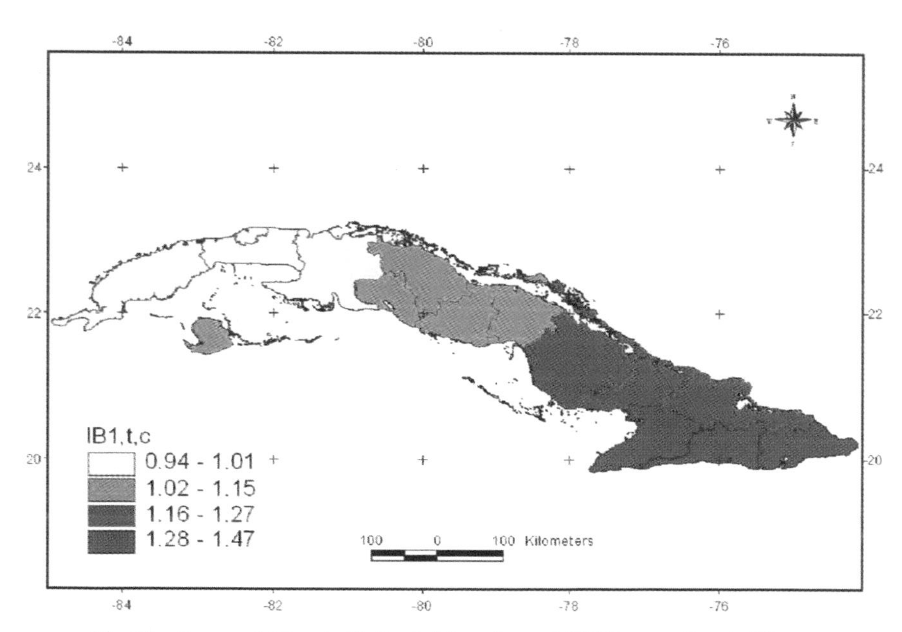

Fig. 14 Seasonal Climate Outlook (May–August/2006) according to $IB_{t,1,C}$.

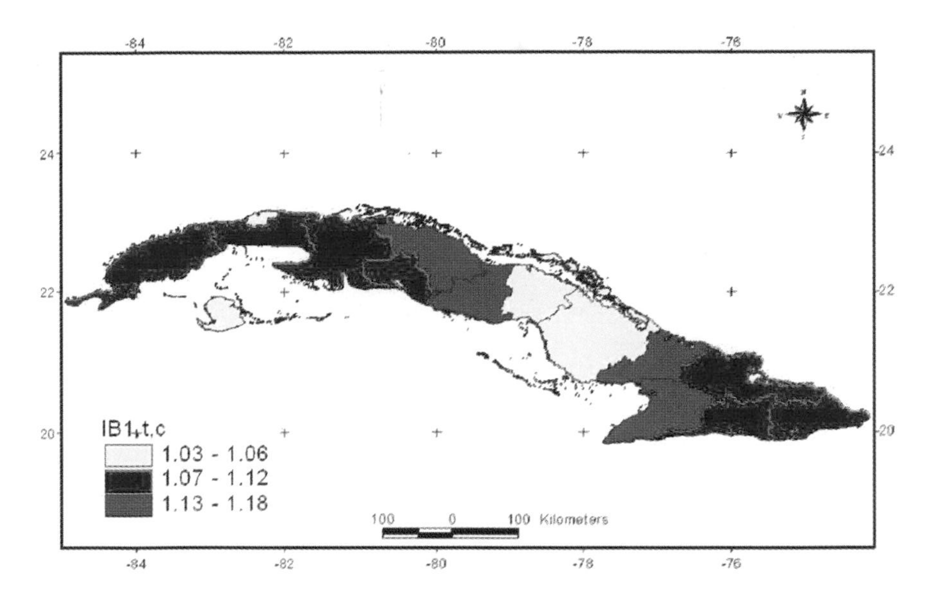

Fig. 15 Climate outlook according to $IB_{t,1,C.}$ August/2006

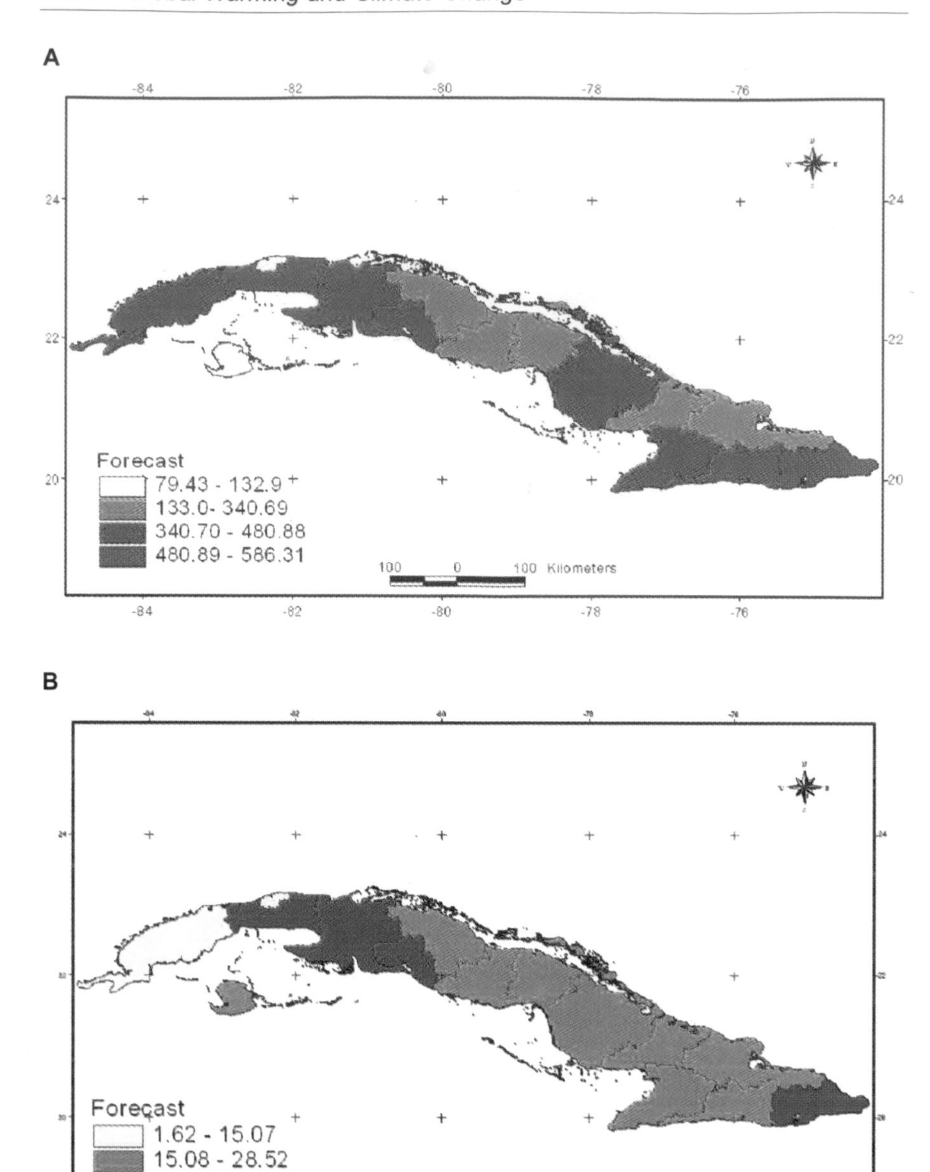

Fig. 16 Rate of per 100,000 habitants, expectation attentions by Acute Diarrhoeal Disease (A) and Acute Respiratory Infections (B) August/2006.

Chapter 44

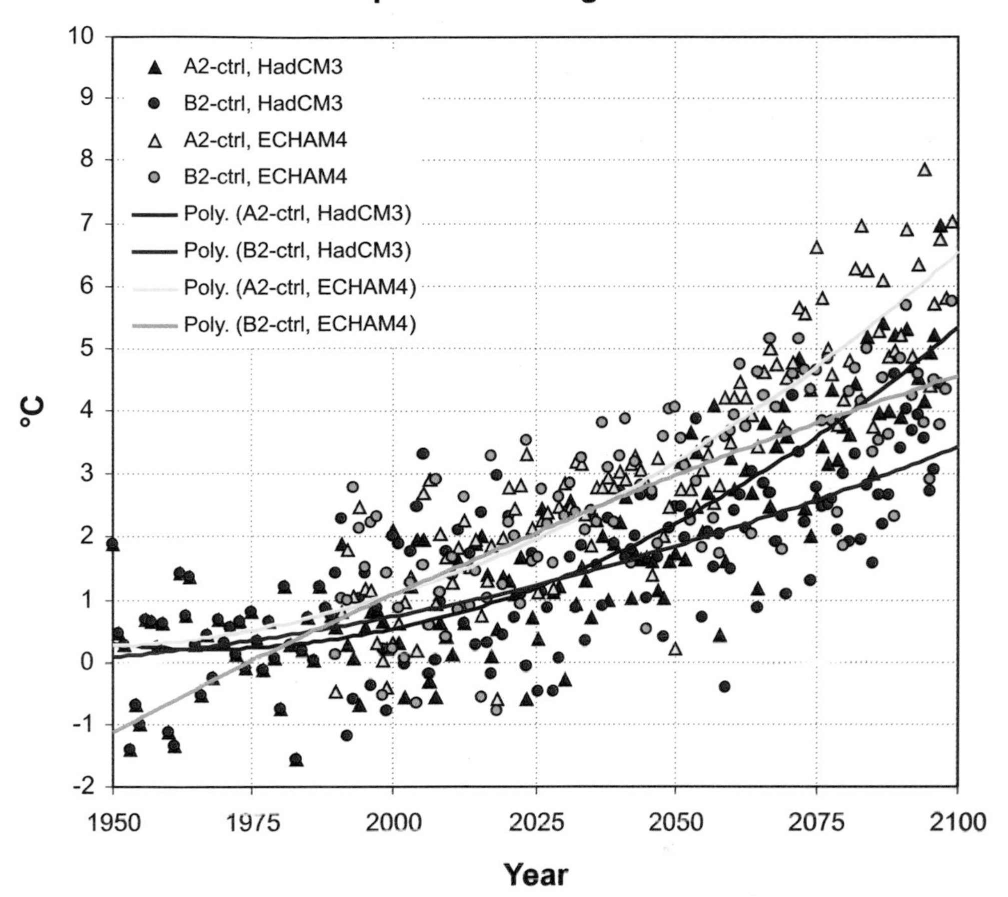

Fig. 3 Shaanxi/Ansai GCM-grid with simulated temperature change by year presented as difference from control scenario. Polynomial regression lines are calculated to indicate trends.

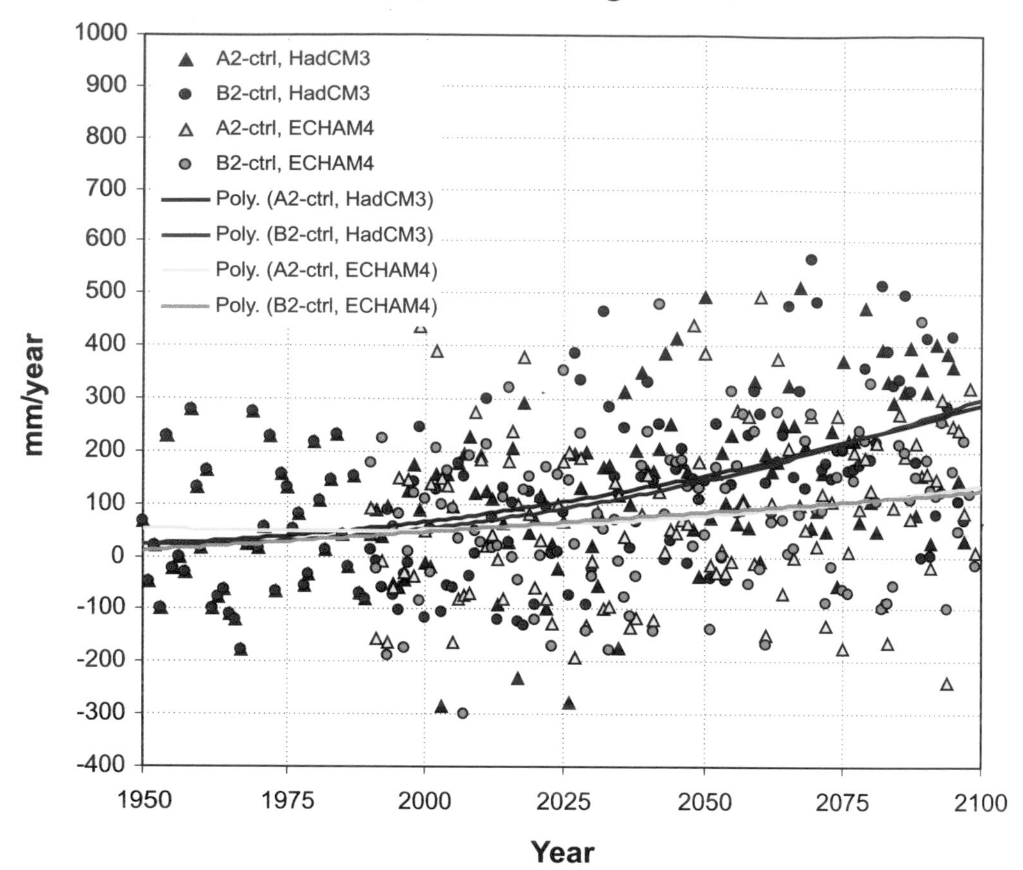

Fig. 4 Shaanxi/Ansai GCM-grid with simulated precipitation change by year presented as difference from control scenario. Polynomial regression lines are calculated to indicate trends.